Graduate Texts in Mathematics 283

Graduate Texts in Mathematics

Graduate Texts in Mathematics bridge the gap between passive study and creative understanding, offering graduate-level introductions to advanced topics in mathematics. The volumes are carefully written as teaching aids and highlight characteristic features of the theory. Although these books are frequently used as textbooks in graduate courses, they are also suitable for individual study.

More information about this series at http://www.springer.com/series/136

Ibrahim Assem • Flávio U. Coelho

Basic Representation Theory of Algebras

Springer

Ibrahim Assem
Département de mathématiques
Université de Sherbrooke
Sherbrooke, QC, Canada

Flávio U. Coelho
IME, University of Sao Paulo
São Paulo, São Paulo, Brazil

ISSN 0072-5285 ISSN 2197-5612 (electronic)
Graduate Texts in Mathematics
ISBN 978-3-030-35117-5 ISBN 978-3-030-35118-2 (eBook)
https://doi.org/10.1007/978-3-030-35118-2

Mathematics Subject Classification: 16G20, 16G60, 16G70, 16G99

This Springer imprint is published by the registered company Springer Nature Switzerland AG.
The registered company address is: Gewerbestrasse 11, 6330 Cham, Switzerland

Preface

This textbook is designed to introduce the reader to the Representation Theory of Algebras as painlessly as possible. It concentrates on the Auslander–Reiten theory, the radical of a module category and related topics. The only prerequisites are some module theory and homological algebra such as are usually taught in a beginner's graduate course, and can be acquired in most of the textbooks in the field.

Representation theory in its broad sense is that part of mathematics that aims at representing abstract mathematical objects as concretely as possible. In this book, we are interested in the Representation Theory of Algebras, by which we mean finite dimensional algebras over a field. Since the works of E. Noether in the 1930s, this is understood as characterising the algebra by means of its module category. That is, it aims at understanding not only the modules, but also the morphisms between them. One is looking for invariants allowing to classify them, but also for algorithms in order to compute them. One may thus view representation theory as an advanced form of linear algebra, in which modern tools such as homological algebra are available.

Since the late 1960s, the theory started growing fast owing to the introduction of almost split sequences by Maurice Auslander and Idun Reiten, and of quivers and their representations by Pierre Gabriel and his school. As the years passed, it became increasingly difficult for beginners to make their way into the field because of the need to master several different results and techniques.

Our book was born out of an unexpected encounter. In 2007, the second author gave a course on the radical of a module category in a Workshop on Representation Theory of Algebras, which took place in Montevideo (Uruguay). In 2013, the first author gave a course on the Auslander–Reiten theory in a CIMPA Research School on the Algebraic and Geometric Aspects of Representation Theory, in Curitiba (Brazil). It was quickly apparent that these courses were complementary and, when put together, would form the socle of a good introductory course in representation theory.

We started writing this book, setting ourselves the following constraints: we wanted to be able to cover the contents completely in a one-semester course, and we also wanted to make it as easy as possible for the student, by avoiding the use of several different techniques or points of view. We have favoured the point of view saying that to understand a module category, one should concentrate on morphisms, and more precisely, morphisms lying in the radical of the module category. Because of these constraints, it was not possible to be encyclopaedic in this work. Perhaps the most obvious omissions are representations of quivers, Gabriel's theorem and covering techniques. We apologise to the reader, pointing out that other books deal with these topics. We do, however, believe that the present volume provides the reader with a solid basis in representation theory, allowing him or her to pursue readings in other directions.

We now briefly describe the contents of the book. Throughout, the word "algebra" stands for a finite dimensional algebra over a commutative field, and the word "module" for a finitely generated right module. The book consists of six chapters. Chapter I is of an introductory nature, it is divided into two sections. In the first section we recall, mostly without proofs, the results of module theory that will be useful in the sequel. Therefore, it can be left out, provided that the reader has the relevant knowledge. The second section, in which all proofs are given, deals with the quiver of a finite dimensional algebra, and classes of examples such as hereditary algebras or Nakayama algebras. In Chapter II, we start by introducing and giving several characterisations of the radical of a module category, which lead us to the definitions of irreducible morphisms and almost split sequences. We prove their existence and then study their relation with the factorisation of a morphism lying in the radical. Chapter III is devoted to construction techniques for almost split sequences and hence irreducible morphisms. With this knowledge, we are able to define and show how to construct the Auslander–Reiten quiver of an algebra (Chapter IV) or at least some of its components. Chapter IV also contains a short discussion on how deep a morphism lies inside the radical of the module category and a description of the module category of the Kronecker algebra. In Chapter V, we discuss the relation between the representation theory of an algebra and that of the endomorphism algebra of a well-chosen module. Auslander's projectivisation technique is presented, as well as a short introduction to tilting theory. The last Chapter VI concentrates on representation-finite algebras. Several characterisations of this class are given, and we end the book with a proof of the so-called Four Terms in the Middle theorem. Throughout the book, several examples are solved in detail. We have included a set of exercises at the end of each section.

At the end of the volume, we have also included a short bibliography divided into three parts: the first part consists of standard textbooks on noncommutative algebra and homological algebra, to which the reader is referred for the results we use from these areas. The second part is a short list of textbooks on (parts of) the representation theory of algebras, and finally the last part is a list of some of the original papers containing the results that are presented here. This bibliography is not complete and we apologise in advance in case some important papers do not appear in it.

The material contained in this textbook is complementary. We believe that it can be covered in a one-semester course. In the case of a shorter course, we believe that the following sections and subsections can be given less emphasis in a first reading:

Chapter II, Subsection II.1.3.
Chapter III, Section III.4.
Chapter IV, Subsection IV.2.3, Sections IV.3 and IV.4.
Chapter V, Section V.2.
Chapter VI, Sections VI.2 and VI.4.

This book has developed from the two courses mentioned above and also from several lectures given to graduate students at the universities of Sherbrooke and São Paulo over a period of several years. It is a pleasure to acknowledge our debt to these students. Their questions, criticisms and suggestions have given us invaluable feedback. We thank in particular Marcia Aguiar, Edson Álvares, Mélissa Barbe-Marcoux, Véronique Bazier-Matte, Guillaume Douville, Marcelo Lanzilotta, Jean-Philippe Morin, Charles Paquette and Sonia Trepode.

We also warmly thank Marion Henry for her precious help in getting our manuscript into shape.

The authors gratefully acknowledge financial support from CNPq and FAPESP, Brazil, and from NSERC, Canada.

Ibrahim Assem — Sherbrooke, QC, Canada
Flávio U. Coelho — São Paulo, São Paulo, Brazil

Contents

I Modules, algebras and quivers ... 1
- I.1 Modules over finite dimensional algebras ... 1
- I.2 Quivers and algebras ... 13

II The radical and almost split sequences ... 41
- II.1 The radical of a module category ... 41
- II.2 Irreducible morphisms and almost split morphisms ... 55
- II.3 The existence of almost split sequences ... 76
- II.4 Factorising radical morphisms ... 89

III Constructing almost split sequences ... 99
- III.1 The Auslander–Reiten translations ... 99
- III.2 The Auslander–Reiten formulae ... 111
- III.3 Examples of constructions of almost split sequences ... 124
- III.4 Almost split sequences over quotient algebras ... 138

IV The Auslander–Reiten quiver of an algebra ... 157
- IV.1 The Auslander–Reiten quiver ... 157
- IV.2 Postprojective and preinjective components ... 192
- IV.3 The depth of a morphism ... 206
- IV.4 Modules over the Kronecker algebra ... 216

V Endomorphism algebras ... 235
- V.1 Projectivisation ... 235
- V.2 Tilting theory ... 246

VI Representation-finite algebras ... 271
- VI.1 The Auslander–Reiten quiver and the radical ... 272
- VI.2 Representation-finiteness using depths ... 278
- VI.3 The Auslander algebra of a representation-finite algebra ... 282
- VI.4 The Four Terms in the Middle theorem ... 298

Bibliography 305
1. Textbooks on noncommutative and homological algebra 305
2. General texts on representations of algebras 305
3. Original papers or surveys related to the contents of the book 306

Index 309

Chapter I
Modules, algebras and quivers

In this book, we assume that the reader has some familiarity with the classical theory of algebras and modules, category theory and homological algebra, such as can be gained from most textbooks in these areas. The first section of this chapter is devoted to recalling, mostly without proofs, some of the fundamental definitions and results from module theory needed later in the book. On the other hand, throughout this book, we shall continuously need to illustrate our results with examples. Therefore, in the second section, we give a concise introduction to the notion of quiver of an algebra, and explain how it can be used to compute examples. We also introduce two classes of algebras that are extensively used later on; namely, Nakayama algebras and hereditary algebras. In this second section, in contrast to the first, complete proofs are given.

I.1 Modules over finite dimensional algebras

I.1.1 Algebras and modules

Because our objects of study are the categories of modules over algebras, it is natural to start by saying what we mean by algebra. Throughout this book, the letter $\mathbf{k}$ denotes a commutative field, and the word "ring" stands for an associative ring with an identity.

A **k-algebra** A is a ring together with a $\mathbf{k}$-vector space structure, in such a way that these structures are compatible, that is, if $a, b \in A$ and $\lambda \in \mathbf{k}$, then

$$a(\lambda b) = \lambda(ab) = (\lambda a)b.$$

I. Assem, F. U. Coelho, *Basic Representation Theory of Algebras*, Graduate Texts in Mathematics 283, https://doi.org/10.1007/978-3-030-35118-2_1

The algebra A is said to be **finite dimensional** if its dimension as a **k**-vector space is finite. Throughout this book, unless otherwise specified, the word algebra always means a finite dimensional **k**-algebra.

Let A, B be algebras. A map $\varphi\colon A \longrightarrow B$ is a **morphism of algebras** if

(a) φ is a morphism of rings; and
(b) φ is a **k**-linear map.

This allows a category to be defined whose objects are the **k**-algebras and whose morphisms are the algebra morphisms.

Example I.1.1. It is easy to verify that the set

$$\begin{pmatrix} \mathbf{k} & 0 & 0 & 0 \\ \mathbf{k} & \mathbf{k} & 0 & 0 \\ \mathbf{k} & 0 & \mathbf{k} & 0 \\ \mathbf{k} & \mathbf{k} & \mathbf{k} & \mathbf{k} \end{pmatrix} = \left\{ \begin{pmatrix} \alpha_{11} & 0 & 0 & 0 \\ \alpha_{21} & \alpha_{22} & 0 & 0 \\ \alpha_{31} & 0 & \alpha_{33} & 0 \\ \alpha_{41} & \alpha_{42} & \alpha_{43} & \alpha_{44} \end{pmatrix} : \alpha_{ij} \in \mathbf{k} \text{ for all } i, j \right\}$$

endowed with the usual addition and multiplication of matrices is a **k**-algebra. This algebra is nine-dimensional.

Example I.1.2. Let V be a finite dimensional **k**-vector space and $\operatorname{End}_{\mathbf{k}} V$ the set of endomorphisms of V, that is, **k**-linear maps from V to V. Then, $\operatorname{End}_{\mathbf{k}} V$, endowed with the usual addition and composition of linear maps, is a **k**-algebra. Its dimension is $(\dim_{\mathbf{k}} V)^2$.

Example I.1.3. Let A be an algebra. We define the **opposite algebra** A^{op} to have as elements those of A, but with the product of $a, b \in A$ defined as follows: $a \times_{op} b = ba$ (where ba denotes the product of b, a in A).

Classically, representing an algebra means to understand its operations by means of matrix operations, and thus to study its structure using properties of matrices. Formally, given an algebra A, a (**k**-linear) **representation** of A is a pair (V, φ), where V is a finite dimensional **k**-vector space and $\varphi\colon A \longrightarrow \operatorname{End}_{\mathbf{k}} V$ is a morphism of algebras. Now, it is easy to see that the data of a representation (V, φ) is equivalent to endowing V with an A-module structure, see Exercise I.1.3.

Definition I.1.4. Let A be a **k**-algebra. A (**right**) A-**module** M is a **k**-vector space M together with a scalar multiplication $M \times A \longrightarrow M$, denoted as $(x, a) \mapsto xa$ (for $x \in M, a \in A$) such that, for every $x, y \in M, a, b \in A$ and $\lambda \in \mathbf{k}$, we have:

(a) $(x + y)a = xa + ya$;
(b) $x(a + b) = xa + xb$;
(c) $x(ab) = (xa)b$;
(d) $(\lambda x)a = \lambda(xa) = x(\lambda a)$;
(e) $x1 = x$, where 1 denotes the identity of A.

The notation M_A indicates that M has an A-module structure (scalar multiplication) on the right. One defines left A-modules similarly. Equivalently, left

A-modules are right modules over the opposite algebra A^{op} of A. Unless otherwise specified, we only deal with right modules.

Given modules M and N, a map $f : M \longrightarrow N$ is called A-**linear**, or a **morphism of** A-**modules**, if the map is $\mathbf{k}$-linear and $f(xa) = f(x)a$ for all $x \in M$, $a \in A$. This defines a category whose objects are the A-modules, and whose morphisms are the A-linear maps.

We denote by $\mathrm{Hom}_A(M, N)$ the $\mathbf{k}$-vector space consisting of all A-linear maps from M to N. If $M = N$, then we write $\mathrm{End}\, M$ or $\mathrm{End}_A\, M$ for this space, which in this case is an algebra, called the **endomorphism algebra** of M.

An A-module M is called **finitely generated** if there exist $d \geq 0$ and an epimorphism $A_A^d \longrightarrow M$. This implies that the images of the vectors of the canonical basis of A^d are generators of M.

Lemma I.1.5. *An A-module is finitely generated if and only if its underlying* $\mathbf{k}$-*vector space is finite dimensional.*

Proof. Suppose that M is a finitely generated A-module. Then there exist $d \geq 0$ and an epimorphism $A_A^d \longrightarrow M$. Because $\dim_{\mathbf{k}} A < \infty$, we have $\dim_{\mathbf{k}} A^d < \infty$ and thus $\dim_{\mathbf{k}} M < \infty$. The converse is obvious. □

We denote by $\mathrm{mod}\, A$ the category whose objects are the finitely generated right A-modules, and whose morphisms are the A-linear maps. Whenever we want to consider finitely generated left A-modules, we consider them to be right modules over A^{op} and denote their category by $\mathrm{mod}\, A^{op}$.

Example I.1.6. Let A be as in Example I.1.1 above and consider the set of row vectors

$$M = (\mathbf{k}\ 0\ \mathbf{k}\ 0) = \{(\lambda\ 0\ \mu\ 0) : \lambda, \mu \in \mathbf{k}\}.$$

It is easily verified that, for every $x \in M$ and $a \in A$, the usual matrix product xa is an element of M. Therefore, M is an A-module, clearly finitely generated, and $\dim_{\mathbf{k}} M = 2$.

I.1.2 The radical and indecomposability

Let A be an algebra and M an A-module. A submodule L of M is **maximal** if $L \neq M$ and, if L' is a proper submodule of M containing L, then $L' = L$. We define the **radical of** M, denoted by $\mathrm{rad}\, M$, to be the intersection of all maximal submodules of M. In particular, $\mathrm{rad}\, A$ is the radical of the module A_A. It can be proved that $\mathrm{rad}\, A$ is a two-sided ideal of A that can be characterised as being the set

$$\{a \in A : 1 - ax \text{ is right invertible, for each } x \in A\}$$

and also the set

$$\{a \in A : 1 - xa \text{ is left invertible, for each } x \in A\}.$$

In addition, for every finitely generated A-module M, we have $\operatorname{rad} M = M \cdot \operatorname{rad} A$.

Because we are considering finite dimensional algebras, we have the following result.

Theorem I.1.7. *Let A be an algebra. Then, its radical is the unique two-sided ideal I of A satisfying the following conditions:*

(a) *I is nil (that is, each element of I is nilpotent); and*
(b) *A/I is a semisimple algebra.* □

Example I.1.8. Let A be the algebra of Example I.1.1 and M the module of Example I.1.6. Consider the set

$$I = \begin{pmatrix} 0 & 0 & 0 & 0 \\ \mathbf{k} & 0 & 0 & 0 \\ \mathbf{k} & 0 & 0 & 0 \\ \mathbf{k} & \mathbf{k} & \mathbf{k} & 0 \end{pmatrix} \subseteq A$$

It is easily verified that I is a two-sided ideal of A. Clearly, I is nil. Because $A/I \cong \mathbf{k}^4$ is semisimple, we get $I = \operatorname{rad} A$. Also,

$$\operatorname{rad} M = M \cdot \operatorname{rad} A = (\mathbf{k}\, 0\, 0\, 0).$$

We have the following useful lemma.

Lemma I.1.9. *Let M, N be modules and $f : M \longrightarrow N$ an epimorphism. Then. $f(\operatorname{rad} M) = \operatorname{rad} N$.*

Proof. Indeed, we have

$$f(\operatorname{rad} M) = f(M \cdot \operatorname{rad} A) = f(M) \cdot \operatorname{rad} A = N \cdot \operatorname{rad} A = \operatorname{rad} N.$$

□

Of particular interest is the following class of algebras.

Definition I.1.10. An algebra is called **local** if it has a unique maximal ideal.

Theorem I.1.11. *The following are equivalent for an algebra A:*

(a) *A is local;*
(b) rad *A is a maximal two-sided ideal;*
(c) *The set of all noninvertible elements of A forms a two-sided ideal;*
(d) *For each $a \in A$, we have that a or $1 - a$ is invertible.* □

If A is local, then the ideal consisting of all noninvertible elements is exactly the radical. Thus, every element of a local algebra is either nilpotent or invertible.

Local algebras yield a criterion allowing us to determine whether or not a given finitely generated module M is indecomposable.

Definition I.1.12. A module M is **indecomposable** if it is nonzero and $M = M' \oplus M''$ implies $M' = 0$ or $M'' = 0$.

Proposition I.1.13. *A finitely generated module M is indecomposable if and only if its endomorphism algebra* End M *is local.* □

The interest in this proposition comes mainly from the following theorem, sometimes called **Unique Decomposition Theorem** and attributed to Remak, Krull, Schmidt and Azumaya.

Theorem I.1.14. *Let M be a finitely generated module. Then:*

(a) *There exists a direct sum decomposition $M = \oplus_{i=1}^m M_i$ with all the M_i indecomposable.*
(b) *This decomposition is unique up to isomorphism: if $M = \oplus_{i=1}^m M_i = \oplus_{j=1}^n N_j$ with the M_i, N_j indecomposable, then $m = n$ and there is a permutation σ of $\{1, \ldots, m\}$ such that $M_i \cong N_{\sigma(i)}$ for all i.* □

Let $f\colon M \longrightarrow N$ be an A-linear map. Decomposing $M = \oplus_{i=1}^m M_i$ and $N = \oplus_{j=1}^n N_j$ with the M_i, N_j indecomposable and letting $q_i\colon M_i \longrightarrow M$ and $p_j\colon N \longrightarrow N_j$ be the injection and the projection associated with these decompositions respectively, we can write f in matrix form as

$$f = (p_j f q_i)_{1 \le i \le m, 1 \le j \le n}$$

with each $p_j f q_i\colon M_i \longrightarrow N_j$ a morphism between indecomposable modules.

For an algebra A, denote by ind A a full subcategory of mod A having as objects a complete set of representatives of the isoclasses (= isomorphism classes) of indecomposable A-modules. Clearly, the subcategory ind A is unique up to equivalence. The above reasoning shows that the category mod A is completely determined by the knowledge of ind A.

Throughout this book, we never distinguish between isomorphic objects; thus, when we speak about "all" modules, we mean all isoclasses of modules.

I.1.3 Idempotents, projectives and injectives

Let A be an algebra. An A-module P is called **projective** if the covariant functor $\mathrm{Hom}_A(P, -) : \mathrm{mod}\, A \longrightarrow \mathrm{mod}\, \mathbf{k}$ is (right) exact, that is, if for every epimorphism $f : M \longrightarrow N$, the induced map $\mathrm{Hom}_A(P, f) : \mathrm{Hom}_A(P, M) \longrightarrow \mathrm{Hom}_A(P, N)$ is surjective. It follows easily from this definition that, if $f : M \longrightarrow P$ is an epimorphism with P projective, then there exists a morphism $g : P \longrightarrow M$ such that $fg = 1_P$. We then say that f is a **retraction** and g is a **section**. A morphism is said to **split** if it is a section or a retraction.

A finitely generated A-module P is projective if and only if it is a direct summand of a finitely generated free A-module. An element $e \in A$ is said to be **idempotent** if $e^2 = e$. Idempotents play an important rôle in understanding algebras and modules. For instance, let us decompose A_A into indecomposable summands

$$A_A = P_1 \oplus \ldots \oplus P_n.$$

In particular, each of the P_i is an indecomposable projective A-module. Then there exists a set of idempotents $\{e_1, \ldots, e_n\}$ such that the e_i are **orthogonal**, that is, $e_i e_j = 0$ for $i \neq j$, **primitive**, that is, $e_i = e_i' + e_i''$ with e_i', e_i'' orthogonal implies $e_i' = 0$ or $e_i'' = 0$, **complete**, that is, $1 = e_1 + e_2 + \ldots + e_n$, and such that $P_i = e_i A$ for each i. Conversely, with every set $\{e_1, \ldots e_n\}$ of complete primitive orthogonal idempotents is associated a decomposition of A_A into indecomposable summands.

Example I.1.15. Let A be as in Example I.1.1. Denoting by e_{ij} the matrix having a coefficient 1 in position (i, j) and 0 in all other positions, it is easy to see that $\{e_{11}, e_{22}, e_{33}, e_{44}\}$ forms a complete set of primitive orthogonal idempotents of A. Thus, the indecomposable projective modules are the $e_{ii}A$, for $i \in \{1, 2, 3, 4\}$. For instance, the module M of Example I.1.6 can be written as $M = e_{33}A$ and, in particular, is indecomposable projective.

The algebra A is called **basic** if, in the above decomposition, $P_i \ncong P_j$ for $i \neq j$. In this case, the P_i form a complete set of representatives of the isoclasses of indecomposable projective A-modules. We can restrict our study to that of basic algebras. Indeed, let A be arbitrary, and P the direct sum of a complete set of representatives of the isoclasses of the indecomposable projective A-modules. Set $B = \operatorname{End} P_A$. Then, B is basic and it follows from the classical Morita theorem that the categories mod A and mod B are equivalent. We may thus assume, from the start, that $A = B$, that is, that A is basic.

Another reduction is possible. A **k**-algebra A is called **connected** if A is nonzero and $A = A_1 \times A_2$ implies $A_1 = 0$ or $A_2 = 0$, that is, A is indecomposable as a ring. Connectedness is characterised by means of idempotents. An idempotent $e \in A$ is called **central** if it belongs to the centre of A, that is, it commutes with every element of A.

For the proof of the next proposition, we refer to Exercise I.1.2.

Proposition I.1.16. *A* **k***-algebra is connected if and only if its only central idempotents are 0 and 1.* □

The following proposition shows that we may, without loss of generality, restrict ourselves to the study of connected algebras.

Proposition I.1.17. *Suppose that* $A = A_1 \times A_2$. *Then the category* mod A *is equivalent to the product category* mod $A_1 \times$ mod A_2. □

One of the most important properties of finite dimensional algebras is the existence of a **duality functor** $\mathrm{D} = \operatorname{Hom}_{\mathbf{k}}(-, \mathbf{k})\colon\ \operatorname{mod} A \longrightarrow \operatorname{mod} A^{op}$. Let M be an A-module, then the underlying **k**-vector space of $\mathrm{D}M = \operatorname{Hom}_{\mathbf{k}}(M, \mathbf{k})$ is the dual space of M and its left A-module structure is defined as follows: if $a \in A$ and $f \in \mathrm{D}M$, then $af \in \mathrm{D}M$ is the linear form defined by

$$(af)(x) = f(xa)$$

for $x \in M$. Similarly, D defines a functor from mod A^{op} to mod A, and it is easily seen that, because we deal with finite dimensional modules, the composition $\mathrm{D}^2 = \mathrm{D} \circ \mathrm{D}$ is functorially isomorphic to the identity functor.

Under the duality D, projective modules correspond to so-called injective modules. An A-module I is called **injective** if the contravariant functor $\mathrm{Hom}_A(-, I)$ is (right) exact, that is, if, for every monomorphism $f : M \longrightarrow N$, the induced map $\mathrm{Hom}_A(f, I) : \mathrm{Hom}_A(N, I) \longrightarrow \mathrm{Hom}_A(M, I)$ is surjective. Thus, an A-module I is injective if and only if its dual $\mathrm{D}I$ is a projective A^{op}-module. If $f : I \longrightarrow M$ is a monomorphism with I injective, then $g : M \longrightarrow I$ exists such that $gf = 1_I$, so that f is a section (and g a retraction).

Let P be an indecomposable projective A-module. Then a primitive idempotent $e \in A$ exists such that $P = eA$. We associate with the idempotent e the A-module $I = \mathrm{D}(Ae)$ (dual of the corresponding indecomposable projective left A-module). Then, I is an indecomposable injective right A-module and every indecomposable injective module is of this form. The modules P and I are related by the fact that the simple module top $P = P/\mathrm{rad}\, P$ of P is isomorphic to the simple socle soc I of I. Then, if $\{e_1 A, \dots, e_n A\}$ is a complete set of representatives of the isoclasses of indecomposable projective A-modules, we get that $\{\mathrm{D}(Ae_1), \dots, \mathrm{D}(Ae_n)\}$ is a complete set of representatives of the isoclasses of the indecomposable injective A-modules. Setting $S_i = e_i A/\mathrm{rad}(e_i A) \cong \mathrm{soc}(\mathrm{D}Ae_i)$, we get that $\{S_1, \dots, S_n\}$ is a complete set of representatives of the isoclasses of simple A-modules.

Further, the indecomposable projective module P_i is a projective cover of S_i. A projective module P is called a **projective cover** of a module M if there exists an epimorphism $f : P \longrightarrow M$ such that, if $f' : P' \longrightarrow M$ is an epimorphism with P' projective, then there exists an epimorphism $g : P' \longrightarrow P$ such that $f' = fg$. The epimorphism p is called a projective cover morphism.

Dually, the indecomposable injective module I_i is an **injective envelope** of S_i, that is, this module verifies the dual property. See Exercise II.2.1 for characterisations of projective covers and injective envelopes.

This correspondence between projectives and injectives is in fact functorial. We define the **Nakayama functor** to be the functor

$$\nu_A = - \otimes_A DA\colon \ \mathrm{mod}\, A \longrightarrow \mathrm{mod}\, A.$$

Clearly, this functor is right exact. Note that $\nu_A \cong \mathrm{D}\,\mathrm{Hom}_A(-, A)$.

Lemma I.1.18. *The Nakayama functor induces an equivalence between the full subcategories of* mod A *consisting of the indecomposable projective and the indecomposable injective modules.*

Proof. Consider the functor $\nu_A^{-1} = \mathrm{Hom}_A(DA, -)\colon \mathrm{mod}A \longrightarrow \mathrm{mod}\, A$, and let $e \in A$ be a primitive idempotent of A. It suffices to prove that, if $P = eA$ and $I = \mathrm{D}(Ae)$, we have $\nu_A P \cong I$ and $\nu_A^{-1} I \cong P$. But this follows from the functorial isomorphisms

$$\nu_A P = eA \otimes_A \mathrm{D}A \cong e(\mathrm{D}A) \cong \mathrm{D}(Ae) = I \quad \text{and}$$

$$\nu_A^{-1} I = \mathrm{Hom}_A(\mathrm{D}A, \mathrm{D}(Ae)) \cong \mathrm{Hom}_{A^{op}}(Ae, A) \cong eA = P.$$

□

The following lemma is extremely useful.

Lemma I.1.19. *Let $e \in A$ be a primitive idempotent, $P = eA$ and $I = \mathrm{D}(Ae)$. For every A-module M, we have isomorphisms of* **k***-vector spaces*

$$Me \cong \mathrm{Hom}_A(P, M) \cong \mathrm{D}\,\mathrm{Hom}_A(M, I).$$

Proof. This follows from the functorial isomorphisms

$$\begin{aligned} \mathrm{D}\,\mathrm{Hom}_A(M, I) &= \mathrm{D}\,\mathrm{Hom}_A(M, \mathrm{D}(Ae)) \cong \mathrm{D}\,\mathrm{Hom}_{A^{op}}(Ae, \mathrm{D}M) \\ &\cong \mathrm{D}(e\mathrm{D}M) \cong (\mathrm{D}^2 M)e \cong Me \\ &\cong \mathrm{Hom}_A(eA, M) = \mathrm{Hom}_A(P, M). \end{aligned}$$

□

I.1.4 The Grothendieck group and composition series

Let A be an algebra and $\mathscr{F}$ the free abelian group having as a basis the set of isoclasses $\widetilde{M}$ of all finitely generated A-modules. Further, let $\mathscr{F}'$ be the subgroup of $\mathscr{F}$ generated by all expressions of the form $\widetilde{L} + \widetilde{N} - \widetilde{M}$, whenever there exists a short exact sequence

$$0 \longrightarrow L \longrightarrow M \longrightarrow N \longrightarrow 0$$

in mod A. The **Grothendieck group** $K_0(A)$ of A is the quotient group $\mathscr{F}/\mathscr{F}'$.

We denote by $[M]$ the image of $\widetilde{M}$ in $K_0(A)$.

Let $\{S_1, \dots, S_n\}$ be a complete set of representatives of the isoclasses of simple A-modules. In this subsection, we prove that $K_0(A)$ is free abelian with basis $\{[S_1], \dots, [S_n]\}$, and thus is isomorphic to $\mathbb{Z}^n$.

A **composition series** of **length** l for an A-module M is a sequence of submodules

$$0 = M_0 \subset M_1 \subset \ldots \subset M_l = M$$

such that, for each i, the quotient M_{i+1}/M_i, called a **composition factor**, is simple. Because M is finite dimensional, such a series always exists, the lengths of all such series are equal, and their common value is called the **composition length** or briefly the **length** of M, denoted as $l(M)$. Also, the number $\mu_i(M)$ of composition factors of M that are isomorphic to some S_i only depends on M and S_i, and not on the particular composition series under consideration. This follows from the classical Jordan–Hölder theorem.

Because finitely generated modules have finite length, we have the following lemma, which generalises a well-known property of finite dimensional vector spaces.

Lemma I.1.20. *Let M be a (finitely generated) A-module and $f \in \operatorname{End} M$. If f is injective or surjective, then f is bijective.* □

We next define a map $\underline{\dim}\colon \mathscr{F} \longrightarrow \mathbb{Z}^n$ by setting, for each module M,

$$\underline{\dim}[M] = (\mu_1(M), \dots, \mu_n(M)).$$

The vector $\underline{\dim}[M]$, which we write simply as $\underline{\dim}\, M$, is called the **dimension vector** of the module M.

Lemma I.1.21. *The map $\underline{\dim}\colon \mathscr{F} \longrightarrow \mathbb{Z}^n$ induces a morphism of groups $\underline{\dim}\colon K_0(A) \longrightarrow \mathbb{Z}^n$.*

Proof. It suffices to show that, if $0 \longrightarrow L \longrightarrow M \longrightarrow N \longrightarrow 0$ is a short exact sequence in $\operatorname{mod} A$, then $\underline{\dim}\, M = \underline{\dim}\, L + \underline{\dim}\, N$, which is equivalent to proving that $\mu_i(M) = \mu_i(L) + \mu_i(N)$ for each i.

We may assume that $L \subseteq M$ and $N = M/L$. Let $0 = L_0 \subset L_1 \subset \dots \subset L_s = L$ and $0 = M_0/L \subset M_1/L \subset \dots \subset M_t/L = M/L$ be composition series for each of L and N respectively. Then,

$$0 = L_0 \subset L_1 \subset \dots \subset L_s = L = M_0 \subset M_1 \subset \dots \subset M_t = M$$

is a composition series for M. The statement follows. □

Theorem I.1.22. *The group $K_0(A)$ is free abelian with basis $\{[S_1], \dots, [S_n]\}$ and the morphism $\underline{\dim}\colon K_0(A) \longrightarrow \mathbb{Z}^n$ is an isomorphism of groups.*

Proof. Let M be an A-module. It follows from the existence of composition series and the definition of $K_0(A)$ that

$$[M] = \sum_{i=1}^{n} \mu_i(M)[S_i].$$

Therefore, $\{[S_1], \dots, [S_n]\}$ is a generating set for the group $K_0(A)$. Because of Lemma I.1.21, the map $\underline{\dim}\colon K_0(A) \longrightarrow \mathbb{Z}^n$ is a morphism of groups. Now, the images of the elements of the generating set $\{[S_1], \dots, [S_n]\}$ are the vectors of the canonical basis of $\mathbb{Z}^n$. Therefore, this generating set is linearly independent, that is, it is a basis of $K_0(A)$. □

There is an important particular case. If A is basic, then the endomorphism algebra $\operatorname{End} S$ of every simple module S is an overfield of $\mathbf{k}$. If $\mathbf{k}$ is algebraically closed, then $\operatorname{End} S = \mathbf{k}$ and $\dim_{\mathbf{k}} S = 1$. In particular, for every A-module M, we have $l(M) = \dim_{\mathbf{k}} M$.

Corollary I.1.23. *Let A be a basic algebra over an algebraically closed field $\mathbf{k}$. Then, for every i with $1 \leq i \leq m$ and every A-module M, we have $\mu_i(M) = \dim_{\mathbf{k}} \operatorname{Hom}_A(P_i, M)$, where P_i is the indecomposable projective A-module such that $P_i / \operatorname{rad} P_i \cong S_i$.*

Proof. For each i, the composition of $\underline{\dim}\colon K_0(A) \longrightarrow \mathbb{Z}^n$ with the ith projection morphism $\mathbb{Z}^n \longrightarrow \mathbb{Z}$ is a morphism of groups mapping $[M] \in K_0(A)$ on $\mu_i(M) \in \mathbb{Z}$. In addition,

$$\mu_i(S_j) = \begin{cases} 0 & i \neq j \\ 1 & i = j. \end{cases}$$

On the other hand, if $0 \longrightarrow L \longrightarrow M \longrightarrow N \longrightarrow 0$ is a short exact sequence, exactness of the functor $\operatorname{Hom}_A(P_i, -)$ yields a short exact sequence

$$0 \longrightarrow \operatorname{Hom}_A(P_i, L) \longrightarrow \operatorname{Hom}_A(P_i, M) \longrightarrow \operatorname{Hom}_A(P_i, N) \longrightarrow 0$$

so that $\dim_{\mathbf{k}} \operatorname{Hom}_A(P_i, -)\colon K_0(A) \longrightarrow \mathbb{Z}$ is also a morphism of groups. Because each simple module is one-dimensional (see above), we have

$$\dim_{\mathbf{k}} \operatorname{Hom}_A(P_i, S_j) = \begin{cases} 0 & i \neq j \\ 1 & i = j. \end{cases}$$

The morphisms μ_i and $\dim_{\mathbf{k}} \operatorname{Hom}_A(P_i, -)$ thus coincide on a basis of the free abelian group $K_0(A)$. Therefore, they are equal. □

Example I.1.24. Let A be as in Example I.1.1. Because $\{e_{11}, e_{22}, e_{33}, e_{44}\}$ is a complete set of primitive orthogonal idempotents, we have four isoclasses of simple modules, so that $K_0(A) \cong \mathbb{Z}^4$. Also, let $M = e_{33}A$ be the indecomposable projective module of Example I.1.6. Its simple top is $S_3 = e_{33}A / \operatorname{rad}(e_{33}A)$. Now, $\operatorname{rad}(e_{33}A)$ is one-dimensional and thus simple, and it is easily seen that right multiplication by the matrix e_{13} yields an isomorphism $S_1 \longrightarrow \operatorname{rad}(e_{33}A)$. Therefore, the (unique) composition series for M is $0 \subset S_1 \subset M$, and $\underline{\dim}\, M = (1, 0, 1, 0)$. In particular, we have $l(M) = \dim_{\mathbf{k}} M = 2$.

Exercises for Section I.1

Exercise I.1.1. Let A be a $\mathbf{k}$-algebra. Prove that the map $\varphi : \mathbf{k} \longrightarrow A$ given by $\lambda \mapsto 1 \cdot \lambda$ (for $\lambda \in \mathbf{k}$) is an injective morphism of rings whose image is contained in the centre of A.

Exercise I.1.2. Let A be a finite dimensional $\mathbf{k}$-algebra and $e \in A$ a central idempotent. Prove the following statements:

(a) eA has a natural algebra structure, with identity e, and
(b) We have an algebra isomorphism $A \cong eA \times (1-e)A$.

Deduce Proposition I.1.16: an algebra is connected if and only if 0 and 1 are its only central idempotents.

Exercise I.1.3. Let A be a finite dimensional $\mathbf{k}$-algebra and $(V, \varphi), (W, \psi)$ be representations of A. A morphism from (V, φ) to (W, ψ) is a $\mathbf{k}$-linear map $f : V \longrightarrow W$ such that $f\varphi(a) = \psi(a)f$ for every $a \in A$. The composition of morphisms is the ordinary composition of $\mathbf{k}$-linear maps. Prove that the resulting category of representations is equivalent to the category $\operatorname{Mod} A$ of all A-modules (not necessarily finitely generated).

Exercise I.1.4. Let A be a finite dimensional $\mathbf{k}$-algebra and $0 \longrightarrow L \longrightarrow M \longrightarrow N \longrightarrow 0$ be an exact sequence of A-modules. Prove that M is finitely generated if and only if L and N are finitely generated.

Exercise I.1.5. Let M_1, M_2, M_3 be submodules of an A-module M. Prove that, if $M_1 \subseteq M_2$, then $M_2 \cap (M_1 + M_3) = M_1 + (M_2 \cap M_3)$. This is the so-called **modular law**.

Exercise I.1.6 (Nakayama's lemma). Let M be a finitely generated A-module, and N a submodule of M. Prove that $N \subseteq \operatorname{rad} M$ if and only if, for every submodule L of M such that $N + L = M$, we have $L = M$.

Exercise I.1.7. Let $A = \mathbf{k}[t]/\langle t^n \rangle$, where $n \geq 1$, $\mathbf{k}[t]$ is the polynomial algebra in one indeterminate t and $\langle t^n \rangle$ is the ideal of $\mathbf{k}[t]$ generated by the nth power t^n. Compute the radical of A and show that A is local.

Exercise I.1.8. Let $e \in A$ be an idempotent. Prove that e is primitive if and only if the algebra eAe is local.

Exercise I.1.9. Let $e \in A$ be an idempotent. Prove that $\operatorname{rad}(eA) = e \cdot \operatorname{rad} A$.

Exercise I.1.10. Let $e_1, e_2 \in A$ be primitive idempotents such that $P_1 = e_1 A$ is not isomorphic to $P_2 = e_2 A$. Prove that $\operatorname{Hom}_A(P_1, P_2) \neq 0$ if and only if $e_2(\operatorname{rad} A)e_1 \neq 0$.

Exercise I.1.11. Prove that an algebra A is local if and only if its radical is equal to the set $\{x \in A : xA \neq A\}$.

Exercise I.1.12. Let $1 = e_1 + \cdots + e_m = f_1 + \cdots + f_n$ be decompositions of the identity 1 of A into primitive orthogonal idempotents. Prove that $m = n$ and that there exists $a \in A$ invertible such that, up to permutation, we have, $f_i = a^{-1}e_i a$, for every i.

Exercise I.1.13. Let A be an algebra, and I an ideal of A. Prove that $\operatorname{rad}(A/I) = (\operatorname{rad} A + I)/I$.

Exercise I.1.14. Let A be an algebra.

(a) Let I, J be ideals of A. Prove that

$$IJ = \{\sum x_i y_i | x_i \in I, y_i \in J\}$$

is an ideal of A contained in $I \cap J$. Prove that, in general, $IJ \neq I \cap J$.

(b) For $n \geq 1$, we define $\operatorname{rad}^{n+1} A$ to be the radical of the module $\operatorname{rad}^n A$. Prove that $\operatorname{rad}^{n+1} A = \operatorname{rad}^n A \cdot \operatorname{rad} A$. Deduce that each $\operatorname{rad}^n A$ is a nilpotent ideal of A.

(c) Let $A = \bigoplus_{i=1}^t P_i$ be a decomposition of A into indecomposable projective A-modules. Prove that, for each $n \geq 1$, we have a decomposition of $A/\operatorname{rad}^n A$ into indecomposable projective A-modules : $A/\operatorname{rad}^n A = \bigoplus_{i=1}^t (P_i/\operatorname{rad}^n P_i)$.

Exercise I.1.15. Let A be the lower triangular matrix algebra

$$A = \begin{pmatrix} \mathbf{k} & 0 & 0 & 0 \\ 0 & \mathbf{k} & 0 & 0 \\ 0 & 0 & \mathbf{k} & 0 \\ \mathbf{k} & \mathbf{k} & \mathbf{k} & \mathbf{k} \end{pmatrix} = \left\{ \begin{pmatrix} \alpha_{11} & 0 & 0 & 0 \\ 0 & \alpha_{22} & 0 & 0 \\ 0 & 0 & \alpha_{33} & 0 \\ \alpha_{41} & \alpha_{42} & \alpha_{43} & \alpha_{44} \end{pmatrix} : \alpha_{ij} \in \mathbf{k} \text{ for all } i, j \right\}$$

with the usual matrix operations. Compute $\operatorname{rad} A$ and give a complete set of representatives of the isoclasses of indecomposable projective and indecomposable injective modules. Compute the dimension vector of each of these modules.

Exercise I.1.16. Let A be the lower triangular matrix algebra

$$A = \begin{pmatrix} \mathbf{k} & 0 & 0 & 0 \\ \mathbf{k} & \mathbf{k} & 0 & 0 \\ \mathbf{k} & \mathbf{k} & \mathbf{k} & 0 \\ \mathbf{k} & \mathbf{k} & \mathbf{k} & \mathbf{k} \end{pmatrix} = \left\{ \begin{pmatrix} \alpha_{11} & 0 & 0 & 0 \\ \alpha_{21} & \alpha_{22} & 0 & 0 \\ \alpha_{31} & \alpha_{32} & \alpha_{33} & 0 \\ \alpha_{41} & \alpha_{42} & \alpha_{43} & \alpha_{44} \end{pmatrix} : \alpha_{ij} \in \mathbf{k} \text{ for all } i, j \right\}$$

and e_{ij} denote the matrix having coefficient 1 in position (i, j), and 0 elsewhere. Prove that

$$e_{ii} \left(\frac{\operatorname{rad} A}{\operatorname{rad}^2 A} \right) e_{jj} = \begin{cases} k \text{ if } (i, j) \in \{(4, 3), (3, 2), (2, 1)\} \\ 0 \text{ otherwise.} \end{cases}$$

I.2 Quivers and algebras

I.2.1 Path algebras and their quotients

Quivers provide an unlimited source of examples of all levels of difficulty. A quiver is a graphical object in which one can encode much of the structural information of an algebra. Viewing an algebra as a quiver (with some relations) not only allows us to visualise the properties of the algebra itself but also gives concrete descriptions of its modules. It is, for instance, particularly easy to describe the simple, projective and injective modules. In this section, we review those properties of quivers that are needed in the sequel. In contrast to Section I.1, all relevant proofs are given.

Definition I.2.1. A **quiver** $Q = (Q_0, Q_1, s, t)$ is a quadruple consisting of two sets: Q_0, whose elements are called **points**, and Q_1, whose elements are called **arrows**, as well as two maps $s, t\colon Q_1 \longrightarrow Q_0$, which associate with each arrow $\alpha \in Q_1$ the points $s(\alpha), t(\alpha) \in Q_0$, called its **source** and its **target** respectively.

An arrow α of source x and target y is denoted by $x \xrightarrow{\alpha} y$ or $\alpha\colon x \longrightarrow y$. The quiver itself is denoted briefly as $Q = (Q_0, Q_1)$ or simply Q. A quiver $Q = (Q_0, Q_1)$ is called **finite** if both Q_0 and Q_1 are finite sets.

A **subquiver** $Q' = (Q'_0, Q'_1, s', t')$ of $Q = (Q_0, Q_1, s, t)$ is a quiver such that $Q'_0 \subseteq Q_0$, $Q'_1 \subseteq Q_1$ and the restrictions of s and t are s' and t' respectively, that is, $s|_{Q'_1} = s'$, $t|_{Q'_1} = t'$ (in other words, if $\alpha\colon x \longrightarrow y$ belongs to Q'_1, then $s'(\alpha) = s(\alpha)$ and $t'(\alpha) = t(\alpha)$). The subquiver Q' is called **full** if $Q'_1 = \{\alpha \in Q_1 \colon s(\alpha), t(\alpha) \in Q'_0\}$. Thus, a full subquiver is completely determined by its set of points.

Let $Q = (Q_0, Q_1, s, t)$ be a quiver. A **path** $\alpha_1\alpha_2\ldots\alpha_l$ of **source** x, **target** y and **length** l in Q is a sequence of l arrows such that $s(\alpha_1) = x$, $t(\alpha_l) = y$ and $t(\alpha_i) = s(\alpha_{i+1})$ for all i such that $1 \leq i < l$. Such a path is represented as:

$$x = x_1 \xrightarrow{\alpha_1} x_2 \xrightarrow{\alpha_2} x_3 \longrightarrow \ldots \xrightarrow{\alpha_l} x_{l+1} = y.$$

We agree to associate with each point $x \in Q_0$ a path ϵ_x of length zero, from x to x, and call it the **stationary** or **trivial** path at x. A quiver Q is called **acyclic** if there is no path in Q of length at least one from one of its points to itself (called **cycle**). Cycles of length exactly one are called **loops**.

With every arrow $\alpha \in Q_1$, we associate a formal inverse α^{-1}, with source $t(\alpha)$ and target $s(\alpha)$. A **walk** in Q of length $l \geq 0$ is a sequence

$$\alpha_1^{\nu_1}\alpha_2^{\nu_2}\ldots\alpha_l^{\nu_l}$$

where for each i, we have $\alpha_i \in Q_1$ and $\nu_i \in \{+1, -1\}$ and for each $i < l$, we have $\{s(\alpha_i), t(\alpha_i)\} \cap \{s(\alpha_{i+1}), t(\alpha_{i+1})\} \neq \varnothing$. A quiver Q is **connected** if for every $x, y \in Q_0$, there exists a walk $\alpha_1^{\nu_1}\alpha_2^{\nu_2}\ldots\alpha_l^{\nu_l}$ such that $x = s(\alpha_1^{\nu_1})$ and $y = t(\alpha_l^{\nu_l})$.

One sometimes calls **underlying graph** of a quiver the structure obtained from it by forgetting the orientation of the arrows (even if such a structure is generally not, strictly speaking, a graph). In this terminology, a quiver is connected if and only if its underlying graph is connected too.

Example I.2.2. Here is an example of a quiver

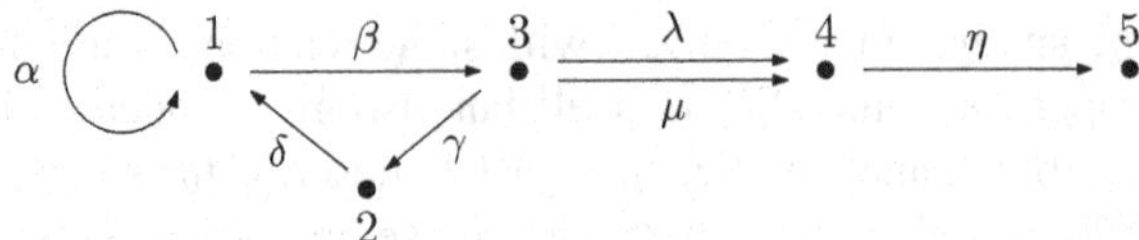

One can see that this quiver contains a path $\beta\gamma\delta\alpha^3\beta\mu$ from 1 to 4 of length 8. The quiver also contains cycles of arbitrary length from 1 to 1, such as the α^i for all $i \geq 1$.

The usual composition of paths in a quiver can be used to define an algebraic structure.

Definition I.2.3. Let Q be a quiver. The **path algebra** $\mathbf{k}Q$ of Q is defined as follows. The underlying $\mathbf{k}$-vector space $\mathbf{k}Q$ has as a basis the set of all paths in Q, including the stationary ones. The product of the basis elements $\alpha_1 \dots \alpha_l$ and $\beta_1 \dots \beta_m$ is defined by:

$$(\alpha_1 \dots \alpha_l)(\beta_1 \dots \beta_m) = \begin{cases} \alpha_1 \dots \alpha_l \beta_1 \dots \beta_m & \text{if } t(\alpha_l) = s(\beta_1) \\ 0 & \text{otherwise.} \end{cases}$$

The product is then extended by distributivity to the whole of $\mathbf{k}Q$.

This defines an associative algebra, in which each stationary path ϵ_x, with $x \in Q_0$, is an idempotent and thus, if Q_0 is finite, then $\sum_{x \in Q_0} \epsilon_x$ is the identity. However, $\mathbf{k}Q$ can be infinite dimensional, as shown in the example below.

Example I.2.4. Let A be the path algebra of the quiver

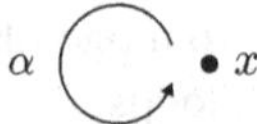

Then A admits as a basis the unique stationary path ϵ_x (which is therefore the identity of A) and all the cycles α^i through x, with $i \geq 1$. Its elements are thus linear combinations of the α^i with coefficients in $\mathbf{k}$ and multiplication is induced from the rule $\alpha^i \alpha^j = \alpha^{i+j}$. Therefore, A is isomorphic to the algebra of polynomials in one indeterminate $\mathbf{k}[t]$.

This example shows that the existence of an oriented cycle in a quiver implies that the path algebra is infinite dimensional. Because, in this book, we are only interested in finite dimensional algebras, we deduce that, starting from the path algebra of a quiver, we need to "kill" all cycles or, more generally, paths that are long enough. We need a notation. In a path algebra $\mathbf{k}Q$, let $\mathbf{k}Q^+$ be the ideal generated by the arrows. Then, $\mathbf{k}Q^+$ contains all paths of positive length. For $m \geq 2$, $\mathbf{k}Q^{+m} = (\mathbf{k}Q^+)^m$ denotes the mth power of $\mathbf{k}Q^+$, that is, the ideal generated by the paths of length m. It contains all paths of length greater than, or equal to m.

Definition I.2.5. Let Q be a finite quiver. An ideal I of $\mathbf{k}Q$ is called **admissible** if there exists $m \geq 2$ such that

$$\mathbf{k}Q^{+m} \subseteq I \subseteq \mathbf{k}Q^{+2}.$$

In this case, the pair (Q, I) is called a **bound quiver** and the algebra $A = \mathbf{k}Q/I$ a **bound quiver algebra**.

The adjective "bound" refers to the verb "to bind".

The finiteness of Q above ensures that $\mathbf{k}Q$ has an identity $\sum\limits_{x \in Q_0} \epsilon_x$. The condition saying that $I \subseteq \mathbf{k}Q^{+2}$ says that I only contains linear combinations of paths of length at least two. Finally, saying that there exists $m \geq 2$ such that $\mathbf{k}Q^{+m} \subseteq I$ amounts to saying that every path of length larger than or equal to m (that is, every path long enough) is contained in I.

Example I.2.6. Let Q be the quiver

$$\beta \circlearrowleft \overset{1}{\bullet} \xleftarrow{\ \alpha\ } \overset{2}{\bullet}$$

The ideal $< \beta^3, \alpha\beta >$ generated by β^3 and $\alpha\beta$ is admissible: indeed, one sees easily that $\mathbf{k}Q^{+3} \subseteq I \subseteq \mathbf{k}Q^{+2}$. On the other hand, neither of the ideals $< \beta >$ nor $< \alpha\beta >$ is admissible: indeed, $< \beta > \not\subseteq \mathbf{k}Q^{+2}$, whereas $\beta^m \notin < \alpha\beta >$ for every $m \geq 2$.

The conditions defining admissibility ensure that the quotient of a path algebra by an admissible ideal is finite dimensional. Actually, we have the following result.

Proposition I.2.7. *Let Q be a finite connected quiver and I an admissible ideal of $\mathbf{k}Q$. Then $A = \mathbf{k}Q/I$ is a basic and connected finite dimensional algebra having $\{e_x = \epsilon_x + I : x \in Q_0\}$ as a complete set of primitive orthogonal idempotents. In addition,* $\operatorname{rad} A = \mathbf{k}Q^+/I$.

Proof. Because Q is finite, the path algebra $\mathbf{k}Q$ is an associative algebra having as identity $1 = \sum_{x \in Q_0} \epsilon_x$. Therefore, A is also associative with identity $1 = \sum_{x \in Q_0} e_x$, where $e_x = \epsilon_x + I$. We claim that A is finite dimensional.

By hypothesis, there exists $m \geq 2$ such that $\mathbf{k}Q^{+m} \subseteq I$. Hence, there exists a surjective morphism of algebras $\mathbf{k}Q/\mathbf{k}Q^{+m} \longrightarrow \mathbf{k}Q/I = A$. Now, $\mathbf{k}Q/\mathbf{k}Q^{+m}$ is spanned as a vector space by (the residual classes of) all paths in the finite quiver Q of length strictly less than m. It is thus finite dimensional. Hence, so is A.

We now prove that $\{e_x : x \in Q_0\}$ forms a complete set of primitive orthogonal idempotents. Clearly, in $\mathbf{k}Q$, $\{\epsilon_x : x \in Q_0\}$ forms a set of orthogonal idempotents. Therefore, in A, $\{e_x : x \in Q_0\}$ also forms a set of orthogonal idempotents. Because $1 = \sum_{x \in Q_0} e_x$, it only remains to show that the e_x are primitive. Now, an idempotent $e \in e_x A e_x$ can be written in the form $e = \lambda\epsilon_x + w + I$ where $\lambda \in \mathbf{k}$, and w is a linear combination of cycles through x. Then, $e^2 = e$ yields $(\lambda^2 - \lambda)\epsilon_x + (2\lambda - 1)w + w^2 \in I$. Because $I \subseteq \mathbf{k}Q^{+2}$, we must have $\lambda^2 - \lambda = 0$ and therefore $\lambda = 0$ or $\lambda = 1$. Assume first that $\lambda = 0$, then $e = w + I$ idempotent gives $w^i - w \in I$ for all i. However, there exists $m \geq 2$ such that $\mathbf{k}Q^{+m} \subseteq I$ and so $w^m \in I$. Then, $w^m - w \in I$ implies $w \in I$ and so $e = 0 + I$ is the zero of $e_x A e_x$. Similarly, if $\lambda = 1$, we get that $e = e_x$.

We next prove that A is connected. If this is not the case, then, because of Proposition I.1.16, A contains a central idempotent $c \neq 0, 1$. We have

$$c = 1 \cdot c \cdot 1 = \left(\sum_{x \in Q_0} e_x\right) \cdot c \cdot \left(\sum_{y \in Q_0} e_y\right) = \sum_{x, y \in Q_0} e_x c e_y = \sum_{x \in Q_0} e_x c,$$

using that the e_x are orthogonal idempotents and that c is central. Because e_x is primitive and $e_x c$ is an idempotent in $e_x A e_x$, we have either $e_x c = e_x$ or $e_x c = 0$. Let $Q'_0 = \{x \in Q_0 \colon e_x c = 0\}$ and $Q''_0 = \{x \in Q_0 \colon e_x c = e_x\}$. Because $c \neq 0, 1$, both Q'_0 and Q''_0 are nonempty, and $Q_0 = Q'_0 \cup Q''_0$ and $Q'_0 \cap Q''_0 = \emptyset$. Because Q is connected, there exist $x \in Q'_0$ and $y \in Q''_0$, which are neighbours in Q. We may even assume without loss of generality that there exists an arrow $\alpha \colon x \longrightarrow y$. But then $e_x A e_y = e_x A e_y c = e_x c A e_y = 0$, whereas $0 \neq \alpha + I = \epsilon_x \alpha \epsilon_y + I = e_x(\alpha + I)e_y \in e_x A e_y$, a contradiction. Thus, A is connected.

To compute the radical of A, we first observe that $\mathbf{k}Q^{+m} \subseteq I$, with m as above, implies that $(\mathbf{k}Q^+/I)^m = 0$ in A, that is, the ideal $\mathbf{k}Q^+/I$ of A is nilpotent (hence nil). In addition,

$$\frac{A}{(\mathbf{k}Q^+/I)} = \frac{(\mathbf{k}Q/I)}{(\mathbf{k}Q^+/I)} \cong \frac{\mathbf{k}Q}{\mathbf{k}Q^+} \cong \mathbf{k}^{|Q_0|}$$

is a product of copies of $\mathbf{k}$. Therefore, A is basic with radical equal to $\mathbf{k}Q^+/I$. □

For instance, if Q is a finite acyclic quiver, then the lengths of paths in Q are bounded; hence, every ideal $I \subseteq \mathbf{k}Q^{+2}$ is admissible. In particular, the zero ideal is admissible and so the path algebra $\mathbf{k}Q$ itself is a basic and connected finite dimensional algebra with radical $\mathbf{k}Q^+$.

Admissible ideals are most commonly defined by means of their generators. A **relation** in a quiver Q is a linear combination of paths of length at least two, all these paths having the same source and the same target. Thus, a relation from x to y in Q is an element of $\mathbf{k}Q$ of the form

$$\rho = \sum_{i=1}^{k} \lambda_i w_i$$

where the λ_i are nonzero scalars, and the w_i paths of length at least two from x to y, say. Thus, every relation is an element of $\mathbf{k}Q^{+2}$. If $k = 1$, then ρ is called a **zero-relation**. If it is of the form $\rho = w_1 - w_2$ (where w_1, w_2 are paths with same end points), then it is a **commutativity relation**.

Proposition I.2.8. *Let Q be a finite connected quiver and I an admissible ideal of $\mathbf{k}Q$. Then there exists a finite set of relations in $\mathbf{k}Q^{+2}$, which generates I as an ideal.*

Proof. We first show that I is finitely generated as an ideal of $\mathbf{k}Q$. Let $m \geq 2$ be such that $\mathbf{k}Q^{+m} \subseteq I$. We have a short exact sequence of $\mathbf{k}Q$-modules

$$0 \longrightarrow \mathbf{k}Q^{+m} \longrightarrow I \longrightarrow \frac{I}{\mathbf{k}Q^{+m}} \longrightarrow 0.$$

Now, $\mathbf{k}Q^{+m}$ is generated, as an ideal, by the paths of length exactly m (thus, it contains all paths of length greater than or equal to m). Because Q is a finite quiver, $\mathbf{k}Q^{+m}$ is thus finitely generated. On the other hand, $I/\mathbf{k}Q^{+m}$ is contained in the finite dimensional vector space $\mathbf{k}Q/\mathbf{k}Q^{+m}$; hence, it is finitely generated. Therefore, I itself is finitely generated.

Let $\{\sigma_1, \ldots, \sigma_t\}$ be a finite set of generators for I. The σ_i are generally not relations. Consider the set

$$\{\epsilon_x \sigma_i \epsilon_y \colon 1 \leq i \leq t, x, y \in Q_0\}.$$

It is finite, because Q is finite, and its nonzero elements are relations that generate I as an ideal because, for each i, we have $\sigma_i = \sum_{x,y \in Q_0} \epsilon_x \sigma_i \epsilon_y$. □

If $A = \mathbf{k}Q/I$ is a bound quiver algebra, where I is generated by the finite set $\{\rho_1, \ldots, \rho_t\}$ of relations, then we say that A is given by the quiver Q bound by the relations $\rho_1 = 0, \ldots, \rho_t = 0$.

I.2.2 Quiver of a finite dimensional algebra

We have seen in Subsection I.2.1 that, given a quiver Q and an admissible ideal I, one can consider the bound quiver algebra $A = \mathbf{k}Q/I$. Conversely, starting with a basic and connected algebra A satisfying an extra condition, one can find a quiver Q_A and an admissible ideal I of $\mathbf{k}Q_A$ such that $A \cong \mathbf{k}Q_A/I$. To see the necessity of an extra condition on A, we recall that, because A is basic, $A/\operatorname{rad} A$ is a product of fields, generally noncommutative. However, as seen in the proof of Proposition I.2.7, if A is a bound quiver algebra, then $A/\operatorname{rad} A$ must be a product of copies of $\mathbf{k}$.

We therefore define an algebra A to be **elementary** if $A/\operatorname{rad} A$ is isomorphic to a product of copies of $\mathbf{k}$. For example, if A is a basic algebra over an algebraically closed field $\mathbf{k}$, then it is automatically elementary: indeed, in this case, $A/\operatorname{rad} A \cong \prod_{i=1}^{n} \mathbf{k}_i$, where the $\mathbf{k}_i$ are (skew) fields that are finite dimensional extensions of $\mathbf{k}$, and $\mathbf{k}$, being algebraically closed, yields that $\mathbf{k}_i = \mathbf{k}$ for each i.

Definition I.2.9. Let A be an elementary algebra, and $\{e_1, \dots, e_n\}$ a complete set of primitive orthogonal idempotents. The **ordinary quiver** or simply **quiver** Q_A of A is defined as follows:

(a) The points $\{1, \dots, n\}$ of Q_A are in bijection with the idempotents $\{e_1, \dots, e_n\}$.
(b) If $x, y \in (Q_A)_0$, then the arrows from x to y are in bijection with the vectors in a basis of the $\mathbf{k}$-vector space

$$e_x \left(\frac{\operatorname{rad} A}{\operatorname{rad}^2 A} \right) e_y.$$

In particular, Q_A is a finite quiver, because A is finite dimensional. We have to show that Q_A is well-defined. Because, for $x, y \in (Q_A)_0$, the number of arrows from x to y depends on the $\mathbf{k}$-dimension of $e_x \left(\frac{\operatorname{rad} A}{\operatorname{rad}^2 A} \right) e_y$ and not on the chosen basis, we only have to prove that this vector space does not depend on the choice of the idempotents.

Lemma I.2.10. *The quiver Q_A does not depend on the choice of a particular complete set of primitive orthogonal idempotents for A.*

Proof. Let $\{e_1, \dots, e_n\}$ and $\{f_1, \dots, f_m\}$ be complete sets of primitive orthogonal idempotents for A. Because

$$A_A = \oplus_{i=1}^{n} (e_i A) = \oplus_{j=1}^{m} (f_j A)$$

and the e_i, f_j are primitive, we get from Theorem I.1.14 that $m = n$ and, up to a permutation, $e_i A \cong f_i A$ for every i. Now, for every i, j, we have isomorphisms of $\mathbf{k}$-vector spaces:

$$\begin{aligned} e_i \left(\frac{\operatorname{rad} A}{\operatorname{rad}^2 A} \right) e_j &\cong \left(\frac{e_i (\operatorname{rad} A)}{e_i (\operatorname{rad}^2 A)} \right) e_j \\ &\cong \left(\frac{\operatorname{rad}(e_i A)}{\operatorname{rad}^2 (e_i A)} \right) e_j \\ &\cong \operatorname{Hom}_A \left(e_j A, \frac{\operatorname{rad}(e_i A)}{\operatorname{rad}^2 (e_i A)} \right). \end{aligned}$$

Then, $e_i A \cong f_i A$ and $e_j A \cong f_j A$ yield an isomorphism of $\mathbf{k}$-vector spaces

$$e_i \left(\frac{\operatorname{rad} A}{\operatorname{rad}^2 A} \right) e_j \cong f_i \left(\frac{\operatorname{rad} A}{\operatorname{rad}^2 A} \right) f_j,$$

as required. □

We next have to show that the definition of Q_A is coherent with that of the corresponding bound quiver algebra.

Lemma I.2.11. *Let $A = \mathbf{k}Q/I$ be a bound quiver algebra, then $Q_A = Q$.*

Proof. Because of Proposition I.2.7, we have that $\{e_x : x \in Q_0\}$ is a complete set of primitive orthogonal idempotents of A and also $\operatorname{rad} A = \mathbf{k}Q^+/I$, which implies

$$\frac{\operatorname{rad} A}{\operatorname{rad}^2 A} = \frac{(\mathbf{k}Q^+/I)}{(\mathbf{k}Q^{+2}/I)} \cong \frac{\mathbf{k}Q^+}{\mathbf{k}Q^{+2}}.$$

The statement follows from the fact that, if $x, y \in Q_0$, then the vector space $e_x\left(\mathbf{k}Q^+/\mathbf{k}Q^{+2}\right)e_y$ has as a basis all paths of length exactly one (that is, all arrows) from x to y. □

We now wish to prove that every elementary finite dimensional algebra is a bound quiver algebra. For this purpose, we start by lifting the arrows to radical elements. Indeed, let x, y be points in Q_A, and $\{\alpha_1, \ldots, \alpha_m\}$ all the arrows from x to y. Then, there exists a set $\{a_{\alpha_1}, \ldots, a_{\alpha_m}\}$ of elements of $e_x(\operatorname{rad} A)e_y$ whose residual classes $\{a_{\alpha_1} + \operatorname{rad}^2 A, \ldots, a_{\alpha_m} + \operatorname{rad}^2 A\}$ form a basis of the space $e_x\left(\operatorname{rad} A/\operatorname{rad}^2 A\right)e_y$.

Lemma I.2.12. *With this notation, the vector space $e_x(\operatorname{rad} A)e_y$ is generated by all products of the form $a_{\alpha_1}\ldots a_{\alpha_l}$, where $\alpha_1\ldots\alpha_l$ is a path from x to y.*

Proof. As a $\mathbf{k}$-vector space, we have

$$e_x(\operatorname{rad} A)e_y = e_x\left(\frac{\operatorname{rad} A}{\operatorname{rad}^2 A}\right)e_y \oplus e_x(\operatorname{rad}^2 A)e_y.$$

Therefore, to every $a \in e_x(\operatorname{rad} A)e_y$ corresponds a linear combination $\sum_{\alpha : x \to y} a_\alpha \lambda_\alpha$, with $\lambda_\alpha \in \mathbf{k}$ such that $a - \sum_{\alpha : x \to y} a_\alpha \lambda_\alpha$ belongs to $e_x(\operatorname{rad}^2 A)e_y$. But now,

$$e_x(\operatorname{rad}^2 A)e_y = \sum_{z \in (Q_A)_0} (e_x(\operatorname{rad} A)e_z) \cdot (e_z(\operatorname{rad} A)e_y).$$

Repeating the above reasoning, we get linear combinations $\sum_{\beta : x \to z} a_\beta \lambda_\beta$ and $\sum_{\gamma : z \to y} a_\gamma \lambda_\gamma$ such that

$$a - \sum_{\alpha : x \to y} a_\alpha \lambda_\alpha - \left(\sum_{\beta : x \to z} a_\beta \lambda_\beta\right)\left(\sum_{\gamma : z \to y} a_\gamma \lambda_\gamma\right) = a - \sum_{\alpha : x \to y} a_\alpha \lambda_\alpha - \sum_{\beta\gamma : x \to y \to z} a_\beta a_\gamma (\lambda_\beta \lambda_\gamma)$$

belongs to $e_x(\operatorname{rad}^3 A)e_y$. Applying induction and using that rad A is nilpotent, we can write a as a linear combination of products of the required form. □

We now prove the main theorem of this subsection.

Theorem I.2.13. *Let A be an elementary finite dimensional* **k***-algebra. Then, there exists a surjective algebra morphism $\varphi\colon \mathbf{k}Q_A \longrightarrow A$ with admissible kernel I. In particular, $A \cong \mathbf{k}Q_A/I$ is a bound quiver algebra.*

Proof. Let a_α be as above and define a morphism of algebras $\varphi\colon \mathbf{k}Q_A \longrightarrow A$ as follows. First, set

$$\begin{aligned} \varphi(\epsilon_x) &= e_x \quad \text{for each } x \in (Q_A)_0, \text{ and} \\ \varphi(\alpha) &= a_\alpha \quad \text{for each } \alpha \in (Q_A)_1. \end{aligned}$$

This defines φ on points and arrows only. To define it on an arbitrary basis vector of $\mathbf{k}Q_A$, that is, a path, we extend this definition by setting

$$\varphi(\alpha_1 \dots \alpha_l) = \varphi(\alpha_1)\dots\varphi(\alpha_l) = a_{\alpha_1}\dots a_{\alpha_l}$$

for each path $\alpha_1 \dots \alpha_l$ in Q_A. Then, φ extends to a **k**-linear map $\mathbf{k}Q_A \longrightarrow A$, which preserves the product of basis vectors, and hence, of every vectors. In addition,

$$\varphi(1) = \varphi\left(\sum_{x\in(Q_A)_0} \epsilon_x\right) = \sum_{x\in(Q_A)_0} \varphi(\epsilon_x) = \sum_{x\in(Q_A)_0} e_x = 1$$

so it preserves the identity as well. Therefore, φ is a morphism of algebras.

Because A is elementary, the elements e_x generate $A/\operatorname{rad} A$ and, because of Lemma I.2.12, the products of the elements a_α generate rad$A$ as a **k**-vector space. Therefore, the set $\{e_x, a_\alpha : x \in (Q_A)_0, \alpha \in (Q_A)_1\}$ generates A as a **k**-algebra. Because these elements lie in the image of φ, we conclude that φ is surjective.

We now prove that $I = \operatorname{Ker}\varphi$ is admissible. Because of the construction of φ, we have $\varphi(\mathbf{k}Q_A^+) \subseteq \operatorname{rad} A$. By induction, we get $\varphi(\mathbf{k}Q_A^{+i}) \subseteq \operatorname{rad}^i A$ for every $i \geq 1$. Because rad$A$ is nilpotent, there exists $m \geq 2$ such that $\varphi(\mathbf{k}Q_A^{+m}) = 0$, that is, $\mathbf{k}Q_A^{+m} \subseteq I$. It remains to show that $I \subseteq \mathbf{k}Q_A^{+2}$. Let $a \in I$. Then there exist linear combinations $\sum_{x\in(Q_A)_0} \epsilon_x\lambda_x$ and $\sum_{\alpha\in(Q_A)_1} \alpha\mu_\alpha$, with $\lambda_x, \mu_\alpha \in \mathbf{k}$ for all x, α such that

$$a - \left(\sum_x \epsilon_x\lambda_x + \sum_\alpha \alpha\mu_\alpha\right) \in \mathbf{k}Q_A^{+2}.$$

Applying φ and using that $\varphi(a) = 0$, we get

$$\sum_x e_x\lambda_x + \sum_\alpha a_\alpha\mu_\alpha \in \varphi(\mathbf{k}Q_A^{+2}) \subseteq \operatorname{rad}^2 A.$$

Because the e_x are orthogonal idempotents, we infer that $\lambda_x = 0$ for all x and therefore $\sum_\alpha a_\alpha \mu_\alpha \in \operatorname{rad}^2 A$, which can be written as $\sum_\alpha (a_\alpha + \operatorname{rad}^2 A)\mu_\alpha = 0$. But the elements of the set $\{a_\alpha + \operatorname{rad}^2 A\}_\alpha$ form, by definition, a basis of the vector space $\operatorname{rad} A/\operatorname{rad}^2 A$. In particular, they are linearly independent and so $\mu_\alpha = 0$ for all α. This shows that $a \in \mathbf{k}Q_A^{+2}$. □

In particular, every basic algebra over an algebraically closed field is a bound quiver algebra. We end this subsection with an example.

Example I.2.14. Consider the algebra in Example I.1.1

$$A = \begin{pmatrix} \mathbf{k} & 0 & 0 & 0 \\ \mathbf{k} & \mathbf{k} & 0 & 0 \\ \mathbf{k} & 0 & \mathbf{k} & 0 \\ \mathbf{k} & \mathbf{k} & \mathbf{k} & \mathbf{k} \end{pmatrix}.$$

Denote as before by e_{ij} the matrix having 1 in position (i, j) and 0 everywhere else. A natural complete set of primitive orthogonal idempotents for A is the set $\{e_{11}, e_{22}, e_{33}, e_{44}\}$. We have proved in Example I.1.8 that

$$\operatorname{rad} A = \begin{pmatrix} 0 & 0 & 0 & 0 \\ \mathbf{k} & 0 & 0 & 0 \\ \mathbf{k} & 0 & 0 & 0 \\ \mathbf{k} & \mathbf{k} & \mathbf{k} & 0 \end{pmatrix}$$

so that $A/\operatorname{rad} A \cong \mathbf{k}^4$. Therefore, A is elementary. In addition,

$$\operatorname{rad}^2 A = \begin{pmatrix} 0 & 0 & 0 & 0 \\ 0 & 0 & 0 & 0 \\ 0 & 0 & 0 & 0 \\ \mathbf{k} & 0 & 0 & 0 \end{pmatrix}.$$

A straightforward calculation gives that each of

$$e_{44}\left(\frac{\operatorname{rad} A}{\operatorname{rad}^2 A}\right)e_{33},\ e_{44}\left(\frac{\operatorname{rad} A}{\operatorname{rad}^2 A}\right)e_{22},\ e_{33}\left(\frac{\operatorname{rad} A}{\operatorname{rad}^2 A}\right)e_{11},\ e_{22}\left(\frac{\operatorname{rad} A}{\operatorname{rad}^2 A}\right)e_{11}$$

is one-dimensional, whereas the rest of the

$$e_{ii}\left(\frac{\operatorname{rad} A}{\operatorname{rad}^2 A}\right)e_{jj}$$

are zero. This gives the quiver Q_A of A

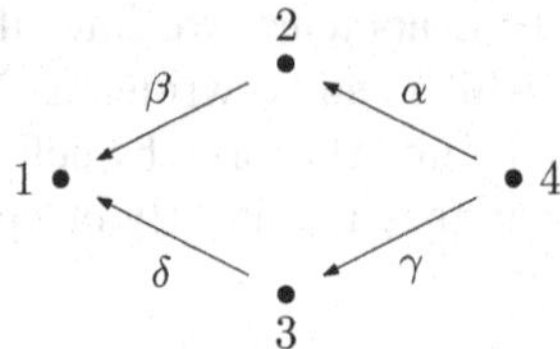

We consider the algebra morphism $\varphi\colon \mathbf{k}Q_A \longrightarrow A$ defined by $\varphi(\epsilon_i) = e_{ii}$ for each $i \in \{1, 2, 3, 4\}$, $\varphi(\alpha) = e_{42}$, $\varphi(\beta) = e_{21}$, $\varphi(\gamma) = e_{31}$, $\varphi(\delta) = e_{43}$. Then, φ is surjective because its image contains a basis of A. We have

$$\varphi(\alpha\beta) = \varphi(\alpha)\varphi(\beta) = e_{42}e_{21} = e_{41} = e_{43}e_{31} = \varphi(\gamma)\varphi(\delta) = \varphi(\gamma\delta).$$

Therefore, $\alpha\beta - \gamma\delta \in \operatorname{Ker}\varphi$. On the other hand, this kernel is one-dimensional because a quick calculation gives $\dim_{\mathbf{k}}\mathbf{k}Q_A = 10$, whereas $\dim_{\mathbf{k}}A = 9$. Thus, if I is the ideal generated by the element $\alpha\beta - \gamma\delta$, we have indeed $A \cong \mathbf{k}Q_A/I$. As mentioned at the end of Subsection I.2.1, we say that A is given by the quiver above bound by the relation $\alpha\beta = \gamma\delta$.

I.2.3 Projective, injective and simple modules

We recall from Subsection I.1.3 that, if A is a finite dimensional algebra, and $\{e_1, \ldots, e_n\}$ is a complete set of primitive orthogonal idempotents, then $\{P_1 = e_1A, \ldots, P_n = e_nA\}$, $\{I_1 = \mathrm{D}(Ae_1), \ldots, I_n = \mathrm{D}(Ae_n)\}$ and $\{S_1 = \operatorname{top} P_1 \cong \operatorname{soc} I_1, \ldots, S_n = \operatorname{top} P_n \cong \operatorname{soc} I_n\}$ are a complete list of representatives of the isoclasses of the indecomposable projective, injective and simple modules respectively. We now show how to construct these modules using bound quivers.

For this purpose, we first need a complete set of primitive orthogonal idempotents. Let $A \cong \mathbf{k}Q/I$ be a bound quiver algebra. Because of Proposition I.2.7, the set $\{e_x = \epsilon_x + I : x \in Q_0\}$ is a complete set of primitive orthogonal idempotents in A. We deduce our first lemma:

Lemma I.2.15. *For each $x \in Q_0$, the indecomposable projective A-module $P_x = e_xA$ is generated, as a* **k***-vector space, by the classes modulo I of all paths in Q starting at x.*

Proof. Indeed, we have

$$P_x = e_xA = e_x\left(\frac{\mathbf{k}Q}{I}\right) \cong \frac{\epsilon_x(\mathbf{k}Q)}{\epsilon_xI}.$$

The statement follows immediately. □

In particular, for every $y \in Q_0$, the vector space $P_x e_y = e_x A e_y = \epsilon_x(\mathbf{k}Q)\epsilon_y/\epsilon_x I \epsilon_y$ is generated by the classes modulo I of all the paths in Q from x to y. This implies that $\mathrm{Hom}_A(P_y, P_x) \cong P_x e_y$ is nonzero if and only if there is a path from x to y not lying in I.

Corollary I.2.16. *For each $x \in Q_0$, the simple A-module $S_x = \operatorname{top} P_x$ is a one-dimensional vector space generated by e_x.*

Proof. Over an elementary algebra, every indecomposable projective module has a one-dimensional top equal to $\mathbf{k}$. Therefore, every simple module is one-dimensional. On the other hand, we have $S_x e_x = \mathrm{Hom}_A(P_x, S_x) \neq 0$ and spanned by e_x. □

A first consequence of the previous results is that a projective module P_x is simple if and only if $x \in Q_0$ is a sink in Q. For every point x, the radical $\operatorname{rad} P_x$ is easy to compute:

$$\operatorname{rad} P_x = P_x \operatorname{rad} A = e_x \operatorname{rad} A = e_x \left(\frac{\mathbf{k}Q^+}{I}\right) \cong \frac{\epsilon_x \mathbf{k}Q^+}{\epsilon_x I}$$

which means that $\operatorname{rad} P_x$ is generated by the classes modulo I of all paths starting at x of length at least one.

Another consequence is as follows. An arrow $\alpha\colon x \longrightarrow y$ in Q corresponds to a nonzero element of $e_x A e_y \cong \mathrm{Hom}_A(e_y A, e_x A)$, and thus to a nonzero morphism $f_\alpha\colon P_y \longrightarrow P_x$. Hence, paths in Q induce sequences of nonzero morphisms between indecomposable projectives.

The next result describes the injective modules in terms of paths in the quiver.

Lemma I.2.17. *For each $x \in Q_0$, the indecomposable injective A-module $\mathrm{D}(Ae_x)$ is isomorphic, as a $\mathbf{k}$-vector space, to the dual of the space generated by all classes modulo I of the paths in Q ending in x.*

Proof. This is similar to Lemma I.2.15 and is thus omitted. □

Also, an injective module I_x is simple if and only if $x \in Q_0$ is a source.

In our examples, we represent these modules in a visually suggestive way, which respects the shape of the quiver and, for a module M, yields immediately its **radical filtration**

$$M \supseteq \operatorname{rad} M \supseteq \operatorname{rad}^2 M \supseteq \ldots \supseteq \operatorname{rad}^t M = 0.$$

The least integer t such that $\operatorname{rad}^t M = 0$ is called the **Loewy length** of M and denoted as $ll(M)$. Thus, if M and N are modules, then $ll(M \oplus N) = \max\{ll(M), ll(N)\}$.

Example I.2.18. Let A be given by the quiver

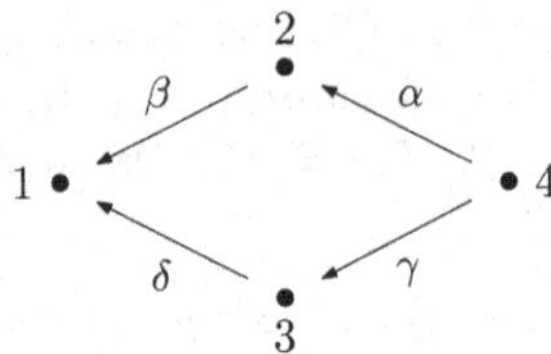

bound by $\alpha\beta = \gamma\delta$.

We first compute the indecomposable projective modules. Because 1 is a sink, we have $P_1 = S_1$ (which we sometimes abbreviate as $P_1 = 1$). Now, P_2, as a **k**-vector space, has a basis $\{e_2, \overline{\beta} = \beta + I\}$. It is then two-dimensional. As seen above, top P_2 is spanned by e_2 and rad P_2 by $\overline{\beta}$. Because P_2 is a submodule of A_A, its scalar multiplication is induced from that of A; thus, it is defined by

$$\begin{array}{ll} e_2 e_2 = e_2 & \\ e_2\overline{\beta} = \overline{\beta} & \\ e_2 u = 0 & \text{for every basis vector } u \neq e_2,\ \overline{\beta} \text{ in } A, \\ \overline{\beta} e_1 = \overline{\beta} & \\ \overline{\beta} v = 0 & \text{for every basis vector } v \neq e_1 \text{ in } A. \end{array}$$

Identifying idempotents to points, and classes of arrows to arrows, we may represent

$$P_2 \quad \text{as} \quad \begin{array}{c} 2 \\ \downarrow{\scriptstyle \beta} \\ 1 \end{array} \quad \text{or, briefly,} \quad P_2 = {\textstyle {2 \atop 1}}\,.$$

This notation clearly suggests that top $P_2 = 2$ and rad $P_2 = 1$. Similarly, $P_3 = {3 \atop 1}$ has top 3 and radical 1. The indecomposable projective module P_4 is generated, as a **k**-vector space, by the classes $\{e_4, \overline{\alpha}, \overline{\alpha}\overline{\beta}, \overline{\gamma}, \overline{\gamma}\overline{\delta}\}$. However, $\overline{\alpha}\overline{\beta} = \overline{\alpha\beta} = \overline{\gamma\delta} = \overline{\gamma}\overline{\delta}$, because $I = <\alpha\beta - \gamma\delta>$. Thus, P_4 is four-dimensional, having as a basis $\{e_4, \overline{\alpha}, \overline{\gamma}, \overline{\alpha\beta}\}$. Using the notation above, P_4 may be represented as

$$\begin{array}{ccccc} & & 4 & & \\ & {\scriptstyle\alpha}\swarrow & & \searrow{\scriptstyle\gamma} & \\ 2 & & & & 3 \\ & {\scriptstyle\beta}\searrow & & \swarrow{\scriptstyle\delta} & \\ & & 1 & & \end{array} \quad \text{or, briefly,} \quad P_4 = \begin{smallmatrix} 4 \\ 2\ 3 \\ 1 \end{smallmatrix}.$$

The reader sees that, if M is a submodule of P_4, then the representation of M corresponds to a subdiagram of P_4 in which all arrows enter (none leaves). Thus, all isoclasses of submodules of P_4 are P_4, rad $P_4 = \begin{smallmatrix} 2\,3 \\ 1 \end{smallmatrix}, \begin{smallmatrix} 2 \\ 1 \end{smallmatrix}, \begin{smallmatrix} 3 \\ 1 \end{smallmatrix}$ and $\operatorname{rad}^2 P_4 = \operatorname{soc} P_4 = 1$. The radical filtration of P_4 is

$$\begin{smallmatrix} 4 \\ 2\ 3 \\ 1 \end{smallmatrix} \supseteq \begin{smallmatrix} 2\,3 \\ 1 \end{smallmatrix} \supseteq 1.$$

One composition series of P_4 is given by

$$\begin{smallmatrix}&4&\\2&&3\\&1&\end{smallmatrix} \supseteq \begin{smallmatrix}2&&3\\&1&\end{smallmatrix} \supseteq \begin{smallmatrix}3\\1\end{smallmatrix} \supseteq 1$$

and the other is obtained by replacing $\begin{smallmatrix}3\\1\end{smallmatrix}$ by $\begin{smallmatrix}2\\1\end{smallmatrix}$. In particular, the Loewy length of P_4 is $ll(P_4) = 3$, whereas its composition length is $l(P_4) = 4$. In the same way, a quotient of P_4 corresponds to a subdiagram in which all arrows leave (and none enters). For instance, $P_4/\operatorname{soc} P_4 \cong \begin{smallmatrix}&4&\\2&&3\end{smallmatrix}$. We may thus write $A_A = 1 \oplus \begin{smallmatrix}2\\1\end{smallmatrix} \oplus \begin{smallmatrix}3\\1\end{smallmatrix} \oplus \begin{smallmatrix}&4&\\2&&3\\&1&\end{smallmatrix}$.

Now for the indecomposable injective modules. Because 4 is a source, $I_4 = S_4 = 4$. On the other hand, I_2 is the dual of the vector space with a basis $\{e_2, \overline{\alpha} = \alpha + I\}$ and its multiplication is also induced from that of A. We can therefore represent I_2 as

$$\begin{array}{c}4\\ \downarrow{\scriptstyle\alpha}\\ 2\end{array} \qquad \text{or, briefly,} \quad \begin{smallmatrix}4\\2\end{smallmatrix}.$$

Similarly, $I_3 = \begin{smallmatrix}4\\3\end{smallmatrix}$. Finally, I_1 is the dual of the vector space spanned by the classes $\{e_1, \overline{\beta}, \overline{\alpha\beta}, \overline{\delta}, \overline{\gamma\delta}\}$. Because $\overline{\alpha\beta} = \overline{\gamma\delta}$, its basis is $\{e_1, \overline{\beta}, \overline{\delta}, \overline{\alpha\beta}\}$. It can be represented as

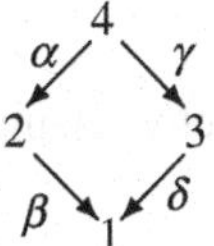

and, in particular, it is isomorphic to P_4. It is then projective–injective and one has

$$(\mathrm{D}A)_A = \begin{smallmatrix}&4&\\2&&3\\&1&\end{smallmatrix} \oplus \begin{smallmatrix}4\\2\end{smallmatrix} \oplus \begin{smallmatrix}4\\3\end{smallmatrix} \oplus 4.$$

One sees that, for instance, P_2 is isomorphic to a submodule of P_4 and the cokernel of the inclusion is just I_3. Thus, we have a short exact sequence

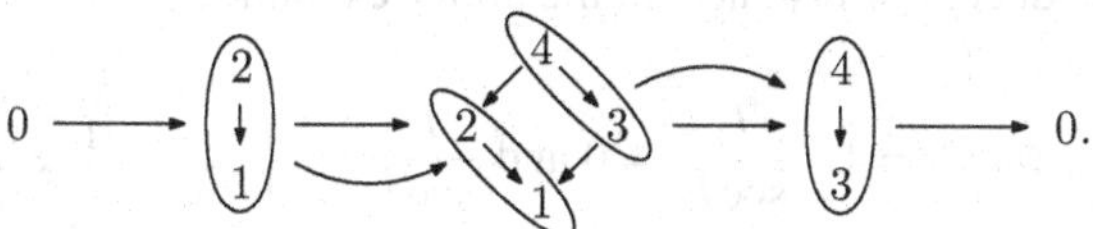

in which the morphisms are represented diagrammatically. It is known that $\begin{smallmatrix}2\\1\end{smallmatrix}$ is a submodule of P_4, and the map $\begin{smallmatrix}2\\1\end{smallmatrix} \longrightarrow P_4$ is the only possible embedding. Factoring out by the image yields $P_4/\begin{smallmatrix}2\\1\end{smallmatrix} \cong \begin{smallmatrix}4\\3\end{smallmatrix}$. The principles used here are that, first, the image of a simple module is only zero or an isomorphic simple (this is the well-known Schur's lemma) and, if $f : L \longrightarrow M$ is a morphism, then $\operatorname{Im} f \cong L/\operatorname{Ker} f$.

Example I.2.19. Let A be given by the quiver

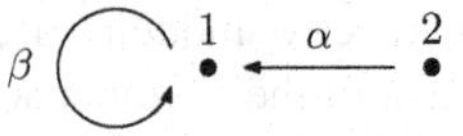

$$\beta \circlearrowright \overset{1}{\bullet} \xleftarrow{\alpha} \overset{2}{\bullet}$$

bound by the relations $\alpha\beta = 0$ and $\beta^3 = 0$. Then, the indecomposable projectives are given by

$$P_1 = \begin{matrix}1\\ \downarrow\beta\\ 1\\ \downarrow\beta\\ 1\end{matrix} = \begin{smallmatrix}1\\1\\1\end{smallmatrix}, \quad P_2 = \begin{matrix}2\\ \downarrow\alpha\\ 1\end{matrix} = \begin{smallmatrix}2\\1\end{smallmatrix}.$$

Thus, $A_A = \begin{smallmatrix}1\\1\\1\end{smallmatrix} \oplus \begin{smallmatrix}2\\1\end{smallmatrix}$.

Similarly, the indecomposable injective A-modules are given by

$$I_1 = \begin{matrix}1 & & \\ \beta\downarrow & & \\ 1 & & 2\\ & \beta\searrow\ \swarrow\alpha & \\ & 1 & \end{matrix} = \begin{smallmatrix}1 & \\ 1 & 2\\ 1 & \end{smallmatrix} \quad \text{and} \quad I_2 = 2$$

so that $(\mathrm{D}A)_A = \begin{smallmatrix}1 & \\ 1 & 2\\ 1 & \end{smallmatrix} \oplus 2$.

This method of representing indecomposable projective and injective modules using their radical filtrations can be extended to other modules. In particular, radicals of indecomposable projectives and quotients of indecomposable injectives by their socles are easy to produce. For instance, in the above example,

$$\operatorname{rad} P_1 = \begin{smallmatrix}1\\1\end{smallmatrix}, \quad \operatorname{rad} P_2 = 1, \quad \frac{I_2}{\operatorname{soc} I_2} = 0 \text{ and } \frac{I_1}{\operatorname{soc} I_1} = \begin{smallmatrix}1 & \\ 1 & 2\end{smallmatrix} = \begin{smallmatrix}1\\1\end{smallmatrix} \oplus 2.$$

This notation has several advantages. For instance, one sees easily that there is a nonzero morphism from P_2 to I_1, which is a monomorphism whose cokernel is rad P_1. This way of visualising modules is generally not very precise and in large examples quickly becomes unpractical. It will however suffice for the small examples we deal with in this book.

There is one class of algebras for which this representation of modules always works very well. These are the Nakayama algebras, which we now consider.

I.2.4 Nakayama algebras

An algebra A is called **representation-finite** if its module category admits only finitely many isoclasses of indecomposable objects. It is called **representation-infinite** if it is not representation-finite. Representation-finite algebras are a particularly nice class to study: indeed, one can classify their indecomposable modules up to isomorphism and, as we shall see later, we know a lot about the morphisms between them. The objective of this subsection is to describe an easy class of representation-finite algebras, that of the so-called Nakayama algebras. Throughout, let A be a finite dimensional **k**-algebra. We start with a definition.

Definition I.2.20. An A-module is called **uniserial** if it admits a unique composition series.

Clearly, every simple module is uniserial. There exist uniserial modules which are not simple (for instance, in Example I.2.19 the modules $P_2 = \begin{smallmatrix} 2 \\ 1 \end{smallmatrix}$ and rad $P_1 = \begin{smallmatrix} 1 \\ 1 \end{smallmatrix}$). A uniserial module has a simple top and a simple socle so, in particular, it is indecomposable. Also, if M is uniserial, then so is every submodule and every quotient of M.

Lemma I.2.21. *An A-module M is uniserial if and only if its radical filtration*

$$M \supseteq \operatorname{rad} M \supseteq \operatorname{rad}^2 M \supseteq \ldots \supseteq \operatorname{rad}^t M = 0$$

is a composition series.

Proof. Assume first that M is uniserial of composition length l. Then, it has a unique maximal submodule, necessarily equal to rad M, whose composition length is equal to $l - 1$. The result follows by induction.

Conversely, let $M \supsetneq M_1 \supsetneq \ldots \supsetneq M_l = 0$ be a composition series for M. The hypothesis says that $M/\operatorname{rad} M$ is simple; thus, M has a unique maximal submodule. Therefore, $M_1 = \operatorname{rad} M$. Inductively, $M_i = \operatorname{rad}^i M$ for all i and thus M has a unique composition series. □

Given a module M, we recall that its Loewy length $ll(M)$ is the least integer t such that $\operatorname{rad}^t M = 0$. If M is now uniserial, then $ll(M)$ equals the composition length $l(M)$ of M because of Lemma I.2.21 above.

Lemma I.2.22. *Every indecomposable projective A-module is uniserial if and only if, for every indecomposable projective P, the module* $\operatorname{rad} P/\operatorname{rad}^2 P$ *is simple or zero.*

Proof. Because necessity follows from Lemma I.2.21, we only need to prove sufficiency. We claim that the hypothesis implies that, for every indecomposable projective P, the radical filtration $P \supseteq \operatorname{rad} P \supseteq \operatorname{rad}^2 P \supseteq \dots$ is a composition series. This implies the uniseriality of P.

We prove the claim. We know that $P/\operatorname{rad} P$ is simple and that $\operatorname{rad} P/\operatorname{rad}^2 P$ is simple or zero. Assume inductively that $\operatorname{rad}^{i-1} P/\operatorname{rad}^i P$ is simple. Let $p\colon P' \longrightarrow \operatorname{rad}^{i-1} P$ be a projective cover. Because $\operatorname{rad}^{i-1} P$ has a simple top, P' is indecomposable. Applying Lemma I.1.9, p induces epimorphisms $\operatorname{rad} P' \longrightarrow \operatorname{rad}^i P$ and $\operatorname{rad}^2 P' \longrightarrow \operatorname{rad}^{i+1} P$. Passing to cokernels, we get an epimorphism $\operatorname{rad} P'/\operatorname{rad}^2 P' \longrightarrow \operatorname{rad}^i P/\operatorname{rad}^{i+1} P$. Because of the hypothesis, $\operatorname{rad} P'/\operatorname{rad}^2 P'$ is simple or zero. Therefore, so is $\operatorname{rad}^i P/\operatorname{rad}^{i+1} P$. This establishes our claim. □

We now define Nakayama algebras.

Definition I.2.23. An algebra A is called a **Nakayama algebra** if all indecomposable projective and all indecomposable injective A-modules are uniserial.

The definition implies immediately a characterisation of Nakayama algebras by means of their quivers.

Theorem I.2.24. *Let A be an elementary algebra. Then, A is a Nakayama algebra if and only if its ordinary quiver* Q_A *is of one of the following two forms:*

(a) 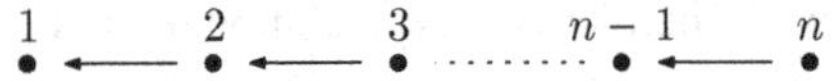

(b) 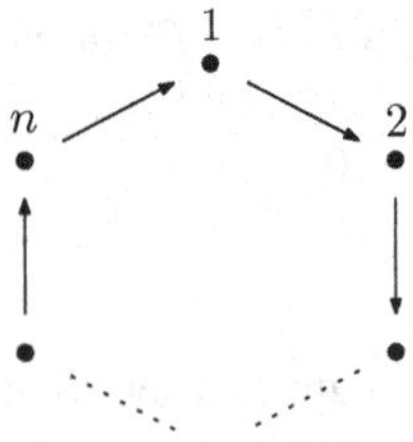

Proof. Because of Lemma I.2.22, every indecomposable projective A-module is uniserial if and only if, for every $x \in (Q_A)_0$, we have

$$\dim_{\mathbf{k}}\left(\frac{\operatorname{rad} P_x}{\operatorname{rad}^2 P_x}\right) = \dim_{\mathbf{k}}\left(e_x\left(\frac{\operatorname{rad} A}{\operatorname{rad}^2 A}\right)\right) \le 1.$$

This occurs if and only if there exists at most one point $y \in (Q_A)_0$ such that $e_x(\operatorname{rad} A/\operatorname{rad}^2 A)e_y \neq 0$ and this vector space is one-dimensional, or, equivalently, there exists at most one arrow starting with x.

Dually, every indecomposable injective A-module is uniserial if and only if, for every x, there is at most one arrow in Q_A ending with x. Thus, A is a Nakayama algebra if and only if, for each $x \in (Q_A)_0$, there is at most one arrow starting at x and one arrow ending at x. This gives the quivers (a) and (b) in the statement. □

In particular, if $A = \mathbf{k}Q/I$ is a Nakayama bound quiver algebra, then the ideal I, on which the theorem imposes no restriction, can only be generated by zero-relations.

We deduce the classification of indecomposable modules over a Nakayama algebra.

Theorem I.2.25. *Let A be a Nakayama algebra and M an indecomposable A-module. Then there exist an indecomposable projective A-module P and a $t \geq 0$ such that $M \cong P/\operatorname{rad}^t P$. In particular, M is uniserial.*

Proof. Let $t = ll(M)$ be the Loewy length of M and set $A' = A/\operatorname{rad}^t A$. Because $0 = \operatorname{rad}^t M = M\cdot\operatorname{rad}^t A$, then M has a natural A'-module structure, and $\operatorname{rad}^{t-1} M \neq 0$ implies $\operatorname{rad}^{t-1} A \neq 0$; therefore, $ll(A') = t$. We claim that A' is Nakayama. If A is elementary, this follows from Theorem I.2.24, but we give an independent proof without this hypothesis. Let $A_A = \oplus_{i=1}^n P_i$ be a decomposition of A into indecomposable projective A-modules. Then,

$$\frac{A}{\operatorname{rad}^t A} = \bigoplus_{i=1}^n \left(\frac{P_i}{P_i \operatorname{rad}^t A}\right) = \bigoplus_{i=1}^n \left(\frac{P_i}{\operatorname{rad}^t P_i}\right).$$

Each of the modules $P_i/\operatorname{rad}^t P_i$ has a simple top and is thus indecomposable. This shows that each indecomposable projective A'-module is isomorphic to some $P_i/\operatorname{rad}^t P_i$. Now, P_i being uniserial, so is $P_i/\operatorname{rad}^t P_i$. This result and its dual imply that A' is Nakayama.

Let $f = (f_1 \dots f_s)\colon \oplus_{j=1}^s P'_j \longrightarrow M$ be a projective cover in $\operatorname{mod} A'$, with all the P'_j indecomposable. Then,

$$t = ll(A') \geq \max\{ll(P'_j)\} \geq ll(M) = t$$

shows that there exists some j such that $ll(P'_j) = t$.

If no $f_j\colon P'_j \longrightarrow M$ with $ll(P'_j) = t$ is injective, then we have $ll(\operatorname{Im} f_j) < t$ for all j, and, because $f = (f_1 \dots f_s)$ is an epimorphism, we get $ll(M) < t$, a contradiction. This proves that there exists j such that $ll(P'_j) = t$ and also $f_j\colon P'_j \longrightarrow M$ is injective. We claim that, in this case, P'_j is also an injective A'-module. Indeed, let $P'_j \longrightarrow I$ be an injective envelope in $\operatorname{mod} A'$. Because $\operatorname{soc} P'_j$ is simple, so is $\operatorname{soc} I$; therefore, I is indecomposable and hence uniserial. In addition, we have

$$t = ll(P'_j) = l(P'_j) \leq l(I) = ll(I) \leq ll(A') = t.$$

Therefore, $l(P'_j) = l(I)$ and so $P'_j \cong I$, which establishes our claim. It implies that the injective morphism $f_j : P'_j \longrightarrow M$ is a section. Because M is indecomposable, we have $M \cong P'_j$. Hence, there exists i such that $M \cong P_i / \operatorname{rad}^t P_i$. □

Because finite dimensional algebras admit only finitely many isoclasses of indecomposable projective modules, each of which has finite (Loewy) length, we infer from the previous theorem that a Nakayama algebra is representation-finite.

Example I.2.26. Let A be given by the quiver

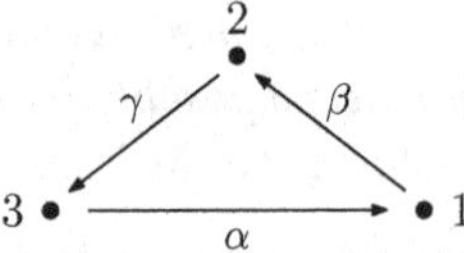

bound by $\alpha\beta\gamma = 0$ and $\gamma\alpha = 0$. Because of Theorem I.2.24, A is a Nakayama algebra. Its indecomposable projective modules are, up to isomorphism,

$$P_1 = \begin{smallmatrix}1\\2\\3\end{smallmatrix}\ , \quad P_2 = \begin{smallmatrix}2\\3\end{smallmatrix}\ , \quad \text{and} \quad P_3 = \begin{smallmatrix}3\\1\\2\end{smallmatrix}\ .$$

We deduce from Theorem I.2.25 the complete list of isoclasses of nonprojective indecomposable A-modules

$$\frac{P_1}{\operatorname{rad} P_1} = 1 \quad \frac{P_2}{\operatorname{rad} P_2} = 2 \quad \frac{P_3}{\operatorname{rad} P_3} = 3$$

$$\frac{P_1}{\operatorname{rad}^2 P_1} = \begin{smallmatrix}1\\2\end{smallmatrix} \quad \frac{P_3}{\operatorname{rad}^2 P_3} = \begin{smallmatrix}3\\1\end{smallmatrix}\ .$$

It is also easy to compute minimal projective resolutions of the simple modules and hence the global dimension of A. For instance, the short exact sequence

$$0 \longrightarrow \begin{smallmatrix}2\\3\end{smallmatrix} \longrightarrow \begin{smallmatrix}1\\2\\3\end{smallmatrix} \longrightarrow 1 \longrightarrow 0$$

shows that $\operatorname{pd} 1 = 1$. Also, there is an exact sequence

$$0 \longrightarrow 3 \longrightarrow \begin{smallmatrix}1\\2\\3\end{smallmatrix} \longrightarrow \begin{smallmatrix}3\\1\\2\end{smallmatrix} \longrightarrow 3 \longrightarrow 0.$$

Splicing infinitely many copies of this sequence shows that $\text{pd}\, 3 = \infty$. The short exact sequence

$$0 \longrightarrow 3 \longrightarrow {\scriptstyle\begin{smallmatrix} 2 \\ 3 \end{smallmatrix}} \longrightarrow 2 \longrightarrow 0$$

shows that $\text{pd}\, 2 = \infty$ as well. Therefore, $\text{gl. dim.}\, A = \infty$.

I.2.5 Hereditary algebras

Hereditary algebras are among the most frequently studied algebras in representation theory. An algebra A is **hereditary** provided that its global dimension is at most one, or, equivalently, every submodule of a projective module is projective. This is equivalent to saying that every quotient of an injective module is injective.

We need the following lemma:

Lemma I.2.27. *Let A be a hereditary algebra, then:*

(a) *A nonzero morphism between indecomposable projectives is a monomorphism.*
(b) *The quiver of A is acyclic.*

Proof.

(a) Let $f : P \longrightarrow P'$ be nonzero, with P and P' indecomposable projectives. Its image $\operatorname{Im} f$ is a submodule of P', and hence is projective. Consequently, the canonical surjection $P \longrightarrow \operatorname{Im} f$ induced by f is a retraction. Because P is indecomposable, we get $P \cong \operatorname{Im} f$.
(b) Assume that the quiver Q_A contains an oriented cycle. Then there exists a sequence of nonzero morphisms between nonisomorphic indecomposable projective A-modules

$$P_0 \xrightarrow{f_1} P_1 \xrightarrow{f_2} \dots \xrightarrow{f_n} P_n = P_0$$

with $n \geq 1$. Because each of the f_i is injective, so is the composition $f_n \dots f_1 : P_0 \longrightarrow P_0$. But then $f_n \dots f_1$ is an isomorphism. Hence, f_n is surjective and therefore an isomorphism between P_n and P_{n-1}. This is a contradiction.

□

We are now able to describe hereditary algebras in terms of bound quivers.

Proposition I.2.28. *A basic, elementary and connected algebra A is hereditary if and only if $A \cong \mathbf{k}Q_A$ with Q_A acyclic.*

Proof. Let S_x be a simple A-module, and P_x its projective cover. The short exact sequence

$$0 \longrightarrow \operatorname{rad} P_x \longrightarrow P_x \longrightarrow S_x \longrightarrow 0$$

says that A is hereditary if and only if, for each x, the radical $\operatorname{rad} P_x$ is projective.

Assume first that $A = \mathbf{k}Q$ and, for $y, z \in Q_0$, let $w(y, z)$ denote the number of paths from y to z in Q. Because A is finite dimensional, implying that Q is finite and acyclic, we have $w(y, z) < \infty$ for all $y, z \in Q_0$. Given $x \in Q_0$, let $y_1, \ldots, y_t$ denote the distinct direct successors of x in Q and assume that, for each i, there are n_i arrows from x to y_i. Then, $\operatorname{top}(\operatorname{rad} P_x) = \oplus_{i=1}^t S_{y_i}^{n_i}$, so that we have a projective cover morphism $p\colon \oplus_{i=1}^t P_{y_i}^{n_i} \longrightarrow \operatorname{rad} P_x$. Let $y \neq x$ be arbitrary in Q, then

$$\begin{aligned}\dim_{\mathbf{k}}(\operatorname{rad} P_x)e_y &= \dim_{\mathbf{k}}(P_x e_y) = \dim_{\mathbf{k}}(e_x A e_y) = w(x, y)\\ &= \textstyle\sum_{i=1}^t n_i w(y_i, y) = \sum_{i=1}^t n_i \dim_{\mathbf{k}}(P_{y_i} e_y)\\ &= \dim_{\mathbf{k}}(\oplus_{i=1}^t P_{y_i}^{n_i})e_y.\end{aligned}$$

Therefore, p is an isomorphism and $\operatorname{rad} P_x$ is projective. Thus, A is hereditary.

Conversely, assume that A is hereditary, and write $A \cong \mathbf{k}Q/I$. Because of Lemma I.2.11, we have $Q = Q_A$. We must prove that $I = 0$. Because of the previous lemma, Q is acyclic, we may number its points so that, if there is an arrow $x \longrightarrow y$ then $x > y$. Assume that $I \neq 0$. Then there is a least x such that $e_x I e_y \neq 0$ for some y. In particular, x is not a sink; thus, $\operatorname{rad} P_x \neq 0$. Because A is hereditary, $\operatorname{rad} P_x$ is projective and in fact $\operatorname{rad} P_x = \oplus_{i=1}^t P_{y_i}^{n_i}$ where, as before, $y_1, \ldots, y_t$ are the direct successors of x in Q_A and n_i the number of arrows from x to y_i. The minimality of x implies that $e_{y_i} I e_y = 0$; thus, $\dim_{\mathbf{k}}(P_{y_i} e_y) = \dim_{\mathbf{k}}(e_{y_i} A e_y) = w(y_i, y)$, where $w(y_i, y)$ denotes, as above, the number of paths from y_i to y. But then

$$\begin{aligned}\dim_{\mathbf{k}}(\operatorname{rad} P_x)e_y &= \textstyle\sum_{i=1}^t n_i \dim_{\mathbf{k}}(P_{y_i} e_y) = \sum_{i=1}^t n_i w(y_i, y) = w(x, y)\\ &> w(x, y) - \dim_{\mathbf{k}}(e_x I e_y) = \dim_{\mathbf{k}}(P_x e_y)\end{aligned}$$

an absurdity. Therefore, $I = 0$ and our claim is established. □

I.2.6 The Kronecker algebra

The Kronecker algebra is a standard example of a representation-infinite algebra and one of the few where it is relatively easy to compute indecomposable modules in detail. It serves to illustrate several of the concepts introduced in these notes, but also opens up new avenues to the reader.

The **Kronecker algebra** is the 2×2 triangular matrix algebra

$$A = \begin{pmatrix} \mathbf{k} & 0 \\ \mathbf{k}^2 & \mathbf{k} \end{pmatrix} = \left\{ \begin{pmatrix} a & 0 \\ (b, c) & d \end{pmatrix} : a, b, c, d \in \mathbf{k} \right\}$$

with the ordinary matrix addition and multiplication, and the operations on the entry below the main diagonal defined componentwise. Thus,

$$\begin{pmatrix} a_1 & 0 \\ (b_1, c_1) & d_1 \end{pmatrix} + \begin{pmatrix} a_2 & 0 \\ (b_2, c_2) & d_2 \end{pmatrix} = \begin{pmatrix} a_1 + a_2 & 0 \\ (b_1 + b_2, c_1 + c_2) & d_1 + d_2 \end{pmatrix}$$

$$\begin{pmatrix} a_1 & 0 \\ (b_1, c_1) & d_1 \end{pmatrix} \cdot \begin{pmatrix} a_2 & 0 \\ (b_2, c_2) & d_2 \end{pmatrix} = \begin{pmatrix} a_1 a_2 & 0 \\ (b_1 a_2 + d_1 b_2, c_1 a_2 + d_1 c_2) & d_1 d_2 \end{pmatrix}.$$

Actually, as we see now, the Kronecker algebra is the path algebra of its ordinary quiver, the so-called **Kronecker quiver** K_2:

$$1 \bullet \underset{\beta}{\overset{\alpha}{\leftleftarrows}} \bullet\, 2$$

Indeed, a natural complete set of primitive orthogonal idempotents of A is provided by the matrix idempotents:

$$e_1 = \begin{pmatrix} 1 & 0 \\ 0 & 0 \end{pmatrix}, \qquad e_2 = \begin{pmatrix} 0 & 0 \\ 0 & 1 \end{pmatrix}.$$

Also, the radical of A is the two-sided ideal

$$\begin{pmatrix} 0 & 0 \\ \mathbf{k}^2 & 0 \end{pmatrix},$$

consisting of the off-diagonal elements: indeed, this ideal is clearly nilpotent and the quotient of A by it is isomorphic to the semisimple algebra $\mathbf{k} \times \mathbf{k}$. Because $\operatorname{rad}^2 A = 0$, it follows easily from the definition of multiplication that $e_2(\operatorname{rad} A / \operatorname{rad}^2 A)e_1 \cong \mathbf{k}^2$, whereas all the other $e_i(\operatorname{rad} A / \operatorname{rad}^2 A)e_j$ vanish. This shows that Q_A is the quiver K_2. We may look at the arrow α as corresponding to the first component of the off-diagonal elements, and the arrow β as corresponding to the second component. Finally, $\dim_{\mathbf{k}} A = \dim_{\mathbf{k}} \mathbf{k}K_2$ implies that $A = \mathbf{k}K_2$. In particular, because of Proposition I.2.28, the Kronecker algebra is hereditary. In the sequel, we give a detailed description of the module category over the Kronecker algebra, In particular, we shall see that it is representation-infinite.

Exercises for Section I.2

Exercise I.2.1. Let Q be a finite connected quiver and I an admissible ideal of $\mathbf{k}Q$. Prove that $\mathbf{k}Q/I$ is local if and only if $|Q_0| = 1$.

Exercise I.2.2. For each of the following bound quiver algebras, give a basis of the algebra, write a multiplication table for this basis, and then give a complete list of the isoclasses of indecomposable projective and indecomposable injective modules.

(a) $1 \xleftarrow{\gamma} 2 \xleftarrow{\beta} 3 \xleftarrow{\alpha} 4$

(b) $1 \xleftarrow{\gamma} 2 \xleftarrow{\beta} 3 \xleftarrow{\alpha} 4$ $\quad \alpha\beta\gamma = 0$

(c) $1 \xleftarrow{\gamma} 2 \xleftarrow{\beta} 3 \xleftarrow{\alpha} 4$ $\quad \alpha\beta = 0$

(d) $1 \xleftarrow{\gamma} 2 \xleftarrow{\beta} 3 \xleftarrow{\alpha} 4$ $\quad \alpha\beta = 0, \beta\gamma = 0$

(e) Quiver with arrows $\beta: 3 \to 1$, $\delta: 3 \to 2$, $\alpha: 4 \to 3$, $\gamma: 5 \to 3$ $\quad \alpha\beta = 0$, $\gamma\delta = 0$

(f) Quiver with arrows $\beta: 3 \to 1$, $\delta: 3 \to 2$, $\alpha: 4 \to 3$, $\gamma: 5 \to 3$ $\quad \alpha\beta = 0$

(g) $1 \xleftarrow{\gamma} 2 \underset{\beta}{\overset{\alpha}{\leftleftarrows}} 3$

(h) $1 \xleftarrow{\gamma} 2 \underset{\beta}{\overset{\alpha}{\leftleftarrows}} 3$ $\quad \alpha\gamma = 0$

(i) β (loop at 1) $\; 1 \xrightarrow{\alpha} 2$ $\quad \beta^2 = 0$

(j) γ (loop at 1) $\; 1 \underset{\beta}{\overset{\alpha}{\rightleftarrows}} 2 \;$ δ (loop at 2) $\quad \gamma^2 = 0, \delta^2 = 0,$ $\gamma\alpha = \alpha\delta, \beta\gamma = \delta\beta$

(k) Quiver with arrows $\varepsilon: 3 \to 1$, $\beta: 3 \to 2$, $\alpha: 5 \to 3$, $\gamma: 5 \to 4$, $\delta: 4 \to 2$ $\quad \alpha\varepsilon = 0, \alpha\beta = \gamma\delta$

Exercise I.2.3. For each of the algebras of Exercise I.2.2, compute the projective resolutions of the simple modules and deduce the global dimension of the algebra.

Exercise I.2.4. Let A be an elementary algebra. Prove that A is connected if and only if Q_A is a connected quiver.

Exercise I.2.5. Let $A = \mathbf{k}Q/I$ be a bound quiver algebra and $x, y \in Q_0$. Prove that:

(a) S_y is a composition factor of P_x if and only if there exists a path w from x to y in Q such that $w \notin I$.
(b) S_y is a composition factor of I_x if and only if there exists a path w from y to x in Q such that $w \notin I$.

Exercise I.2.6. Let $A = \mathbf{k}Q$ be a hereditary algebra and $x \in Q_0$. Prove that:

(a) If x is a sink, then $P_x \cong S_x$ and, if x is not a sink, then

$$\operatorname{rad} P_x = \bigoplus_{\alpha : x \to y} P_y .$$

(b) If x is a source, then $I_x \cong S_x$ and, if x is not a source, then

$$\frac{I_x}{\operatorname{soc} I_x} = \bigoplus_{\alpha : y \to x} I_y .$$

Exercise I.2.7. Prove that an A-module M is uniserial if and only if $l(M) = ll(M)$.

Exercise I.2.8. Let I be an ideal in a Nakayama algebra A. Prove that A/I is a Nakayama algebra.

Exercise I.2.9. Let A be a Nakayama algebra and P_A an indecomposable projective module such that $ll(P) = ll(A_A)$. Prove that P is also injective.

Exercise I.2.10. Let $0 \longrightarrow L \longrightarrow M \longrightarrow N \longrightarrow 0$ be a short exact sequence. Prove that $\max(ll(L), ll(N)) \le ll(M) \le ll(L) + ll(N)$.

Exercise I.2.11. An algebra A is called **selfinjective** provided that the A-module A_A is injective. Let A be an elementary Nakayama algebra. Prove that A is selfinjective if and only if A is given by the quiver.

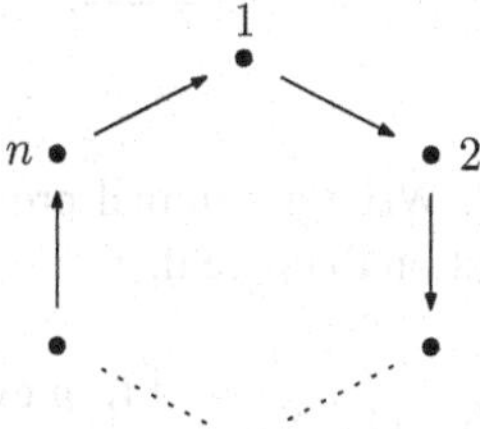

bound by $\operatorname{rad}^i A = 0$ for some $i \ge 2$.

Exercise I.2.12. For each of the following Nakayama algebras, give a complete list of all indecomposable modules up to isomorphism.

(a) $1 \xleftarrow{\varepsilon} 2 \xleftarrow{\delta} 3 \xleftarrow{\gamma} 4 \xleftarrow{\beta} 5 \xleftarrow{\alpha} 6$ $\qquad \alpha\beta = \beta\gamma\delta\varepsilon = 0$

(b) $1 \xleftarrow{\varepsilon} 2 \xleftarrow{\delta} 3 \xleftarrow{\gamma} 4 \xleftarrow{\beta} 5 \xleftarrow{\alpha} 6$ $\qquad \alpha\beta\gamma\delta\varepsilon = 0$

(c) $1 \xrightarrow{\gamma} 3,\ 3 \xrightarrow{\alpha} 2,\ 2 \xrightarrow{\beta} 1$ $\qquad \alpha\beta = 0$

(d) $1 \xrightarrow{\gamma} 3,\ 3 \xrightarrow{\alpha} 2,\ 2 \xrightarrow{\beta} 1$ $\qquad \alpha\beta\gamma = 0$

(e) $1 \underset{\beta}{\overset{\alpha}{\rightleftarrows}} 2$ $\qquad \alpha\beta\alpha\beta = 0$

(f) $1 \underset{\beta}{\overset{\alpha}{\rightleftarrows}} 2$ $\qquad \alpha\beta\alpha = 0$

(g) $1 \xrightarrow{\alpha} 2,\ 2 \xrightarrow{\beta} 3,\ 3 \xrightarrow{\gamma} 4,\ 4 \xrightarrow{\delta} 1$ $\qquad \alpha\beta = \beta\gamma = 0$

(h) $1 \xrightarrow{\alpha} 2,\ 2 \xrightarrow{\beta} 3,\ 3 \xrightarrow{\gamma} 4,\ 4 \xrightarrow{\delta} 1$ $\qquad \alpha\beta\gamma = 0$

Exercise I.2.13. Let A be given by the quiver

$$1 \underset{\beta}{\overset{\alpha}{\rightleftarrows}} 2$$

bound by $\alpha\beta = 0$, $\beta\alpha = 0$. Write a minimal projective resolution of the simple A-module S_1. Use this resolution to prove that

$$\operatorname{Ext}_A^n(S_1, S_1) = \begin{cases} \mathbf{k} & n \text{ even} \\ 0 & n \text{ odd}. \end{cases}$$

Exercise I.2.14. Let A be given by the quiver

$$1 \xleftarrow{\delta} 2 \xleftarrow{\gamma} 3 \xleftarrow{\beta} 4 \xleftarrow{\alpha} 5$$

bound by $\alpha\beta\gamma = 0, \gamma\delta = 0$.

(a) Compute the global dimension of A.
(b) Prove that $\mathrm{Ext}_A^3(5, 1) = \mathbf{k}$.
(c) Compute $\mathrm{Ext}_A^1\left(2 \oplus \begin{smallmatrix}3\\2\end{smallmatrix}, 2 \oplus \begin{smallmatrix}3\\2\end{smallmatrix}\right)$.
(d) Prove that $\mathrm{id}\left(\begin{smallmatrix}3\\2\end{smallmatrix}\right) > 1$.

Exercise I.2.15. Prove that the Kronecker algebra is isomorphic to the algebra of all triangular 3×3 matrices of the form

$$\begin{pmatrix} a & 0 & 0 \\ b & c & 0 \\ d & 0 & c \end{pmatrix}$$

with $a, b, c, d \in \mathbf{k}$, with ordinary matrix operations.

Exercise I.2.16. Let P be an indecomposable projective module over a path algebra A. Prove that $\mathrm{End}\, P_A = \mathbf{k}$.

Exercise I.2.17. For each of the following lower triangular matrix algebras, construct the ordinary quiver and deduce that the given algebra is hereditary.

(a) $A = \begin{pmatrix} \mathbf{k} & 0 & 0 \\ 0 & \mathbf{k} & 0 \\ \mathbf{k} & \mathbf{k} & \mathbf{k} \end{pmatrix} = \left\{ \begin{pmatrix} \alpha_{11} & 0 & 0 \\ 0 & \alpha_{22} & 0 \\ \alpha_{31} & \alpha_{32} & \alpha_{33} \end{pmatrix} \mid \alpha_{ij} \in \mathbf{k} \text{ for all } i, j \right\}$

(b) $A = \begin{pmatrix} \mathbf{k} & 0 & 0 & 0 \\ 0 & \mathbf{k} & 0 & 0 \\ \mathbf{k} & \mathbf{k} & \mathbf{k} & 0 \\ \mathbf{k} & \mathbf{k} & \mathbf{k} & \mathbf{k} \end{pmatrix} = \left\{ \begin{pmatrix} \alpha_{11} & 0 & 0 & 0 \\ 0 & \alpha_{22} & 0 & 0 \\ \alpha_{31} & \alpha_{32} & \alpha_{33} & 0 \\ \alpha_{41} & \alpha_{42} & \alpha_{43} & \alpha_{44} \end{pmatrix} \mid \alpha_{ij} \in \mathbf{k} \text{ for all } i, j \right\}$

Exercise I.2.18. Prove that the following conditions are equivalent for an algebra A:

(a) A is hereditary,
(b) $\mathrm{rad}\, A$ is a projective A-module,
(c) $\mathrm{pd}\,(A/\mathrm{rad}\, A) \leq 1$ where $A/\mathrm{rad}\, A$ is considered an A-module,
(d) $\mathrm{Ext}_A^1(M, -)$ is right exact, for every A-module M,
(e) $\mathrm{Ext}_A^1(S, -)$ is right exact, for every simple A-module S.

Exercise I.2.19. Let $A = \mathbf{k}Q/I$ be a bound quiver algebra and $x, y \in Q_0$.

(a) Applying $\mathrm{Hom}_A(-, S_y)$ to the exact sequence $0 \longrightarrow \mathrm{rad}\, P_x \longrightarrow P_x \longrightarrow S_x \longrightarrow 0$, prove that $\mathrm{Ext}_A^1(S_x, S_y) \cong \mathrm{Hom}_A(\frac{\mathrm{rad}\, P_x}{\mathrm{rad}^2\, P_x}, S_y)$.
(b) Deduce that $\dim_{\mathbf{k}} \mathrm{Ext}_A^1(S_x, S_y)$ equals the number of arrows from x to y in Q.

Exercise I.2.20 (Triangular matrix algebras). Let A, B be algebras and ${}_BM_A$ a $B-A$-bimodule. The set R of all matrices

$$R = \begin{pmatrix} A & 0 \\ M & B \end{pmatrix} = \left\{ \begin{pmatrix} a & 0 \\ x & b \end{pmatrix} \mid a \in A, x \in M, b \in B \right\}$$

becomes an algebra when endowed with the ordinary matrix addition and the multiplication induced from the bimodule structure of M.

(a) Prove that $\operatorname{rad} R = \begin{pmatrix} \operatorname{rad} A & 0 \\ M & \operatorname{rad} B \end{pmatrix}$.

(b) Let $\mathscr{C}$ be the category whose objects are the triples (X, Y, ϕ) where X is an A-module, Y a B-module and $\phi : Y \otimes_B M \longrightarrow X$ an A-linear map. A morphism $(u, v) : (X, Y, \phi) \longrightarrow (X', Y', \phi')$ is a pair consisting of an A-linear map $u : X \longrightarrow X'$ and a B-linear map $v : Y \longrightarrow Y'$ such that $u\phi = \phi'(v \otimes M)$. Composition is induced from the usual composition of morphisms. Prove that $\mathscr{C} \cong \operatorname{mod} R$.

(c) Prove that the module category over the Kronecker algebra $A = \begin{pmatrix} \mathbf{k} & 0 \\ \mathbf{k}^2 & \mathbf{k} \end{pmatrix}$ is equivalent to the category $\mathscr{C}$ whose objects are quadruples (X, Y, f, g), where X, Y are $\mathbf{k}$-vector spaces, and $f, g : X \longrightarrow Y$ are $\mathbf{k}$-linear maps. A morphism is a pair of maps $(u, v) : (X, Y, f, g) \longrightarrow (X', Y', f', g')$ such that $u : X \longrightarrow X'$ and $v : Y \longrightarrow Y'$ verify $uf = vf'$, $ug = g'v$ and the composition of morphisms is induced from the usual composition of $\mathbf{k}$-linear maps.

Exercise I.2.21 (One-point extensions). Let A be an algebra, M an A-module and

$$B = \begin{pmatrix} A & 0 \\ M & \mathbf{k} \end{pmatrix} = \left\{ \begin{pmatrix} a & 0 \\ x & \lambda \end{pmatrix} \mid a \in A, x \in M, \lambda \in \mathbf{k} \right\}$$

be equipped with the usual matrix addition and the multiplication induced from the module structure of M. Thus, B is an algebra, called the one-point extension of A by M and denoted as $A[M]$. Prove the following facts:

(a) $\operatorname{rad} B = \operatorname{rad} A \oplus M$, as vector spaces;

(b) The quiver Q_B contains Q_A as a full subquiver and there is exactly one additional point x, which is a source. In addition, there is an additional arrow $x \to y$ each time S_y appears as a summand in $\operatorname{top} M$ and these are all additional arrows;

(c) Every indecomposable projective A-module remains indecomposable in $\operatorname{mod} B$ and there is exactly one additional indecomposable projective B-module whose radical equals M;

(d) $\operatorname{gl.dim.} B = \max \{\operatorname{gl.dim.} A, 1 + \operatorname{pd} M\}$;

(e) B is hereditary if and only if A is hereditary and M is projective;

(f) Let A be the hereditary algebra given by the quiver.

$$1 \longleftarrow 2 \longleftarrow 3$$

Compute the bound quiver of $A[M]$, where M equals each of the following A-modules:

(i) $M = \begin{smallmatrix} 3 \\ 2 \\ 1 \end{smallmatrix}$,

(ii) $M = \begin{smallmatrix} 3 \\ 2 \end{smallmatrix}$,

(iii) $M = 3$,

(iv) $M = \begin{smallmatrix} 3 \\ 2 \\ 1 \end{smallmatrix} \oplus 3$,

(v) $M = \begin{smallmatrix} 3 \\ 2 \\ 1 \end{smallmatrix} \oplus 3 \oplus 3$,

(vi) $M = 3 \oplus \begin{smallmatrix} 2 \\ 1 \end{smallmatrix}$.

Exercise I.2.22 (Representations of quivers). Let $A = \mathbf{k}Q/I$ be a bound quiver algebra. We define a category $\text{rep}(Q, I)$, called the category of representations of the bound quiver (Q, I); an object M in $\text{rep}(Q, I)$ is defined by the following data:

1) With each $x \in Q_0$ is associated a finite dimensional $\mathbf{k}$-vector space $M(x)$.
2) With each arrow $\alpha : x \longrightarrow y$ in Q_1 is associated a $\mathbf{k}$-linear map $M(\alpha) : M(x) \longrightarrow M(y)$. This is extended to a path $\alpha_1 \dots \alpha_t$ by setting $M(\alpha_1 \dots \alpha_t) = M(\alpha_t) \dots M(\alpha_1)$, and $M(e_x) = 1_{M(x)}$ for each $x \in Q_0$.
3) If $\rho = \Sigma_i \lambda_i w_i$ is a relation in I, then $M(\rho) = \Sigma_i \lambda_i M(w_i) = 0$.

A morphism $f : M \longrightarrow N$ in $\text{rep}(Q, I)$ is a family of $\mathbf{k}$-linear maps $f = (f_x : M(x) \longrightarrow N(x))$ such that, for each arrow $\alpha : x \longrightarrow y$, we have $N(\alpha) f_x = f_y M(\alpha)$, that is, the following square commutes:

$$\begin{array}{ccc} M(x) & \xrightarrow{f_x} & N(x) \\ {\scriptstyle M(\alpha)}\downarrow & & \downarrow{\scriptstyle N(\alpha)} \\ M(y) & \xrightarrow{f_y} & N(y). \end{array}$$

The composition of $f : L \longrightarrow M$, $g : M \longrightarrow N$ is defined in the obvious way: $(gf)_x = g_x f_x$ for each $x \in Q_0$.

(a) Prove that $\text{rep}(Q, I)$ is an abelian category.
(b) With each A-module M, we associate an object $M' = F(M)$ of $\text{rep}(Q, I)$ as follows. For $x \in Q_0$, we set $M'(x) = Me_x$ and, for $\alpha : x \longrightarrow y$ in Q_1, we let $M'(\alpha) : M'(x) \longrightarrow M'(y)$ be given by the right multiplication $me_x \mapsto m(\alpha + I) = m(\alpha + I)e_y$ (for $m \in M$). Prove that this extends to a functor $F : \text{mod}\, A \longrightarrow \text{rep}(Q, I)$.
(c) Conversely, with an object M' of $\text{rep}(Q, I)$ we associate a module $M = G(M')$ as follows. As a $\mathbf{k}$-vector space, set $M = \bigoplus_{x \in Q_0} M'(x)$. For a path w from x to y in Q and $m = (m_x)_{x \in Q_0} \in M$, set

$$m(w + I) = M'(w)(m_x).$$

Prove that M is indeed a $\mathbf{k}Q$-module annihilated by I, thus an A-module, then prove that this assignment extends to a functor $G : \text{rep}(Q, I) \longrightarrow \text{mod}\, A$.

(d) Prove that F and G are quasi-inverse functors so that $\text{mod}\, A \cong \text{rep}(Q, I)$.

Exercise I.2.23. Prove that in each of the following bound quivers $\text{Ext}^2_A(I_x, P_y)$ is one-dimensional.

(a) $y \bullet \xleftarrow{\ \gamma\ } \bullet \xleftarrow{\ \beta\ } \bullet \xleftarrow{\ \alpha\ } \bullet\, x$ $\qquad \alpha\beta\gamma = 0$

(b) $y \bullet$ (paths $y \xleftarrow{\beta} \bullet \xleftarrow{\alpha} x$ and $y \xleftarrow{\delta} \bullet \xleftarrow{\gamma} x$) $\bullet\, x$ $\qquad \alpha\beta = \gamma\delta$

Exercise I.2.24. Prove that for the following bound quiver algebra A:

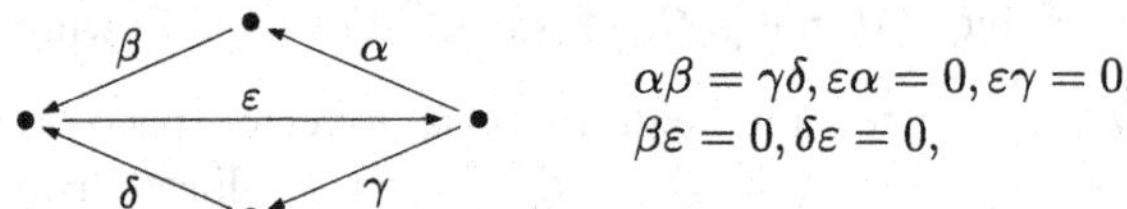

$$\alpha\beta = \gamma\delta, \varepsilon\alpha = 0, \varepsilon\gamma = 0,$$
$$\beta\varepsilon = 0, \delta\varepsilon = 0,$$

we have $\text{id}\, A_A \leq 1$ and $\text{pd}(DA)_A \leq 1$, while gl. dim. $A = \infty$.

Chapter II
The radical and almost split sequences

As in Chapter I, we let **k** be an arbitrary (commutative) field. Our algebras are finite dimensional **k**-algebras, associative and with an identity. The main working tool in this book is the notion of almost split sequences. It arose from an attempt to understand the morphisms lying in the radical of a module category. From this attempt, Auslander and Reiten extracted the notions of irreducible morphisms and almost split sequences, which allow all irreducible morphisms to be arranged in a neat way. We start our discussion in Section II.1 with a short description of the radical of a module category. We define and study irreducible morphisms and almost split sequences in Section II.2. We prove in Section II.3 the existence theorem for almost split sequences and we proceed to apply these sequences to the study of the radical in Section II.4.

II.1 The radical of a module category

II.1.1 Categorical framework

In several places, we use a categorical language. For this reason, it is convenient to fix the terminology and recall a few results.

Definition II.1.1. Let **k** be a field. A category $\mathscr{C}$ is called a **k-category** if it satisfies the following conditions:

(a) For every pair of objects X, Y in $\mathscr{C}$, the set $\mathrm{Hom}_{\mathscr{C}}(X, Y)$ of morphisms from X to Y is a **k**-vector space.
(b) The composition of morphisms is **k**-bilinear, that is, if $f, f_1, f_2 : X \longrightarrow Y$ and $g, g_1, g_2 : Y \longrightarrow Z$ are morphisms while $\lambda_1, \lambda_2, \mu_1, \mu_2$ are scalars, then we have

I. Assem, F. U. Coelho, *Basic Representation Theory of Algebras*, Graduate Texts in Mathematics 283, https://doi.org/10.1007/978-3-030-35118-2_2

$$g \circ (\lambda_1 f_1 + \lambda_2 f_2) = \lambda_1 (g \circ f_1) + \lambda_2 (g \circ f_2)$$

and

$$(\mu_1 g_1 + \mu_2 g_2) \circ f = \mu_1 (g_1 \circ f) + \mu_2 (g_2 \circ f).$$

In general, we do not need to assume that the Hom-spaces of a **k**-category are finite dimensional **k**-vector spaces, though, in all of our examples, that will be the case.

Clearly, if $\mathscr{C}$ is a **k**-category and X an object in $\mathscr{C}$, then the **k**-vector space $\operatorname{End}_{\mathscr{C}} X$, of all morphisms from X to itself, has a natural **k**-algebra structure. It is in general infinite dimensional.

Often, one needs to consider not only the objects of a **k**-category, but also their finite direct sums and products. This leads to the following definition.

Definition II.1.2. A **k**-category $\mathscr{C}$ is **k-linear** if it is additive, that is, if every finite family of objects in $\mathscr{C}$ admits a direct sum and a direct product.

It is well-known that, given a finite family $\{X_1, \cdots, X_n\}$ of objects in a **k**-linear category $\mathscr{C}$, then its direct sum $\oplus_{i=1}^n X_i$ and its direct product $\prod_{i=1}^n X_i$ are isomorphic. In particular, the empty sum and the empty product are isomorphic and called the **zero object**. Predictably, the zero object is denoted by 0.

Example II.1.3. Examples of **k**-categories abound. Let, for instance, A be a finite dimensional **k**-algebra. Then the category $\operatorname{mod} A$ of all finitely generated right A-modules is a **k**-linear category. Also, the full subcategories $\operatorname{proj} A$ of projective objects and $\operatorname{inj} A$ of injective objects in $\operatorname{mod} A$, are **k**-linear.

Let, as before, $\operatorname{ind} A$ denote a full subcategory of $\operatorname{mod} A$ whose objects form a complete set of representatives of the isoclasses of indecomposable A-modules. Then, $\operatorname{ind} A$, and actually every full subcategory of $\operatorname{ind} A$, is a **k**-category, but not a **k**-linear category.

Another class of examples is as follows: let M be an A-module, not necessarily indecomposable and $\operatorname{add} M$ denote the full subcategory of $\operatorname{mod} A$ where objects are all direct sums of indecomposable direct summands of M. Then, $\operatorname{add} M$ is a **k**-linear category, see Exercise II.1.1.

The appropriate notion of functor between **k**-categories is that of **k**-functor.

Definition II.1.4. Let $\mathscr{C}, \mathscr{D}$ be **k**-categories. A (covariant or contravariant) functor $F : \mathscr{C} \longrightarrow \mathscr{D}$ is a **k-functor** if, for each pair of morphisms $f, g : X \longrightarrow Y$ in $\mathscr{C}$ and each pair of scalars λ, μ in **k**, we have

$$F(\lambda f + \mu g) = \lambda F(f) + \mu F(g).$$

Reformulating, F is a **k**-functor whenever, for each pair of objects X, Y in $\mathscr{C}$ the mapping $f \mapsto Ff$ induced by F on the Hom-spaces is a **k**-linear map.

For instance, for every object X in $\mathscr{C}$, the functors $\mathrm{Hom}_{\mathscr{C}}(-, X)$ and $\mathrm{Hom}_{\mathscr{C}}(X, -)$ from $\mathscr{C}$ to the category $\mathrm{Mod}\,\mathbf{k}$ of all (not necessarily finite dimensional) **k**-vector spaces are **k**-functors. Unless otherwise specified, all functors we deal with are **k**-functors.

We need the definition of a subfunctor. Let $\mathscr{C}, \mathscr{D}$ be **k**-categories and $F, G : \mathscr{C} \longrightarrow \mathscr{D}$ **k**-functors. We say that F is a **subfunctor** of G (and we write $F \subseteq G$) if there exists a functorial monomorphism $\varphi : F \longrightarrow G$, that is, for every object X in $\mathscr{C}$ there exists a monomorphism $\varphi_X : FX \longrightarrow GX$, which is compatible with the morphisms in $\mathscr{C}$. If, for instance, F and G are covariant, this means that, for every morphism $f : X \longrightarrow Y$ in $\mathscr{C}$, we have a commutative square:

$$\begin{array}{ccc} FX & \xrightarrow{\varphi_X} & GX \\ {\scriptstyle Ff}\downarrow & & \downarrow{\scriptstyle Gf} \\ FY & \xrightarrow{\varphi_Y} & GY \end{array}$$

where φ_X and φ_Y are monomorphisms. The definition is similar if F and G are contravariant.

We recall the notion of an ideal in a **k**-category.

Definition II.1.5. An **ideal** $\mathscr{I}$ in a **k**-category $\mathscr{C}$ is defined by the following data: for each pair of objects X, Y in $\mathscr{C}$, there exists a **k**-subspace $\mathscr{I}(X, Y)$ of $\mathrm{Hom}_{\mathscr{C}}(X, Y)$ such that:

(a) $f \in \mathscr{I}(X, Y)$ and $h \in \mathrm{Hom}_{\mathscr{C}}(W, X)$ imply $fh \in \mathscr{I}(W, Y)$, and
(b) $f \in \mathscr{I}(X, Y)$ and $g \in \mathrm{Hom}_{\mathscr{C}}(Y, Z)$ imply $gf \in \mathscr{I}(X, Z)$.

In other words, an ideal $\mathscr{I}$ is a family $\{\mathscr{I}(X, Y)\}_{X,Y}$ of **k**-subspaces of the Hom-spaces, which is stable under left and right compositions with arbitrary morphisms in $\mathscr{C}$.

For instance, let $\mathscr{C}, \mathscr{D}$ be **k**-categories and $F : \mathscr{C} \longrightarrow \mathscr{D}$ a **k**-functor. Its **kernel** $\mathscr{K} = \mathrm{Ker}\, F$ is defined by assigning to each pair of objects X, Y in $\mathscr{C}$ the set

$$\mathscr{K}(X, Y) = \{f \in \mathrm{Hom}_{\mathscr{C}}(X, Y) : Ff = 0\},$$

which is clearly a **k**-subspace of $\mathrm{Hom}_{\mathscr{C}}(X, Y)$. It is easily verified that these data define an ideal $\mathscr{K}$ of $\mathscr{C}$.

Given an ideal $\mathscr{I}$ in a **k**-category $\mathscr{C}$, one can define the **quotient category** $\mathscr{C}/\mathscr{I}$. This is the category having the same class of objects as $\mathscr{C}$ and the set of morphisms from the object X to the object Y is the quotient space:

$$\mathrm{Hom}_{\mathscr{C}/\mathscr{I}}(X, Y) = \frac{\mathrm{Hom}_{\mathscr{C}}(X, Y)}{\mathscr{I}(X, Y)}.$$

To define the composition, let $f : X \longrightarrow Y$ and $g : Y \longrightarrow Z$ be morphisms in $\mathscr{C}$ and set:

$$(g + \mathscr{I}(Y, Z)) \circ (f + \mathscr{I}(X, Y)) = (g \circ f) + \mathscr{I}(X, Z).$$

Because of the definition of ideal, this yields a well-defined operation

$$\mathrm{Hom}_{\mathscr{C}/\mathscr{I}}(Y, Z) \times \mathrm{Hom}_{\mathscr{C}/\mathscr{I}}(X, Y) \longrightarrow \mathrm{Hom}_{\mathscr{C}/\mathscr{I}}(X, Z),$$

which is the required composition.

With this definition, $\mathscr{C}/\mathscr{I}$ inherits from $\mathscr{C}$ the structure of a **k**-category. Further, if $\mathscr{C}$ is **k**-linear, then it is easily seen that so is $\mathscr{C}/\mathscr{I}$. In addition, if the Hom-spaces in $\mathscr{C}$ are finite dimensional vector spaces, then the Hom-spaces in $\mathscr{C}/\mathscr{I}$ are also finite dimensional vector spaces. There is a natural functor from $\mathscr{C}$ to $\mathscr{C}/\mathscr{I}$, mapping each object to itself, and each morphism $f \in \mathrm{Hom}_{\mathscr{C}}(X, Y)$ to its residual class $f + \mathscr{I}(X, Y) \in \mathrm{Hom}_{\mathscr{C}/\mathscr{I}}(X, Y)$. This functor is called the **projection functor** from $\mathscr{C}$ to $\mathscr{C}/\mathscr{I}$. It is clearly full and dense, and its kernel is precisely the ideal $\mathscr{I}$.

II.1.2 Defining the radical of mod *A*

Motivated by the analogy between categories and algebras, we expect that all the "significant" information of mod A is contained in its radical, in such a way that the quotient of mod A by its radical is semisimple. Following a familiar strategy, we start by defining the radical on ind A and then extend this definition to the whole of mod A. In addition, the radical has to be an ideal in mod A, exactly as the radical of an algebra is an ideal in it. Thus, with each pair M, N of indecomposable A-modules, we wish to associate a subspace $\mathrm{rad}_A(M, N)$ of $\mathrm{Hom}_A(M, N)$, stable under left and right composition by arbitrary morphisms.

A natural requirement is that, if $M = N$, then the radical $\mathrm{rad}_A(M, N)$ should coincide with the radical $\mathrm{rad}\,\mathrm{End}_A M$ of the endomorphism algebra $\mathrm{End}\, M$ of M. Because we are assuming that M is indecomposable, the algebra $\mathrm{End}\, M$ is local; thus, its radical consists of all noninvertible elements, that is, all nonisomorphisms from M to itself. Generalising this observation to the case where M is perhaps not equal to N, we are led to define the subspace $\mathrm{rad}_A(M, N)$ to consist of all nonisomorphisms from M to N.

One way to extend this definition to decomposable modules is as follows. Let M, N be arbitrary modules and $M = \oplus_{i=1}^m M_i$, $N = \oplus_{j=1}^n N_j$ direct sum decompositions, with all M_i, N_j indecomposable. To these decompositions, we associate the projections $p_j : N \longrightarrow N_j$ and injections $q_i : M_i \longrightarrow M$. Because we want the radical to be an ideal of mod A, it is reasonable to require that $f : M \longrightarrow N$ belongs to $\mathrm{rad}_A(M, N)$ if and only if, for all i and j, the morphism $p_j f q_i : M_i \longrightarrow N_j$ belongs to $\mathrm{rad}_A(M_i, N_j)$, that is, it is not an isomorphism. It turns out that this requirement suffices to define an ideal of mod A. We recall from Subsection II.1.3

that a morphism $f: X \longrightarrow Y$ in a category is a section (or a retraction) if there exists $f': Y \longrightarrow X$ such that $f'f = 1_X$ (or $ff' = 1_Y$ respectively).

Lemma II.1.6. *There exists a unique ideal* rad_A *of* $\operatorname{mod} A$ *such that, if* M, N *are* A*-modules, then* $\operatorname{rad}_A(M, N)$ *consists of all morphisms* $f: M \longrightarrow N$ *such that, for every section* $q : M' \longrightarrow M$ *and retraction* $p: N \longrightarrow N'$*, the composition* $pfq : M' \longrightarrow N'$ *is not an isomorphism.*

Proof. We first show that we can assume that M' and N' are indecomposable. Namely, we claim that a morphism $f : M \longrightarrow N$ belongs to $\operatorname{rad}_A(M, N)$ if and only if for every section $q : M' \longrightarrow M$ and retraction $p : N \longrightarrow N'$, with M', N' indecomposable, the composition $pfq : M' \longrightarrow N'$ is not an isomorphism. Indeed, sufficiency is trivial; therefore, let us prove necessity. Assume on the contrary that there exist a section $q : M' \longrightarrow M$ and a retraction $p : N \longrightarrow N'$ such that the composition $g = pfq : M' \longrightarrow N'$ is an isomorphism. Then, $g^{-1}pfq = 1_{M'}$. Let X be an indecomposable summand of M'; then, for the injection $v : X \longrightarrow M'$ and the projection $u : M' \longrightarrow X$, we have $uv = 1_X$. But then

$$1_X = uv = ug^{-1}pfqv = (ug^{-1}p)f(qv)$$

is an isomorphism, with $ug^{-1}p : N \longrightarrow X$ a retraction, $qv : X \longrightarrow M$ a section and X indecomposable. This completes the proof of our claim.

Clearly, if M, N are given, then the property in the statement uniquely defines a subset $\operatorname{rad}_A(M, N)$ of $\operatorname{Hom}_A(M, N)$. We thus have to prove that these data define an ideal in $\operatorname{mod} A$.

Let $f, g \in \operatorname{rad}_A(M, N)$ and $\lambda, \mu \in \mathbf{k}$. Then, for every section $q : M' \longrightarrow M$ and retraction $p : N \longrightarrow N'$, with M', N' indecomposable, the composition $pfq: M' \longrightarrow N'$ is not an isomorphism. We have two cases to consider. If $M' \ncong N'$, then

$$p\,(\lambda f + \mu g)\,q = \lambda\,(pfq) + \mu(pgq)$$

is clearly not an isomorphism. On the other hand, if $M' \cong N'$, then the above linear combination belongs to $\operatorname{Hom}_A\left(M', N'\right) \cong \operatorname{End} M'$. Because the latter is a local algebra, the sum of two noninvertible elements (radical elements) is noninvertible (thus, it belongs to the radical). This shows that $\operatorname{rad}_A(M, N)$ is a subspace of $\operatorname{Hom}_A(M, N)$.

Let now $f \in \operatorname{rad}_A(M, N)$ and $g \in \operatorname{Hom}_A(L, M)$. We claim that $fg \in \operatorname{rad}_A(L, N)$. If this is not the case, then there exist a section $q : L' \longrightarrow L$ and a retraction $p : N \longrightarrow N'$, with L', N' indecomposable, such that the composition $p(fg)q : M' \longrightarrow N'$ is an isomorphism. But now $p(fg)q = (pf)(gq)$. Hence, gq is a section and the invertibility of $pf(gq)$ contradicts the hypothesis that f belongs to $\operatorname{rad}_A(M, N)$. This establishes our claim. The proof that the radical is stable under left composition by arbitrary morphisms is similar. □

The previous lemma justifies the definition of the radical of the module category.

Definition II.1.7. The **radical** rad_A of $\operatorname{mod} A$ is the unique ideal such that, if M, N are A-modules, then $\operatorname{rad}_A(M, N)$ consists of all morphisms $f : M \longrightarrow N$ such that, for every section $q : M' \longrightarrow M$ and retraction $p : N \longrightarrow N'$, the composition $pfq : M' \longrightarrow N'$ is not an isomorphism. The morphisms lying in some $\operatorname{rad}_A(M, N)$ are called **radical morphisms**.

From this definition, we have an immediate consequence:

Corollary II.1.8. *Let M, N be indecomposable A-modules.*

(a) *If $M \ncong N$, then* $\operatorname{rad}_A(M, N) = \operatorname{Hom}_A(M, N)$.
(b) *If $M \cong N$, then* $\operatorname{rad}_A(M, N) \cong \operatorname{rad} \operatorname{End} M$ *consists of all nonisomorphisms, that is, of the nilpotent endomorphisms.* □

Also, it is easily seen that

$$\operatorname{rad}_A\left(\oplus_{i=1}^m M_i, \oplus_{j=1}^n N_j\right) = \oplus_{i=1}^m \oplus_{j=1}^n \operatorname{rad}_A\left(M_i, N_j\right),$$

see Exercise II.1.2.

We now prove a first characterisation of the radical, which is sometimes used as a definition.

Corollary II.1.9. *Let M, N be A-modules. A morphism $f : M \longrightarrow N$ is radical if and only if, for every indecomposable module X and morphisms $u : X \longrightarrow M$, $v : N \longrightarrow X$, the composition vfu is not an isomorphism.*

Proof. Let X be an indecomposable module, and $u : X \longrightarrow M$, $v : N \longrightarrow X$ morphisms. If vfu is an isomorphism, then v is a retraction and u is a section. Thus, $f \notin \operatorname{rad}_A(M, N)$. The converse is obvious. □

In the case of one of the modules M, N being indecomposable, the definition of $\operatorname{rad}_A(M, N)$ becomes simpler.

Corollary II.1.10. *Let $f : M \longrightarrow N$ be a morphism of A-modules.*

(a) *If M is indecomposable, then f is radical if and only if f is not a section.*
(b) *If N is indecomposable, then f is radical if and only if f is not a retraction.*

Proof. We only prove (a), because the proof of (b) is dual.

Assume $f \notin \operatorname{rad}_A(M, N)$. Then there exist a section $q : M' \longrightarrow M$ and a retraction $p : N \longrightarrow N'$ such that $pfq : M' \longrightarrow N'$ is an isomorphism. However, the indecomposability of M implies that q is an isomorphism. Hence, so is pf. Therefore, f is a section. Conversely, if f is a section, then there exists a retraction f' such that $f'f = 1_M$. But then $f \notin \operatorname{rad}_A(M, N)$. □

Example II.1.11. Let A be given by the quiver:

$$1 \bullet \underset{\beta}{\overset{\alpha}{\rightleftarrows}} \bullet 2$$

bound by the relation $\beta\alpha = 0$. The indecomposable projective A-modules

$$P_1 = \begin{smallmatrix} 1 \\ 2 \\ 1 \end{smallmatrix} \qquad \text{and} \qquad P_2 = \begin{smallmatrix} 2 \\ 1 \end{smallmatrix}$$

are uniserial. Clearly, $\operatorname{rad}_A(P_1, P_1) = \operatorname{rad}\operatorname{End} P_1$ is one-dimensional and generated by the morphism $f : P_1 \longrightarrow P_1$ having as an image the simple socle of P_1 (and as a kernel its radical, which is isomorphic to P_2). Observe that $f^2 = 0$, that is, f is nilpotent. On the other hand, $\operatorname{rad}_A(P_2, P_2) = \operatorname{rad}\operatorname{End} P_2 = 0$, whereas $\operatorname{rad}_A(P_2, P_1) = \operatorname{Hom}_A(P_2, P_1)$ is one-dimensional, generated by the inclusion of P_2 as radical of P_1. Finally, $\operatorname{rad}_A(P_1, P_2) = \operatorname{Hom}_A(P_1, P_2)$ is one-dimensional, generated by the morphism $P_1 \longrightarrow P_2$ having as an image the simple socle of P_2.

We finish this subsection by proving that, as expected, the quotient of $\operatorname{mod} A$ by its radical is a semisimple category. We recall the definition of the latter. Let Λ be a set and $(\mathscr{C}_\lambda)_{\lambda\in\Lambda}$ a collection of **k**-linear categories. The **direct sum** of the $\mathscr{C}_\lambda$ is the full subcategory $\oplus_\lambda \mathscr{C}_\lambda$ of $\Pi_\lambda \mathscr{C}_\lambda$ consisting of all the objects $(X_\lambda)_{\lambda\in\Lambda}$ such that $X_\lambda = 0$ for all but at most finitely many $\lambda \in \Lambda$ with the obvious morphisms. A **k**-linear category is called **semisimple** if it is equivalent to the direct sum of categories of the form $\operatorname{mod} K$, with K a skew field containing **k**.

Corollary II.1.12. *If A is an algebra, then the category* $\operatorname{mod} A/\operatorname{rad} A$ *is semisimple.*

Proof. Let $(M_\lambda)_{\lambda\in\Lambda}$ denote a complete set of representatives of the isoclasses of indecomposable A-modules. Clearly, Λ is in general an infinite set. For each $\lambda \in \Lambda$, the algebra $\operatorname{End} M_\lambda$ is local and therefore

$$K_\lambda = \frac{\operatorname{End} M_\lambda}{\operatorname{rad}(\operatorname{End} M_\lambda)}$$

is a skew overfield of **k**. We consider the functor

$$F\colon \operatorname{mod} A \longrightarrow \Pi_\lambda \operatorname{mod} K_\lambda$$

defined on the objects as follows: for each X in $\operatorname{mod} A$, we set

$$F(X) = \left(\frac{\operatorname{Hom}_A(M_\lambda, X)}{\operatorname{rad}_A(M_\lambda, X)}\right)_{\lambda\in\Lambda}$$

and in the obvious way on the morphisms.

We claim that the essential image of the functor F is the full subcategory $\oplus_\lambda \operatorname{mod} K_\lambda$ of $\Pi_\lambda \operatorname{mod} K_\lambda$.

Indeed, let M be an A-module. Because of the Krull–Schmidt theorem, we can write $M = \oplus_\lambda M_\lambda^{m_\lambda}$, where the M_λ are equal to zero for all but at most finitely many values of λ. Then,

$$F(M) = \left(\frac{\mathrm{Hom}_A(M_\lambda, M_\lambda^{m_\lambda})}{\mathrm{rad}_A(M_\lambda, M_\lambda^{m_\lambda})} \right)_{\lambda \in \Lambda} = (K_\lambda^{m_\lambda})_{\lambda \in \Lambda}$$

which is indeed an object in $\oplus_\lambda \bmod K_\lambda$.

This shows that the essential image of F lies inside $\oplus_\lambda \bmod K_\lambda$, so that F is actually a functor from $\bmod A$ to $\oplus_\lambda \bmod K_\lambda$. As such, this functor is dense: indeed, let X be an object in $\oplus_\lambda \bmod K_\lambda$, then $X \cong (K_\lambda^{m_\lambda})_{\lambda \in \Lambda}$ where the m_λ equal zero for all but at most finitely many values of λ. Setting $M = \oplus_\lambda M_\lambda^{m_\lambda}$, we see that $F(M) \cong X$. This proves the density of $F\colon \bmod A \longrightarrow \oplus_\lambda \bmod K_\lambda$. Fullness is proved in exactly the same way.

Finally, it is easily seen that the kernel of F is precisely the ideal rad_A. The statement now follows. □

II.1.3 Characterisations of the radical

The radical of an algebra is commonly defined as being the intersection of all maximal right ideals, and then it equals the intersection of all maximal left ideals. As stated in Subsection I.1.2, it can also be seen as the set of all elements a in the algebra such that, for every x, the element $1-ax$ is right invertible, and then it equals the set of all elements a such that, for every x, the element $1 - xa$ is left invertible. The purpose of the present subsection is to provide similar characterisations for the radical of a module category.

Our first observation is that the radical of $\bmod A$ defines a subbifunctor $\mathrm{rad}_A(-, ?)$ of $\mathrm{Hom}_A(-, ?)$. Indeed, let N be an A-module. We define a subfunctor $\mathrm{rad}_A(-, N)$ of the contravariant Hom-functor $\mathrm{Hom}_A(-, N)$ by setting

$$\mathrm{rad}_A(-, N)(M) = \mathrm{rad}_A(M, N)$$

and, for a morphism $f : M' \longrightarrow M$,

$$\mathrm{rad}_A(-, N)(f) = \mathrm{Hom}_A(f, N) : \mathrm{rad}_A(M, N) \longrightarrow \mathrm{rad}_A(M', N).$$

Indeed, it follows from the definition of ideal in a category that, if $v \in \mathrm{rad}_A(M, N)$, then $\mathrm{Hom}_A(f, N)(v) = vf$ belongs to $\mathrm{rad}_A\big(M', N\big)$. Thus, $\mathrm{Hom}_A(f, N)$ is indeed a map from $\mathrm{rad}_A(M, N)$ to $\mathrm{rad}_A\big(M', N\big)$.

In exactly the same way, for a fixed module M, we define a subfunctor $\mathrm{rad}_A(M, -)$ of the covariant Hom-functor $\mathrm{Hom}_A(M, -)$. As required, this defines a subbifunctor $\mathrm{rad}_A(-, ?)$ of $\mathrm{Hom}_A(-, ?)$.

For our first lemma, we need one more definition. A proper subfunctor F of a functor G is **maximal** if, whenever F' is a subfunctor of G such that $F \subseteq F'$ then $F' = F$ or $F' = G$.

Lemma II.1.13. *Let $\mathscr{C}$ be a linear category and X an object in $\mathscr{C}$. Then, there exist bijections between:*

(a) *The maximal right ideals of* $\operatorname{End}_{\mathscr{C}} X$ *and the maximal subfunctors of* $\operatorname{Hom}_{\mathscr{C}}(-, X)$.
(b) *The maximal left ideals of* $\operatorname{End}_{\mathscr{C}} X$ *and the maximal subfunctors of* $\operatorname{Hom}_{\mathscr{C}}(X, -)$.

Proof. We construct the bijection in (a) because the construction in (b) is similar.

Let I be a maximal right ideal of $\operatorname{End}_{\mathscr{C}} X$. We define a corresponding subfunctor F_I of $\operatorname{Hom}_{\mathscr{C}}(-, X)$. For an object Y in $\mathscr{C}$, we let $F_I(Y)$ be the subset of $\operatorname{Hom}_{\mathscr{C}}(Y, X)$ defined by

$$F_I(Y) = \{f \in \operatorname{Hom}_A(Y, X) : fg \in I \text{ for every } g : X \longrightarrow Y\}.$$

Clearly, $F_I(Y)$ is a subspace of $\operatorname{Hom}_A(Y, X)$. Also, F_I is made into a functor by setting, for $u : Y' \longrightarrow Y$,

$$F_I(u) = \operatorname{Hom}_A(u, X).$$

Indeed, if $f \in F_I(Y)$ and $g : X \longrightarrow Y'$ is arbitrary, then $(fu)g = f(ug) \in I$ because $ug : X \longrightarrow Y$. This shows that $F_I(u)$ is well-defined as a map from $F_I(Y)$ to $F_I(Y')$, and, therefore, that F_I is a subfunctor of $\operatorname{Hom}_{\mathscr{C}}(-, X)$. In addition, it is apparent that $F_I(X) = I$.

We have to prove that F_I is a maximal subfunctor. Assume that F is a functor such that $F_I \subseteq F \subseteq \operatorname{Hom}_{\mathscr{C}}(-, X)$. In particular, $I = F_I(X) \subseteq F(X) \subseteq \operatorname{End}_{\mathscr{C}} X$. Additionally, $F(X)$ is actually a right ideal in $\operatorname{End}_{\mathscr{C}} X$: this indeed follows from the fact that, because of the definition of a subfunctor, we have a commutative square

$$\begin{array}{ccc} FX & \longrightarrow & \operatorname{Hom}_{\mathscr{C}}(X,X) \\ {\scriptstyle F(u)}\downarrow & & \downarrow{\scriptstyle \operatorname{Hom}_{\mathscr{C}}(u,X)} \\ FX & \longrightarrow & \operatorname{Hom}_{\mathscr{C}}(X,X) \end{array}$$

where $u : X \longrightarrow X$. Then, the maximality of I implies that we have one of the following two cases. In the first case, $F(X) = \operatorname{End}_{\mathscr{C}} X$ and then $f \in \operatorname{Hom}_{\mathscr{C}}(Y, X)$ and the commutative square

$$\begin{array}{ccc} 1_X \in FX & \longrightarrow & \operatorname{End}_{\mathscr{C}}(X) \ni 1_X \\ \downarrow & & \downarrow{\scriptstyle \operatorname{Hom}_{\mathscr{C}}(f,X)} \\ F(f)(1_X) \in FY & \longrightarrow & \operatorname{Hom}_{\mathscr{C}}(Y,X) \ni f \end{array}$$

give $F(f)(1_X) = f \in F(Y)$ so that $\operatorname{Hom}_{\mathscr{C}}(Y, X) = F(Y)$ for every object Y. Consequently, $F = \operatorname{Hom}_{\mathscr{C}}(-, X)$. In the second case, $F(X) = I$ and then $f \in$

$F(Y)$ and $g \in \mathrm{Hom}_{\mathscr{C}}(X, Y)$ together with the commutative square

$$\begin{array}{ccc} FY & \longrightarrow & \mathrm{Hom}_{\mathscr{C}}(Y,X) \\ {\scriptstyle F(g)}\downarrow & & \downarrow{\scriptstyle \mathrm{Hom}_{\mathscr{C}}(g,X)} \\ I = FX & \longrightarrow & \mathrm{End}_{\mathscr{C}}(X) \end{array}$$

give $F(g)(f) = fg \in F(X) = I$ and so $f \in F_I(Y)$ by definition. Hence, $F = F_I$. This completes the proof of the maximality of F_I.

Conversely, let F be a maximal subfunctor of $\mathrm{Hom}_{\mathscr{C}}(-, X)$. We claim that there exists a unique maximal right ideal I of $\mathrm{End}_{\mathscr{C}} X$ such that $F = F_I$. Let us set $I = F(X)$. Then, I is certainly a subspace of $\mathrm{End}_{\mathscr{C}} X$. The fact that it is a right ideal follows from the fact that F is a subfunctor, as seen before. It remains to prove that I is maximal. Certainly, there exists a maximal right ideal J of $\mathrm{End}_{\mathscr{C}} X$ such that $I \subseteq J$. Let $f \in F(Y)$ and $g : Y \longrightarrow X$. Then

$$F(g)(f) = fg \in F(X) = I \subseteq J$$

gives $f \in F_J(Y)$. Because $F(X) = I \subseteq J = F_J(X)$, we have $F \subseteq F_J$. Then, the maximality of F implies that $F = F_J$ and so $I = F(X) = F_J(X) = J$. This shows the maximality of I. Its uniqueness being obvious, the proof is complete. □

Corollary II.1.14. *Let $\mathscr{C}$ be a linear category and $f : X \longrightarrow Y$ a morphism in $\mathscr{C}$. Then:*

(a) *$f \in F(X)$ for every maximal subfunctor F of $\mathrm{Hom}_{\mathscr{C}}(-, Y)$ if and only if $1_Y - fg$ is invertible for every $g : Y \longrightarrow X$.*
(b) *$f \in F(Y)$ for every maximal subfunctor F of $\mathrm{Hom}_{\mathscr{C}}(X, -)$ if and only if $1_X - gf$ is invertible for every $g : Y \longrightarrow X$.*

Proof. We only prove (a), because the proof of (b) is similar.

Because of Lemma II.1.13 above, $f \in F(X)$ for every maximal subfunctor F of $\mathrm{Hom}_{\mathscr{C}}(-, Y)$ if and only if $fg \in I$ for every maximal right ideal I of $\mathrm{End}_{\mathscr{C}} Y$ and every morphism $g : Y \longrightarrow X$. This is the case if and only if $fg \in \mathrm{rad}\,\mathrm{End}_{\mathscr{C}} Y$ for every $g : Y \longrightarrow X$. A well-known property of the radical of an algebra implies that $1_Y - fg$ is invertible. Conversely, assume that $1_Y - fg$ is invertible for every $g : Y \longrightarrow X$, and let $h \in \mathrm{End}_{\mathscr{C}} Y$. Our condition implies that $1_Y - f(gh) = 1_Y - (fg)h$ is invertible for every such h. This shows that $fg \in \mathrm{rad}\,\mathrm{End}_{\mathscr{C}} Y$ and completes the proof. □

We can relax a bit the second condition of the previous corollary.

Lemma II.1.15. *Let $\mathscr{C}$ be a linear category and $f : X \longrightarrow Y$ a morphism in $\mathscr{C}$. Then:*

(a) *The morphism $1_Y - fg$ is invertible for every $g : Y \longrightarrow X$ if and only if it is right invertible for every $g : Y \longrightarrow X$.*

(b) *The morphism* $1_X - gf$ *is invertible for every* $g : Y \longrightarrow X$ *if and only if it is left invertible for every* $g : Y \longrightarrow X$.

Proof. We only prove (a), because the proof of (b) is similar.

Because the necessity is obvious, we show the sufficiency. If $1_Y - fg$ is right invertible, there exists h such that $(1_Y - fg)h = 1_Y$. Then, $h = 1_Y + fgh$ also admits a right inverse l, because of the hypothesis. But then $1_Y = hl = (1_Y + fgh)\,l = l + fg$ yields $l = 1_Y - fg$ and so h is also a left inverse of $1_Y - fg$, which is thus invertible. □

We now prove the equivalence of the conditions stated in (a) and (b) of Lemma II.1.15 (and thus of Corollary II.1.14).

Lemma II.1.16. *Let* $\mathscr{C}$ *be a linear category and* $f : X \longrightarrow Y$ *a morphism in* $\mathscr{C}$. *Then,* $1_X - gf$ *is invertible for every* $g : Y \longrightarrow X$ *if and only if* $1_Y - fg$ *is invertible for every* $g : Y \longrightarrow X$.

Proof. Assume that $1_X - gf$ is invertible and let h be its inverse. Then, $h\,(1_X - gf) = 1_X$ yields $h = 1_X + hgf$ and we have

$$(1_Y + fhg)\,(1_Y - fg) = 1_Y - fg + fhg - fhgfg = 1_Y - f\,(1_X - h + hgf)\,g = 1_Y.$$

Similarly, $(1_X - gf)\,h = 1_X$ yields $(1_Y - fg)\,(1_Y + fhg) = 1_Y$. Thus, $1_Y - fg$ is invertible. The converse is proven in exactly the same way. □

The reader should be aware that we use in this subsection the terminology "invertible, right invertible, left invertible" (instead of the more familiar "isomorphism, retraction, section" respectively) to underline the analogy between the radical of a category and that of an algebra.

We are now able to prove the main result of this subsection, which gives various equivalent characterisations of radical morphisms.

Theorem II.1.17. *Let* A *be a finite dimensional* **k**-*algebra and* $f : M \longrightarrow N$ *a morphism of* A-*modules. The following conditions are equivalent:*

(a) $f \in \mathrm{rad}_A\,(M, N)$.
(b) $f \in F(M)$ *for every maximal subfunctor* F *of* $\mathrm{Hom}_A(-, N)$.
(c) $f \in F(N)$ *for every maximal subfunctor* F *of* $\mathrm{Hom}_A(M, -)$.
(d) $1_N - fg$ *is invertible for every* $g : N \longrightarrow M$.
(e) $1_M - gf$ *is invertible for every* $g : N \longrightarrow M$.
(f) $1_N - fg$ *is right invertible for every* $g : N \longrightarrow M$.
(g) $1_M - gf$ *is left invertible for every* $g : N \longrightarrow M$.

Proof. We have proved the equivalence of conditions (b) to (g). It thus suffices to prove the equivalence of (f) with (a).

Let $\mathscr{R}\,(M, N)$ be the set

$$\{f \in \mathrm{Hom}_A(M, N) \mid 1_N - fg \text{ is right invertible for every } g : N \longrightarrow M\}.$$

(i) We first prove that, if M and N are indecomposable, then $\mathscr{R}(M, N)$ equals $\operatorname{rad}_A(M, N)$.

Indeed, assume $f \in \mathscr{R}(M, N)$, then f cannot be an isomorphism because, if it were, then $1_N - ff^{-1} = 0$ would be invertible, an absurdity. Thus, $\mathscr{R}(M, N) \subseteq \operatorname{rad}_A(M, N)$. Conversely, assume that $f : M \longrightarrow N$ is not an isomorphism. Then, for every $g : N \longrightarrow M$, the composition $fg : N \longrightarrow N$ is not an isomorphism either: for, if it were, then f would be a retraction, and hence an isomorphism owing to the indecomposability of M, and this is a contradiction. But then, because $\operatorname{End} N$ is local, the morphism $1_N - fg$ is (right) invertible.

(ii) We next prove that $\mathscr{R}$ defines an ideal of mod A.

Clearly, $0 \in \mathscr{R}(M, N)$. Also, $f \in \mathscr{R}(M, N)$ and $\lambda \in \mathbf{k}$ imply $\lambda f \in \mathscr{R}(M, N)$. We now show that, if $f_1, f_2 \in \mathscr{R}(M, N)$, then $f_1 + f_2 \in \mathscr{R}(M, N)$.

Let $g : N \longrightarrow M$ be arbitrary. Then, $1_N - f_1 g$ has a right inverse h_1 and $1_N - f_2 g h_1$ has a right inverse h_2. We claim that $h_1 h_2$ is a right inverse of $1_N - (f_1 + f_2)g$.

We first observe that $(1_N - f_1 g)\, h_1 = 1_N$ gives $h_1 - 1_N = f_1 g h_1$ whereas $(1_N - f_2 g h_1)\, h_2 = 1_N$ gives $h_2 - 1_N = f_2 g h_1 h_2$. Hence,

$$\begin{aligned}(1_N - (f_1 + f_2)g)h_1h_2 &= h_1h_2 - f_1gh_1h_2 - f_2gh_1h_2\\ &= h_1h_2 - (h_1 - 1_N)\, h_2 - (h_2 - 1_N)\\ &= 1_N\end{aligned}$$

This shows that $\mathscr{R}(M, N)$ is a $\mathbf{k}$-subspace of $\operatorname{Hom}_A(M, N)$.

Let now $f \in \mathscr{R}(M, N)$ and $u \in \operatorname{Hom}_A(L, M)$. Then, for every morphism $g : N \longrightarrow M$, the morphism $1_N - (fu)g = 1_N - f\,(ug)$ is right invertible. Therefore, $fu \in \mathscr{R}(M, N)$. Let $v \in \operatorname{Hom}_A(N, L)$ and $g : L \longrightarrow M$ be arbitrary. Then, because of the hypothesis, $1_N - f\,(gv)$ has a right inverse h, and $(1_N - fgv)\, h = 1_N$ yields $h - 1_N = fgvh$. Hence,

$$\begin{aligned}(1_L - vfg)\,(1_L + vhfg) &= 1_L + vhfg - vfg - vfgvhfg\\ &= 1_L + vhfg - vfg - v\,(h - 1_N)\, fg\\ &= 1_L\end{aligned}$$

Therefore, $vf \in \mathscr{R}(M, L)$. This completes the proof of (ii).

(iii) Let M, N be arbitrary modules. We prove that $f \in \mathscr{R}(M, N)$ if and only if, for every section $q : M' \longrightarrow M$ and retraction $p : N \longrightarrow N'$ with M', N' indecomposable, we have $pfq \in \mathscr{R}\left(M', N'\right)$.

Because necessity follows from (ii), we prove sufficiency. Assume that $M = \oplus_{i=1}^m M_i$, $N = \oplus_{j=1}^n N_j$ are direct sum decompositions with all M_i, N_j indecomposable. Associate with these decompositions the projections $p_i : M \longrightarrow M_i$, $p'_j : N \longrightarrow N_j$ and injections $q_i : M_i \longrightarrow M$, $q'_j : N_j \longrightarrow N$.

The hypothesis asserts that, for each pair (i, j), we have $p'_j f q_i \in \mathscr{R}(M_i, N_j)$. But then, using again that $\mathscr{R}$ is an ideal, we get that:

$$f = 1_N f 1_M = \left(\sum_j q'_j p'_j\right) f \left(\sum_i q_i p_i\right) = \sum_{i,j} q'_j \left(p'_j f q_i\right) p_i \in \mathscr{R}(M, N).$$

The statement of the Theorem then follows from (i), (ii), (iii) and Lemma II.1.6. □

As a consequence, we obtain one further characterisation of the radical (which the reader can compare with Corollary II.1.14 above).

Corollary II.1.18. *Let $f : M \longrightarrow N$ be a morphism of A-modules. The following conditions are equivalent:*

(a) $f \in \operatorname{rad}_A(M, N)$.
(b) *gf is nilpotent for every morphism $g : N \longrightarrow M$.*
(c) *fg is nilpotent for every morphism $g : N \longrightarrow M$.*

Proof. We only prove the equivalence of (a) and (b), because the equivalence of (a) and (c) is similar.

Assume first that gf is nilpotent for every morphism $g : N \longrightarrow M$. Let $n > 0$ be such that $(gf)^n = 0$ but $(gf)^{n-1} \neq 0$. Then, $1_M + (gf) + \ldots + (gf)^{n-1}$ is an inverse for $1_M - gf$, which is therefore invertible.

Conversely, assume that $1_M - gf$ is invertible for every morphism $g : N \longrightarrow M$. Then we have $gf \in \operatorname{rad}\operatorname{End} M$. But every element in $\operatorname{rad}\operatorname{End} M$ is nilpotent, which gives the result. □

Exercises for Section II.1

Exercise II.1.1. Let $\mathscr{C}$ be a **k**-category. Prove that there exists a **k**-linear category add $\mathscr{C}$, unique up to isomorphism (called the **linearisation** of $\mathscr{C}$) such that:

(a) $\mathscr{C}$ is a full subcategory of add $\mathscr{C}$.
(b) If $\mathscr{D}$ is a **k**-linear category and $F : \mathscr{C} \longrightarrow \mathscr{D}$ a **k**-functor, then there exists a unique **k**-functor $F' : \operatorname{add}\mathscr{C} \longrightarrow \mathscr{D}$ whose restriction to $\mathscr{C}$ equals F.

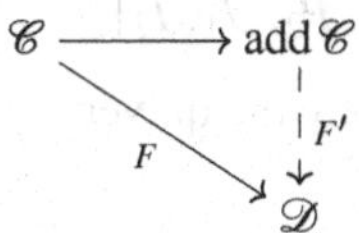

Exercise II.1.2. Let A be an algebra and $\mathscr{I}$ an ideal in mod A. Assume that $M = \bigoplus_{i=1}^m M_i$, $M = \bigoplus_{j=1}^n N_i$ are A-modules with the M_i, N_j indecomposable, with associated inclusions and projections $q_i : M_i \longrightarrow M$, $p'_j : N \longrightarrow N_j$.

(a) Let $f : M \longrightarrow N$ be a morphism. Prove that $f \in \mathscr{I}(M, N)$ if and only if $p'_j f q_i \in \mathscr{I}(M_i, N_j)$ for all i, j.
(b) Deduce that $\mathscr{I}(M, N) \cong \bigoplus_{i=1}^m \bigoplus_{j=1}^n \mathscr{I}(M_i, N_j)$.

Exercise II.1.3. Let $\mathscr{C}$ be a **k**-linear category and $\mathscr{I}$, $\mathscr{J}$ ideals in $\mathscr{C}$.

(a) We define the product $\mathscr{I}\mathscr{J}$ as follows: for each pair of objects X, Y in $\mathscr{C}$, we let $\mathscr{I}\mathscr{J}(X, Y)$ be the set of all sums $\sum_i g_i f_i$ where $g_i \in \mathscr{I}(Z_i, Y)$, $f_i \in \mathscr{J}(X, Z_i)$ for some objects Z_i in $\mathscr{C}$. Prove that these data define an ideal in $\mathscr{C}$.
(b) We define inductively, for $m \geq 1$, $\mathscr{I}^m = \mathscr{I}^{m-1}.\mathscr{I}$ and $\mathscr{I}^\infty = \cap_{m\geq 1}\mathscr{I}^m$. Prove that these data define ideals in $\mathscr{C}$.

Exercise II.1.4. Let $\mathscr{C}$ be a **k**-linear category and $\mathscr{I}$, $\mathscr{J}$ ideals in $\mathscr{C}$. We define $\mathscr{I} \cap \mathscr{J}$ by:

$$\left(\mathscr{I} \cap \mathscr{J}\right)(X, Y) = \mathscr{I}(X, Y) \cap \mathscr{J}(X, Y)$$

for all X, Y in $\mathscr{C}$. Prove that $\mathscr{I} \cap \mathscr{J}$ is an ideal in $\mathscr{C}$ and that it contains the product ideal $\mathscr{I}\mathscr{J}$.

Exercise II.1.5. Let A be an algebra, $\mathscr{I}$ an ideal of mod A and P : mod $A \longrightarrow$ (mod $A)/\mathscr{I}$ the canonical projection. Prove that $\mathscr{I} \subseteq \text{rad}_A$ if and only if a morphism f in mod A is such that $P(f)$ is an isomorphism, then f is an isomorphism.

Exercise II.1.6. Let $\mathscr{C}$ be a **k**-linear category and $\mathscr{I}$, $\mathscr{J}$ ideals in $\mathscr{C}$. Prove that, if $\mathscr{I}(X, X) \subseteq \mathscr{J}(X, X)$ for all objects X in $\mathscr{C}$, then $\mathscr{I} \subseteq \mathscr{J}$.

Exercise II.1.7. Let A be given by the quiver

$$\alpha \circlearrowright 1 \xrightarrow{\ \beta\ } 2$$

bound by $\alpha^3 = 0, \alpha^2\beta = 0$.Let P_1, P_2 and I_1, I_2 be the indecomposable projective and injective modules corresponding to the points 1 and 2 respectively. Compute $\text{rad}_A(M, N)$ for all $M, N \in \{P_1, P_2, I_1, I_2\}$.

Exercise II.1.8. Let A be given by the quiver

$$1 \underset{\beta}{\overset{\alpha}{\rightleftarrows}} 2$$

bound by $\beta\alpha\beta\alpha = 0$. Compute $\text{rad}_A(M, N)$ for all indecomposable modules M, N.

Exercise II.1.9. Let A be a finite dimensional algebra and M, N modules. Prove that, for each $m \geq 2$, $\mathrm{rad}_A^m(M, N)$ is a **k**-subspace of $\mathrm{rad}_A^{m-1}(M, N)$.

Exercise II.1.10. Let A be a finite dimensional algebra and M an A-module. As in Example II.1.3, we denote by add M the full subcategory of mod A consisting of all direct sums of direct summands of M.

(a) Prove that add M is the linearisation of the full subcategory of mod A consisting of all the indecomposable summands of M, see Exercise II.1.1.
(b) Denote by $\langle \mathrm{add}\, M \rangle$ the set of all morphisms in mod A which factor through an object in add M. That is, $f : X_A \longrightarrow Y_A$ lies in $\langle \mathrm{add}\, M \rangle$ whenever there exist M_0 in add M and morphisms $g : X \longrightarrow M_0$, $h : M_0 \longrightarrow Y$ such that $f = hg$. Prove that these data define an ideal in mod A.

II.2 Irreducible morphisms and almost split morphisms

II.2.1 Irreducible morphisms

If one admits that the relevant information about mod A (at least about indecomposable modules) lies in its radical, then it is reasonable to ask which morphisms generate all radical morphisms by successive compositions and linear combinations. Clearly, these are those radical morphisms between indecomposable modules that cannot be further factored as sums of compositions of other radical morphisms.

Now, let $f : L \longrightarrow M$, $g : M \longrightarrow N$ be radical morphisms. Their composition lies in the product of the ideal rad_A of mod A with itself, namely rad_A^2, see Exercise II.1.3. Given modules L, N, the **radical square** $\mathrm{rad}_A^2(L, N)$ is defined as the set of all sums of the form $\sum_i g_i f_i$ where each f_i is a radical morphism from L to some A-module M_i, and each g_i is a radical morphism from M_i to N.

Setting $M = \oplus_{i=1}^m M_i$, this may be rewritten as

$$\mathrm{rad}_A^2(L, N) = \{gf : \text{ for some } M \text{ in mod} A,\ f \in \mathrm{rad}_A(L, M),\ g \in \mathrm{rad}_A(M, N)\}.$$

In view of that, we are interested in exactly those morphisms that belong to the radical but not to the radical square. Dropping the assumption that these are morphisms between indecomposable modules, we get to the following definition. As usual, all modules are assumed to be finitely generated right modules over a finite dimensional **k**-algebra A.

Definition II.2.1. Let L, M be modules (not necessarily indecomposable). A morphism $f : L \longrightarrow M$ is called **irreducible** if:

(a) f is neither a section nor a retraction, and
(b) whenever $f = gh$, then h is a section or g is a retraction:

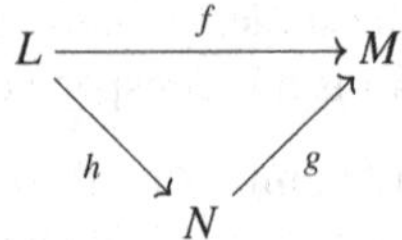

Clearly, this notion is self-dual, that is, $f: L \longrightarrow M$ is irreducible in $\operatorname{mod} A$ if and only if $\mathrm{D}f: \mathrm{D}M \longrightarrow \mathrm{D}L$ is irreducible in $\operatorname{mod} A^{op}$.

Before giving examples, we justify this definition by considering the case where L or M or both are indecomposable.

Lemma II.2.2. *Let $f: L \longrightarrow M$ be a morphism in* $\operatorname{mod} A$.

(a) *If L or M is indecomposable, and f is irreducible, then f is radical.*
(b) *If both L and M are indecomposable, then f is irreducible if and only if it belongs to* $\operatorname{rad}_A(L, M) \setminus \operatorname{rad}^2_A(L, M)$.

Proof.

(a) Assume that L is indecomposable, and $f: L \longrightarrow M$ is irreducible. Then f is not a section. Because of Corollary II.1.10, it is radical. The proof is similar if M is indecomposable.
(b) Assume that L and M are both indecomposable, then, again because of Corollary II.1.10, $f \in \operatorname{rad}_A(L, M)$ if and only if it is neither a section nor a retraction. In addition, $f \notin \operatorname{rad}^2_A(L, M)$ if and only if for every decomposition $f = gh$ with $h: L \longrightarrow X$ and $g: X \longrightarrow M$, we have $h \notin \operatorname{rad}_A(L, X)$ or $g \notin \operatorname{rad}_A(X, M)$. Invoking Corollary II.1.10 again, we see that this is the case if and only if h is a section or g is a retraction.

□

Another property of irreducible morphisms is that they are either injective or surjective.

Lemma II.2.3. *Every irreducible morphism is a monomorphism or an epimorphism.*

Proof. Let $f: L \longrightarrow M$ be irreducible and $f = jp$ its canonical factorisation through its image, with $p: L \longrightarrow \operatorname{Im} f$ surjective and $j: \operatorname{Im} f \longrightarrow M$ injective. Because f is irreducible, p is a section or j is a retraction. In the first case, p is an isomorphism and f a monomorphism, and, in the second case, j is an isomorphism and f an epimorphism. □

As a consequence, there are no irreducible morphisms from a module to itself because a monomorphism (or an epimorphism) $f: M \longrightarrow M$ is necessarily an isomorphism, see Lemma I.1.20.

We now give examples.

Example II.2.4. Let P be an indecomposable projective A-module. We claim that the inclusion morphism $j: \operatorname{rad} P \longrightarrow P$ is irreducible. Indeed, j is evidently neither a section nor a retraction. Assume $j = gh$ with $g: X \longrightarrow P$ and

$h\colon \operatorname{rad} P \longrightarrow X$. Suppose that g is not a retraction. Because P is projective, then g is not surjective. Therefore, $\operatorname{Im} g \subseteq \operatorname{rad} P$, that is, there exists $g'\colon X \longrightarrow \operatorname{rad} P$ such that $g = jg'$. But then $j = gh = jg'h$, which implies $g'h = 1_{\operatorname{rad} P}$, because j is a monomorphism. This proves that h is a section.
Dually, if I is indecomposable injective, then the projection morphism $I \longrightarrow I/\operatorname{soc} I$ is irreducible.

Example II.2.5. Let A be the path algebra of the quiver

$$1 \bullet \xleftarrow{\ \alpha\ } \bullet\, 2$$

Consider the indecomposable projective A-module $P_2 = \begin{smallmatrix}2\\1\end{smallmatrix}$ at the point 2. Then we have $\dim_{\mathbf{k}} \operatorname{Hom}_A(P_2, S_2) = 1$ and every nonzero morphism $f\colon P_2 \longrightarrow S_2$ is surjective with kernel S_1. We claim that every such morphism is irreducible.

Clearly, f is not a section, because it is a proper surjection, and not a retraction, because $P_2 \not\cong S_2$. Assume that there exists a factorisation $f = gh$, with $h\colon P_2 \longrightarrow X$ and $g\colon X \longrightarrow S_2$. Then, g is surjective. Hence, S_2 is a direct summand of the top of X. But A is a Nakayama algebra so, up to isomorphism, there are only two indecomposable A-modules having S_2 in their top, namely P_2 and S_2. Thus, $X \cong S_2 \oplus X'$ or $X \cong P_2 \oplus X'$. In addition, in the first case, the restriction $g_{|S_2}$ of g to S_2 is an isomorphism (so g is a retraction) and in the second case, the restriction $g_{|P_2}$ of g to P_2 is a scalar multiple of f (so h is a section).

Example II.2.6. The statement in Lemma II.2.2(b) ceases to be true if we stop assuming that both L and M are indecomposable. Indeed, assume that $f\colon L_1 \longrightarrow M$ is an irreducible morphism with both L_1 and M indecomposable. Let L_2 be an indecomposable module that is isomorphic to neither L_1 nor M and such that $\operatorname{rad}(\operatorname{End} L_2) \neq 0$. Let u be a nonzero morphism in $\operatorname{rad}(\operatorname{End} L_2)$ and $v\colon L_1 \longrightarrow L_2$ be arbitrary (maybe zero). Then the morphism $(f, 0)\colon L_1 \oplus L_2 \longrightarrow M$ is not irreducible. Indeed, it is certainly neither a section nor a retraction, but it admits the factorisation

$$(f, 0) = (f, 0)\begin{pmatrix} 1_{L_1} & 0 \\ v & u \end{pmatrix}$$

that is, the following diagram commutes:

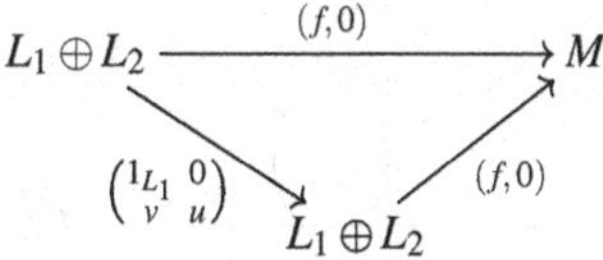

Assume that $\begin{pmatrix} 1_{L_1} & 0 \\ v & u \end{pmatrix}$ is a retraction. Then there exists a matrix $\begin{pmatrix} g & h \\ g' & h' \end{pmatrix}$ such that

$$\begin{pmatrix} g & h \\ vg + ug' & vh + uh' \end{pmatrix} = \begin{pmatrix} 1_{L_1} & 0 \\ v & u \end{pmatrix} \begin{pmatrix} g & h \\ g' & h' \end{pmatrix} = \begin{pmatrix} 1_{L_1} & 0 \\ 0 & 1_{L_2} \end{pmatrix}.$$

Thus, $h = 0$ and so $uh' = 1_{L_2}$, that is, u is a retraction, a contradiction.
On the other hand, it is clear that $(f, 0)$ belongs to the radical but not to the radical square, because $f \in \mathrm{rad}_A(L_1, M) \setminus \mathrm{rad}^2_A(L_1, M)$.

Explicit versions of this example are easy to construct. Consider for instance the algebra given by the quiver

$$1 \bullet \underset{\beta}{\overset{\alpha}{\rightleftarrows}} \bullet\, 2$$

bound by $\beta\alpha = 0$. As seen in Example II.2.4 above, the inclusion $\iota\colon S_1 \longrightarrow P_2$ of the radical S_1 in the indecomposable projective module P_2 is irreducible. Consider now the morphism $u\colon P_1 \longrightarrow P_1$ mapping the top of P_1 onto its socle and the inclusion $v\colon S_1 \longrightarrow P_1$. Then the morphism $(\iota, 0)\colon S_1 \oplus P_1 \longrightarrow P_2$ admits the factorisation $(\iota, 0) = (\iota, 0) \begin{pmatrix} 1_{S_1} & 0 \\ v & u \end{pmatrix}$.

Example II.2.7. Similarly, the statement in Lemma II.2.2(a) ceases to be true if we stop assuming that L or M is indecomposable. We show an example of an irreducible morphism between decomposable modules, which is not a radical morphism. Let $f\colon L \longrightarrow M$ be irreducible, with both L and M indecomposable. Let N be an indecomposable which is neither comparable to L nor to M in the sense that $\mathrm{Hom}_A(L, N) = 0$, $\mathrm{Hom}_A(N, L) = 0$, $\mathrm{Hom}_A(M, N) = 0$ and $\mathrm{Hom}_A(N, M) = 0$. We claim that the morphism $\begin{pmatrix} f & 0 \\ 0 & 1_N \end{pmatrix} : L \oplus N \longrightarrow M \oplus N$ is irreducible.

Indeed, it is not a section, because if it were, then there would exist a morphism $\begin{pmatrix} u & v \\ u' & v' \end{pmatrix} : M \oplus N \longrightarrow L \oplus N$ such that

$$\begin{pmatrix} u & v \\ u' & v' \end{pmatrix} \begin{pmatrix} f & 0 \\ 0 & 1_N \end{pmatrix} = \begin{pmatrix} 1_L & 0 \\ 0 & 1_N \end{pmatrix}.$$

But this implies that $uf = 1_L$ so that f is a section, a contradiction. In exactly the same way, we prove that it is not a retraction. So, assume that we have a factorisation

$$\begin{array}{ccc} L\oplus N & \xrightarrow{\begin{pmatrix} f & 0 \\ 0 & 1 \end{pmatrix}} & M\oplus N \\ & {\scriptstyle (u',v')} \searrow \quad \nearrow {\scriptstyle \begin{pmatrix} u \\ v \end{pmatrix}} & \\ & X & \end{array}$$

then we have

$$\begin{pmatrix} f & 0 \\ 0 & 1_N \end{pmatrix} = \begin{pmatrix} u \\ v \end{pmatrix} (u' \ v') = \begin{pmatrix} uu' & uv' \\ vu' & vv' \end{pmatrix} = \begin{pmatrix} uu' & 0 \\ 0 & vv' \end{pmatrix}$$

because $uv' \in \mathrm{Hom}_A(N, M) = 0$ and $vu \in \mathrm{Hom}_A(L, N) = 0$. Therefore, $f = uu'$ and $1_N = vv'$. In particular, v is a retraction and v' is a section. Additionally, because f is irreducible, u is a retraction or u' is a section.
If u is a retraction, then there exists $u''\colon M \longrightarrow X$ such that $uu'' = 1_M$. Then,

$$\begin{pmatrix} u \\ v \end{pmatrix} (u'' \ v') = \begin{pmatrix} uu'' & uv' \\ vu'' & vv' \end{pmatrix} = \begin{pmatrix} 1_M & 0 \\ 0 & 1_N \end{pmatrix}.$$

because $uv' = 0$ as said before, whereas $vu'' \in \mathrm{Hom}_A(M, N) = 0$. Therefore, $\begin{pmatrix} u \\ v \end{pmatrix}$ is a retraction.
If, on the other hand, u' is a section, then there exists $u''\colon X \longrightarrow L$ such that $u''u' = 1_L$. Then,

$$\begin{pmatrix} u'' \\ v \end{pmatrix} (u' \ v') = \begin{pmatrix} u''u' & u''v' \\ vu' & vv' \end{pmatrix} = \begin{pmatrix} 1_L & 0 \\ 0 & 1_N \end{pmatrix}$$

because $vu' = 0$ as said before, whereas $u''v \in \mathrm{Hom}_A(N, L) = 0$. Therefore, $(u' \ v')$ is a section.

This completes the proof that $\begin{pmatrix} f & 0 \\ 0 & 1_N \end{pmatrix}$ is irreducible. On the other hand, it is certainly not radical, because the composition

$$(0 \ 1_N) \begin{pmatrix} f & 0 \\ 0 & 1_N \end{pmatrix} \begin{pmatrix} 0 \\ 1_N \end{pmatrix} = 1_N$$

is an isomorphism.

Again, explicit versions of this example are easy to construct. Assume for instance that A is the path algebra of the quiver

$$\overset{1}{\bullet} \longleftarrow \overset{2}{\bullet} \longleftarrow \overset{3}{\bullet}$$

Then, the inclusion $i\colon S_1 \longrightarrow P_2$ of the radical S_1 into the indecomposable projective P_2 is irreducible. In addition, the simple module S_3 is certainly neither comparable to S_1 nor to P_2. Therefore, the morphism $\begin{pmatrix} i & 0 \\ 0 & 1_{S_3} \end{pmatrix} : S_1 \oplus S_3 \longrightarrow P_2 \oplus S_3$ is irreducible but not radical.

The previous examples show that the irreducible morphisms are proper generalisations of the radical morphisms that are not in the radical square. In these notes, however, we are exclusively concerned with morphisms having source or target (or both) indecomposable.

In the sequel, we are particularly interested in the situation where we have a short exact sequence

$$0 \longrightarrow L \xrightarrow{f} M \xrightarrow{g} N \longrightarrow 0$$

with f and/or g irreducible. It turns out that, in this situation, L and/or N is indecomposable; namely, we prove that the kernel (or cokernel) of an irreducible epimorphism (or monomorphism respectively) is indecomposable. We need a lemma.

Lemma II.2.8. *Let* $0 \longrightarrow L \xrightarrow{f} M \xrightarrow{g} N \longrightarrow 0$ *be a nonsplit exact sequence.*

(a) *The morphism f is irreducible if and only if for every $v\colon V \longrightarrow N$, there exists $v_1\colon V \longrightarrow M$ such that $v = gv_1$ or there exists $v_2\colon M \longrightarrow V$ such that $g = vv_2$.*
(b) *The morphism g is irreducible if and only if for every $u\colon L \longrightarrow U$, there exists $u_1\colon M \longrightarrow U$ such that $u = u_1 f$ or there exists $u_2\colon U \longrightarrow M$ such that $f = u_2 u$.*

Proof. It suffices to prove (a), because the proof of (b) is dual.

Necessity. A morphism $v\colon V \longrightarrow N$ induces a commutative diagram with exact rows

$$\begin{array}{ccccccccc} 0 & \longrightarrow & L & \xrightarrow{f'} & E & \xrightarrow{g'} & V & \longrightarrow & 0 \\ & & \| & & \downarrow{\scriptstyle u} & & \downarrow{\scriptstyle v} & & \\ 0 & \longrightarrow & L & \xrightarrow{f} & M & \xrightarrow{g} & N & \longrightarrow & 0 \end{array}$$

where E is the fibered product of g and v. Because f is irreducible, f' is a section or u is a retraction. In the first case, g' is a retraction; thus, there exists $g''\colon V \longrightarrow E$ such that $g'g'' = 1_V$. Setting $v_1 = ug''$ we get $gv_1 = g(ug'') = vg'g'' = v$. In the second case, there exists $u'\colon M \longrightarrow E$ such that $uu' = 1_M$ and so, setting $v_2 = g'u'$ yields $vv_2 = v(g'u') = guu' = g$ as required.

Sufficiency. Because the given sequence is not split, f is neither a section nor a retraction. Assume that $f = f_1 f_2$ with $f_2\colon L \longrightarrow X$, $f_1\colon X \longrightarrow M$. Because f is injective, so is f_2 and we get a commutative diagram with exact rows

$$\begin{array}{ccccccccc} 0 & \longrightarrow & L & \xrightarrow{f_2} & X & \longrightarrow & \operatorname{Coker} f_2 & \longrightarrow & 0 \\ & & \| & & \downarrow{\scriptstyle f_1} & & \downarrow{\scriptstyle v} & & \\ 0 & \longrightarrow & L & \xrightarrow{f} & M & \xrightarrow{g} & N & \longrightarrow & 0 \end{array}$$

where v is deduced by passing to cokernels. In particular, X is isomorphic to the fibered product of g and v. If there exists $v_1\colon \operatorname{Coker} f_2 \longrightarrow M$ such that $v = gv_1$, then the universal property of the fibered product implies that the upper sequence splits and so f_2 is a section. If there exists $v_2\colon M \longrightarrow \operatorname{Coker} f_2$ such that $g = vv_2$, then the same universal property yields that f_1 is a retraction. □

Corollary II.2.9.

(a) *The cokernel of an irreducible monomorphism is indecomposable.*
(b) *The kernel of an irreducible epimorphism is indecomposable.*

Proof. It suffices to prove (a) because the proof of (b) is dual.

Let $f\colon L \longrightarrow M$ be an irreducible monomorphism, $N = \operatorname{Coker} f$ and $g\colon M \longrightarrow N$ the surjection. Assume $N = N_1 \oplus N_2$ with $N_1 \neq 0$ and $N_2 \neq 0$. Then, the inclusions $q_1\colon N_1 \longrightarrow N$ and $q_2\colon N_2 \longrightarrow N$ are both proper monomorphisms. Apply Lemma II.2.8. If there exists $u_1\colon M \longrightarrow N_1$ such that $g = q_1u_1$, then q_1 would be surjective, a contradiction. Hence, there exists $v_1\colon N_1 \longrightarrow M$ such that $gv_1 = q_1$. Similarly, there exists $v_2\colon N_2 \longrightarrow M$ such that $gv_2 = q_2$. But then $g(v_1, v_2) = (q_1, q_2) = 1_N$ and g is a retraction, which implies that f is a section, a contradiction. □

II.2.2 Almost split and minimal morphisms

As stated in the introduction to Subsection II.2.1, the consideration of irreducible morphisms came from the need to identify building blocks for radical morphisms, so that other radical morphisms could be obtained from the irreducible ones by successive compositions and linear combinations. Therefore, the next step is to study the factorisation behaviour of radical morphisms. This leads to the following definition.

Definition II.2.10.

(a) A radical morphism $f\colon L \longrightarrow M$ with L indecomposable is called **left almost split** if, for every radical morphism $u\colon L \longrightarrow U$, there exists $u'\colon M \longrightarrow U$ such that $u = u'f$.

$$\begin{array}{ccc} L & \xrightarrow{f} & M \\ {\scriptstyle u}\downarrow & \swarrow{\scriptstyle u'} & \\ U & & \end{array}$$

(b) A radical morphism $g\colon M \longrightarrow N$ with N indecomposable is called **right almost split** if, for every radical morphism $v\colon V \longrightarrow N$ there exists $v'\colon V \longrightarrow M$ such that $v = gv'$.

$$\begin{array}{ccc} & & V \\ & {}^{v'}\swarrow & \downarrow{\scriptstyle v} \\ M & \xrightarrow{g} & N \end{array}$$

These notions are evidently dual to each other, that is, $f: L \longrightarrow M$ is left almost split if and only if $\mathrm{D}f: \mathrm{D}M \longrightarrow \mathrm{D}L$ is right almost split.

Saying in (a) that $u: L \longrightarrow U$ with L indecomposable is a radical morphism amounts to saying that u is not a section, because of Corollary II.1.10. In addition, if U itself is indecomposable, then this amounts to saying that u is a nonisomorphism. Dually, in (b), $v: V \longrightarrow N$ with N indecomposable is radical if and only if it is not a retraction. If V is also indecomposable, then this is the case if and only if it is a nonisomorphism.

The characterisation of almost split morphisms given in the lemma below is their original definition.

Lemma II.2.11.

(a) *A morphism $f: L \longrightarrow M$ is left almost split if and only if:*

 (i) *f is not a section; and*
 (ii) *if $u: L \longrightarrow U$ is not a section, then there exists $u': M \longrightarrow U$ such that $u = u'f$.*

(b) *A morphism $g: M \longrightarrow N$ is right almost split if and only if:*

 (i) *g is not a retraction; and*
 (ii) *if $v: V \longrightarrow N$ is not a retraction, then there exists $v': V \longrightarrow M$ such that $v = gv'$.*

Proof. We only prove (a), because the proof of (b) is dual.

We first assume that conditions (i) and (ii) hold and prove that this implies that L is indecomposable. Indeed, if this is not the case, then $L = L_1 \oplus L_2$ with L_1, L_2 nonzero. Hence, the projection morphisms $p_1: L \longrightarrow L_1$ and $p_2: L \longrightarrow L_2$ are proper epimorphisms, and in particular are not sections. Because of condition (ii), there exist $p_1': M \longrightarrow L_1$ and $p_2': M \longrightarrow L_2$ such that $p_1 = p_1'f$ and $p_2 = p_2'f$. Now,

$$\begin{pmatrix} p_1' \\ p_2' \end{pmatrix} f = \begin{pmatrix} p_1 \\ p_2 \end{pmatrix} = 1_L$$

and f is a section, a contradiction that establishes the indecomposability of L. But then, a morphism with source L is radical if and only if it is not a section. □

Example II.2.12. Let P be an indecomposable projective module. The inclusion $j: \operatorname{rad} P \longrightarrow P$ is right almost split. Indeed, a radical morphism $v: V \longrightarrow P$ is a nonretraction, and hence a nonsurjection (because P is projective). Therefore, $\operatorname{Im} v \subseteq \operatorname{rad} P$ and so there exists $v': V \longrightarrow \operatorname{rad} P$ such that $v = jv'$.
Dually, if I is indecomposable injective, then the projection $I \longrightarrow I/\operatorname{soc} I$ is left almost split.

Example II.2.13. Let A be the path algebra of the quiver

$$\overset{1}{\bullet} \longleftarrow \overset{2}{\bullet} \longleftarrow \overset{3}{\bullet}$$

There exist an epimorphism $p : P_3 = \begin{smallmatrix} 3 \\ 2 \\ 1 \end{smallmatrix} \longrightarrow I_2 = \begin{smallmatrix} 3 \\ 2 \end{smallmatrix}$ and a monomorphism $j \colon S_2 = 2 \longrightarrow I_2 = \begin{smallmatrix} 3 \\ 2 \end{smallmatrix}$, both unique up to scalar multiples, because $\dim_{\mathbf{k}} \operatorname{Hom}(P_3, I_2) = \dim_{\mathbf{k}} \operatorname{Hom}(S_2, I_2) = 1$. We claim that the morphism

$$\begin{pmatrix} p & j \end{pmatrix} : P_3 \oplus S_2 \longrightarrow I_2$$

is right almost split. Indeed, it is not a retraction, because I_2 is isomorphic to neither P_3 nor S_2. Let $v : V \longrightarrow I_2$ be a radical morphism. One sees that the only indecomposable modules that map nontrivially to I_2 are S_2, P_2 and P_3. Therefore, $V \cong V_1 \oplus V_2$, where V_1 is one of these three indecomposable modules. If $V_1 = S_2$, then the restriction $v|_{V_1}$ is equal to (a scalar multiple of) j. If $V_1 = P_3$, then $v|_{V_1}$ is equal to (a scalar multiple of) p. Finally, if $V_1 = P_2$, then $v|_{V_1}$ obviously factors through p or j. In any case, v factors through $\begin{pmatrix} p & j \end{pmatrix}$.
Similarly, one proves that the obvious morphism $P_2 \longrightarrow S_2 \oplus P_3$ is left almost split.

Example II.2.14. Knowing one almost split morphism, it is easy to construct a lot. Indeed, let $f : L \longrightarrow M$ be left almost split and $f' : L \longrightarrow M'$ be radical. We claim that the morphism

$$\begin{pmatrix} f \\ f' \end{pmatrix} : L \longrightarrow M \oplus M'$$

is also left almost split. Indeed, neither f nor f' is a section, hence the indecomposable module L is isomorphic to no direct summand of $M \oplus M'$. Therefore, $\begin{pmatrix} f \\ f' \end{pmatrix}$ is not a section either. Let $u : L \longrightarrow U$ be radical and $u' \colon M \longrightarrow U$ be such that $u = u'f$. Then we have the factorisation

$$u = \begin{pmatrix} u' & 0 \end{pmatrix} \begin{pmatrix} f \\ f' \end{pmatrix}.$$

This establishes our claim. Dually, if $g : M \longrightarrow N$ is right almost split and $g' \colon M' \longrightarrow N$ is radical, then $\begin{pmatrix} g \\ g' \end{pmatrix} : M \oplus M' \longrightarrow N$ is also right almost split.

The previous example tends to suggest that the "good" almost split morphisms should satisfy a minimality condition, namely the target of a left almost split morphism, or the source of a right almost split morphism, should be as small as possible. This brings us to the definition of minimal morphisms.

Definition II.2.15.

(a) A morphism $f : L \longrightarrow M$ is **left minimal** if every $h \in \operatorname{End} M$ such that $hf = f$ is an automorphism.
(b) A morphism $g : M \longrightarrow N$ is **right minimal** if every $h \in \operatorname{End} M$ such that $gh = g$ is an automorphism.

Again, these notions are dual to each other: $f : L \to M$ is left minimal if and only if $\mathrm{D}f : \mathrm{D}M \to \mathrm{D}L$ is right minimal.

Example II.2.16. Clearly, every epimorphism is left minimal and every monomorphism is right minimal. In addition, if P is an indecomposable projective module, then the inclusion morphism $\operatorname{rad} P \to P$ is right minimal. Dually, if I is indecomposable injective, then the projection morphism $I \to I/\operatorname{soc} I$ is left minimal.

Example II.2.17. An epimorphism $g : M \longrightarrow N$ is called **superfluous** if, for every morphism $h : L \longrightarrow M$ such that $gh : L \longrightarrow N$ is an epimorphism, we have that h itself is an epimorphism. Typical superfluous epimorphisms are the projective covers, see Subsection I.1.3. Now, we claim that every superfluous epimorphism is right minimal. Indeed, let $g : M \longrightarrow N$ be a superfluous epimorphism and $h : M \longrightarrow M$ be such that $gh = g$. In particular, gh is an epimorphism. Hence, so is h. But M has finite length; hence, applying Lemma I.1.20, we get that h is an automorphism.
The dual notion is that of an essential monomorphism. A monomorphism $f : L \longrightarrow M$ is called **essential** if, for every morphism $h : M \longrightarrow N$ such that $hf : L \longrightarrow N$ is a monomorphism, we have that h itself is a monomorphism. Typical essential monomorphisms are injective envelopes. Just as above, ones proves that every essential monomorphism is left minimal.

Lemma II.2.18. *Every irreducible morphism is both left and right minimal.*

Proof. Let $f : L \to M$ be irreducible and $h \in \operatorname{End} M$ be such that $hf = f$. Because f is not a section, then h must be a retraction, and in particular an epimorphism. But then h is an automorphism, because M has finite length. This proves left minimality. The proof of right minimality is similar. □

As a consequence, it follows from Example II.2.4 that, if P is indecomposable projective, then the inclusion $\operatorname{rad} P \to P$ is left and right minimal, and, if I is indecomposable injective, then the projection $I \to I/\operatorname{soc} I$ is left and right minimal.

Now, we make explicit the meaning of minimality for almost split morphisms. A left almost split morphism $f : L \to M$ will turn out to be left minimal if and only if its target M has least composition length $l(M)$ among the targets of left almost split morphisms of source L. This means that, if $f' : L \to M'$ is also left almost split, then $l(M) \leq l(M')$. Clearly, the dual statement holds true for right almost split morphisms.

Proposition II.2.19.

(a) *Let $f : L \to M$ be a left almost split morphism. Then, f is left minimal if and only if its target M has least length among the targets of left almost split morphisms having source L. In addition, this condition uniquely determines f up to isomorphism.*
(b) *Let $g : M \to N$ be a right almost split morphism. Then, g is right minimal if and only if its source M has least length among the sources of right almost split morphisms having target N. In addition, this condition uniquely determines g up to isomorphism.*

Proof. We only prove (a), because the proof of (b) is dual.

Sufficiency. Let $f : L \to M$ be left almost split, with $l(M)$ minimal among the lengths of the targets of left almost split morphisms of source L. Let $h \in \operatorname{End} M$ be such that $hf = f$. Let $h = jp$ be the canonical factorisation of h through its image $M' = \operatorname{Im} h$. We claim that $pf : L \to M'$ is left almost split. Clearly, pf is not a section, because f is not. Because L is indecomposable, this implies that pf is a radical morphism. Let $u : L \to U$ be a radical morphism. Then there exists $u' : M \to U$ such that $u = u'f$. But then, $u = u'f = u'hf = u'jpf$ factors through M'. This establishes our claim that pf is left almost split. Then, by hypothesis, $l(M) \leq l\left(M'\right)$. On the other hand, $M' \subseteq M$ implies that $l\left(M'\right) \leq l(M)$. Hence, $l(M) = l\left(M'\right)$; therefore, $M = M'$, and so h is surjective. Now M has finite length; hence, h is an automorphism. This shows that f is left minimal.

Necessity. Assume now that the left almost split morphism $f : L \longrightarrow M$ is left minimal, and let $f' : L \longrightarrow M'$ be also left almost split. There exist $h : M \longrightarrow M'$ such that $f' = hf$ and $h' : M' \longrightarrow M$ such that $f = h'f'$.

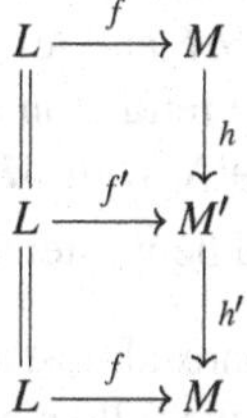

But then $f = h'hf$ and left minimality of f imply that $h'h$ is an automorphism. Therefore, h is injective and we have $l(M) \leq l\left(M'\right)$ as required.

Uniqueness. This is proved using the same argument. Indeed, assume that, the left almost split morphism $f' : L \longrightarrow M'$ is also left minimal. Then, as above, $f' = hh'f'$ gives that hh' is an automorphism. Similarly, $h'h$ is an automorphism. Hence, h and h' are isomorphisms (in particular, $M \cong M'$). □

Definition II.2.20.

(a) A morphism is called **left minimal almost split** if it is at the same time left minimal and left almost split.

(b) A morphism is called **right minimal almost split** if it is at the same time right minimal and right almost split.

It is an easy consequence of Proposition II.2.19 that, given an indecomposable module L, there exists a left minimal almost split morphism with source L, provided that there exists a left almost split morphism of source L. Dually, the existence of right almost split morphisms with given indecomposable target N implies the existence of right minimal almost split morphisms of target N.

Example II.2.21. As we have seen, for every indecomposable projective module P, the inclusion morphism $\operatorname{rad} P \to P$ is right minimal almost split. Dually, if I is an indecomposable injective module, then the projection $I \longrightarrow I/\operatorname{soc} I$ is left minimal almost split.

As a corollary to Proposition II.2.19, we prove that typical almost split morphisms are exactly as in Example II.2.14.

Corollary II.2.22.

(a) *Let $f' : L \longrightarrow M'$ be left almost split. Then, there exists a decomposition $M' = M \oplus X$ such that $f' = \binom{f}{0}$ with $f : L \longrightarrow M$ left minimal almost split.*
(b) *Let $g' : M' \longrightarrow N$ be right almost split. Then, there exists a decomposition $M' = M \oplus Y$ such that $g' = (g\ 0)$ with $g : M \longrightarrow N$ right minimal almost split.*

Proof. We only prove (a), because the proof of (b) is dual.

Let $f' : L \longrightarrow M'$ be left almost split. Because of Proposition II.2.19, there exists a left minimal almost split morphism $f : L \longrightarrow M$, having source L. As before, we find morphisms $h : M \longrightarrow M'$ such that $f' = hf$ and $h' \colon M' \longrightarrow M$ such that $f = h'f'$. But then $f = h'hf$ and left minimality of f implies that $h'h$ is an automorphism. Therefore, h' is a retraction and h is a section. Identifying $h'h$ with the identity, we get $M' = M \oplus X$ with $M = \operatorname{Im} h$ and $X = \operatorname{Ker} h'$. Then, $f' : L \longrightarrow M' = M \oplus X$ may indeed be written in the form $f' = \binom{f}{0}$. □

Our present objective is to compare almost split morphisms with irreducible ones. The first step in this direction is the following lemma, which the reader should compare with Lemma II.2.18.

Lemma II.2.23. *Every (left or right) minimal almost split morphism is irreducible.*

Proof. We only prove the statement for left minimal almost split morphisms, the other case being dual. Let $f : L \longrightarrow M$ be left minimal almost split.

Because f is a radical morphism with an indecomposable source, then it is not a section. It is not a retraction either, because otherwise the indecomposability of L would imply that it is an isomorphism. Assume thus that $f = f_1 f_2$ with $f_2 : L \longrightarrow X$ and $f_1 : X \longrightarrow M$. Suppose that f_2 is not a section. Because f is left almost split, there exists $f_2' : M \longrightarrow X$ such that $f_2 = f_2' f$.

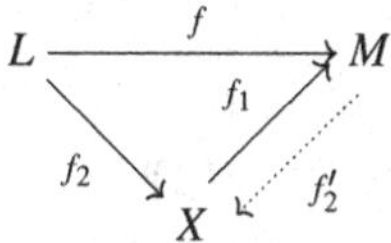

But then $f = f_1 f_2 = f_1 f_2' f$. Left minimality of f yields that $f_1 f_2'$ is an automorphism. Hence, f_1 is a retraction. □

We are now able to prove our structure theorem for irreducible morphisms. It says that irreducible morphisms with a given indecomposable source (or target) are exactly those morphisms that can be completed to a minimal almost split morphism having the same source (or target respectively).

Theorem II.2.24.

(a) *Let $f : L \longrightarrow M$ be left minimal almost split. Then, $f' : L \longrightarrow M'$ is irreducible if and only if $M' \neq 0$ and there exist a decomposition $M = M' \oplus M''$ and a morphism $f'' : L \longrightarrow M''$ such that $\begin{pmatrix} f' \\ f'' \end{pmatrix} : L \longrightarrow M$ is left minimal almost split.*

(b) *Let $g : M \longrightarrow N$ be right minimal almost split. Then, $g' : M' \longrightarrow N$ is irreducible if and only if $M' \neq 0$ and there exist a decomposition $M = M' \oplus M''$ and a morphism $g'' : M'' \longrightarrow N$ such that $\begin{pmatrix} g' & g'' \end{pmatrix} : M \longrightarrow N$ is right minimal almost split.*

Proof. We only prove (a), because the proof of (b) is dual.

Necessity. Assume f' is irreducible. Then, clearly, $M' \neq 0$. Because f is left almost split, there exists $h : M \longrightarrow M'$ such that $f' = hf$. But now f is not a section. Hence, h is a retraction. This implies the statement.

Sufficiency. Assume $f = \begin{pmatrix} f' \\ f'' \end{pmatrix} : L \longrightarrow M = M' \oplus M''$ is left minimal almost split. We claim that f' is irreducible. To prove this statement, we first assume that f' is a section. Let $h : M' \longrightarrow L$ be such that $hf' = 1_L$. Then,

$$\begin{pmatrix} h & 0 \end{pmatrix} \begin{pmatrix} f' \\ f'' \end{pmatrix} = 1_L$$

implies that f itself is a section, a contradiction. Thus, f' is not a section. Assume now that f' is a retraction. Because L is indecomposable, then f' would be an isomorphism and therefore a section, and we have seen that this leads to a contradiction. Thus, f' is not a retraction either.

Suppose now that $f' = f_1 f_2$ with $f_2 : L \longrightarrow X$ and $f_1 : X \longrightarrow M'$. If f_2 is not a section then, because f is left almost split, there exists a morphism $\begin{pmatrix} h' & h'' \end{pmatrix} : M' \oplus M'' \longrightarrow X$ such that

$$\begin{pmatrix} h' & h'' \end{pmatrix} \begin{pmatrix} f' \\ f'' \end{pmatrix} = f_2.$$

We deduce the following commutative diagram:

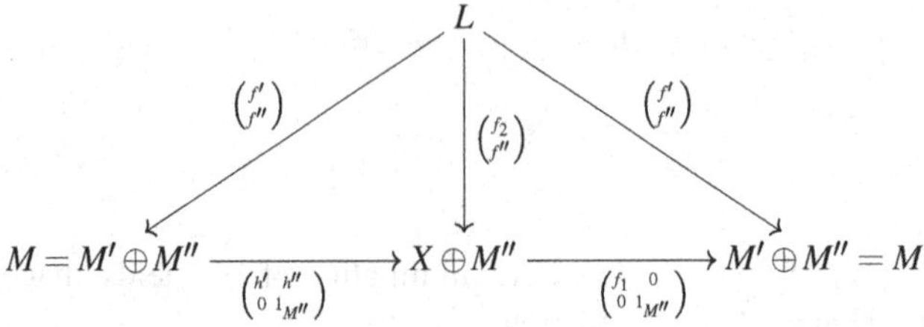

The minimality of f implies that

$$\begin{pmatrix} f_1 & 0 \\ 0 & 1_{M''} \end{pmatrix} \begin{pmatrix} h' & h'' \\ 0 & 1_{M''} \end{pmatrix} = \begin{pmatrix} f_1 h' & f_1 h'' \\ 0 & 1_{M''} \end{pmatrix}$$

is an automorphism. Hence, $f_1 h'$ is an isomorphism and so f_1 is a retraction. □

We finish this subsection with the following consequence of the above theorem.

Corollary II.2.25.

(a) *Let $f : L \longrightarrow M$ be left minimal almost split and $p : M \longrightarrow M'$ a retraction. Then, $pf : L \longrightarrow M'$ is irreducible.*
(b) *Let $g : M \longrightarrow N$ be right minimal almost split and $j : M' \longrightarrow M$ a section. Then, $gj : M' \longrightarrow N$ is irreducible.* □

II.2.3 Almost split sequences

We are now ready to define almost split sequences. In the previous subsection, we considered those radical morphisms through which other radical morphisms factor. These are the almost split morphisms. Because of Corollary II.2.22, we may even assume that we deal with minimal almost split morphisms. And then, these morphisms are irreducible because of Lemma II.2.23; hence, if both their source and target are indecomposable, they belong to the radical but not to the radical square of the module category. The question therefore naturally arises whether there exist sufficiently many minimal almost split morphisms inside the module category. Our objective in this subsection and the next is to prove that this is indeed the case. We start by showing that composable irreducible morphisms can be arranged in a neat way, giving rise to minimal almost split morphisms.

Definition II.2.26. A short exact sequence $0 \longrightarrow L \xrightarrow{f} M \xrightarrow{g} N \longrightarrow 0$ is an **almost split sequence** (or an **Auslander–Reiten sequence**) if both of the morphisms f and g are irreducible.

Remark II.2.27.

(a) Clearly, this is a self-dual concept, that is, a short exact sequence $0 \longrightarrow L \xrightarrow{f} M \xrightarrow{g} N \longrightarrow 0$ is almost split in $\operatorname{mod} A$ if and only if $0 \longrightarrow DN \xrightarrow{Dg} DM \xrightarrow{Df} DL \longrightarrow 0$ is an almost split sequence in $\operatorname{mod} A^{op}$.
(b) Because irreducible morphisms never split, an almost split sequence never splits.
(c) Because of Corollary II.2.9, f irreducible implies N indecomposable, and g irreducible implies L indecomposable: an almost split sequence always has indecomposable end terms.
(d) Because of Lemma II.2.18, both morphisms f and g are left and right minimal.

We give a first example of an almost split sequence.

Example II.2.28. Let $A = \mathbf{k}[t] / \langle t^2 \rangle$. Then, A is local and so has A_A a unique indecomposable projective module, up to isomorphism. Now $\dim_{\mathbf{k}} A = 2$ and $\operatorname{rad} A$ is equal to the simple module $S = \langle t \rangle / \langle t^2 \rangle$. As seen in Example II.2.4, the inclusion $j : S \hookrightarrow A$ is irreducible. Its cokernel is the morphism $p : A \longrightarrow S$ induced by the multiplication by t. But A_A is also the unique indecomposable injective module, up to isomorphism, and $p : A \longrightarrow S$ is the projection of A onto its quotient by its socle. In particular, p is irreducible and we have an almost split sequence:

$$0 \longrightarrow S \xrightarrow{j} A \xrightarrow{p} S \longrightarrow 0.$$

Our objective in this subsection is to prove that almost split sequences may also be defined via minimal almost split morphisms. We start with the following lemma.

Lemma II.2.29.

(a) *Let*

$$\begin{array}{ccccccccc} 0 & \longrightarrow & L & \xrightarrow{f} & M & \xrightarrow{g} & N & \longrightarrow & 0 \\ & & \| & & \downarrow u & & \downarrow v & & \\ 0 & \longrightarrow & L & \xrightarrow{f} & M & \xrightarrow{g} & N & \longrightarrow & 0 \end{array}$$

be a commutative diagram with exact nonsplit rows and N indecomposable. Then, u and v are automorphisms.

(b) *Let*

$$\begin{array}{ccccccccc} 0 & \longrightarrow & L & \xrightarrow{f} & M & \xrightarrow{g} & N & \longrightarrow & 0 \\ & & \downarrow u & & \downarrow v & & \| & & \\ 0 & \longrightarrow & L & \xrightarrow{f} & M & \xrightarrow{g} & N & \longrightarrow & 0 \end{array}$$

be a commutative diagram with exact nonsplit rows and L indecomposable. Then, u and v are automorphisms.

Proof. We only prove (a), because the proof of (b) is dual.

If v is not an automorphism, then the indecomposability of N implies that v is nilpotent. Let $m > 0$ be such that $v^m = 0$. Then, $gu^m = v^m g = 0$ implies that u^m factors through L: there exists $h : M \longrightarrow L$ such that $fh = u^m$. But then $fhf = u^m f = f$ implies $hf = 1_L$, because f is a monomorphism. Hence, f is a section, a contradiction. This shows that v is an automorphism. Therefore, so is u. □

As a consequence of this lemma, the indecomposability of the end terms in a nonsplit short exact sequence implies minimality of the morphisms.

Corollary II.2.30. *Let* $0 \longrightarrow L \xrightarrow{f} M \xrightarrow{g} N \longrightarrow 0$ *be a nonsplit short exact sequence.*

(a) *If N is indecomposable, then f is left minimal.*
(b) *If L is indecomposable, then g is right minimal.*

Proof. We only prove (a), because the proof of (b) is dual.

Assume $hf = f$ for some $h \in \operatorname{End} M$. We have a commutative diagram with exact nonsplit rows:

$$\begin{array}{ccccccccc} 0 & \longrightarrow & L & \xrightarrow{f} & M & \xrightarrow{g} & N & \longrightarrow & 0 \\ & & \| & & \downarrow{\scriptstyle h} & & \downarrow{\scriptstyle h'} & & \\ 0 & \longrightarrow & L & \xrightarrow{f} & M & \xrightarrow{g} & N & \longrightarrow & 0 \end{array}$$

where h' is deduced by passing to cokernels. Applying Lemma II.2.29 completes the proof. □

We are ready to prove our structure theorem for almost split sequences. It shows that an almost split sequence is characterised by any of its two nonzero morphisms.

Theorem II.2.31. *Let* $0 \longrightarrow L \xrightarrow{f} M \xrightarrow{g} N \longrightarrow 0$ *be a short exact sequence. The following conditions are equivalent.*

(a) *The sequence is almost split.*
(b) *The morphism f is left minimal almost split.*
(c) *The morphism g is right minimal almost split.*

Proof. We first show that (a) implies (b). Because of Lemma II.2.18, f is left minimal. We prove it is left almost split. Because f is irreducible and L is indecomposable, it is a radical morphism. Let $u : L \longrightarrow U$ be a radical morphism. We may assume that U itself is indecomposable (then u is a nonisomorphism). Because g is irreducible, it follows from Lemma II.2.8 that there exist $u_1 : M \longrightarrow U$ such that $u = u_1 f$ (in which case we have finished) or there exists $u_2 : U \longrightarrow M$ such that $f = u_2 u$. In the latter case, the irreducibility of f and the fact that u is not a section imply that u_2 is a retraction. Because U is indecomposable, this implies that u_2 is an isomorphism and we get $u = u_2^{-1} f$. Thus, we have finished in this case as well.

We now prove that (b) implies (c). First, g is not a retraction, because f is not a section. Assume that $v : V \longrightarrow N$ is a radical morphism. Corollary II.2.9 implies that $N = \operatorname{Coker} f$ is indecomposable. Then, v is not a retraction. We have a commutative diagram with exact rows.

$$\begin{array}{ccccccccc} 0 & \longrightarrow & L & \xrightarrow{h} & E & \xrightarrow{k} & V & \longrightarrow & 0 \\ & & \| & & \downarrow{\scriptstyle u} & & \downarrow{\scriptstyle v} & & \\ 0 & \longrightarrow & L & \xrightarrow{f} & M & \xrightarrow{g} & N & \longrightarrow & 0 \end{array}$$

where E is the fibered product of the morphisms g and v. We claim that the upper sequence splits, that is, the morphism k is a retraction. If this is not the case, then h is not a section and, because f is left almost split, there exists $u' : M \longrightarrow E$ such that $u'f = k$. We deduce a larger commutative diagram with exact rows.

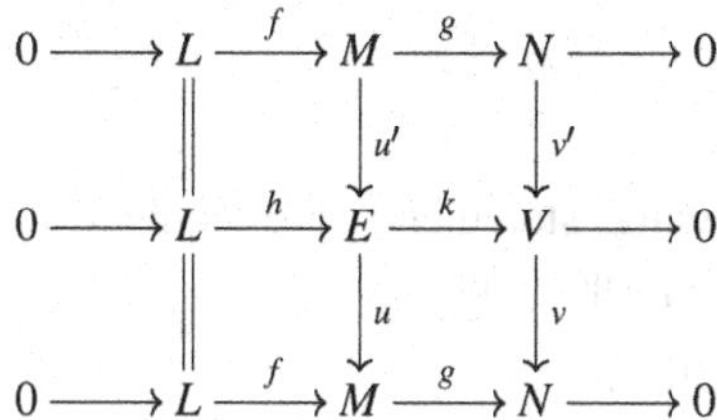

where v' is deduced by passing to cokernels. Because f is not a section, the lower (= upper) sequence does not split. Applying Lemma II.2.29, we get that vv' is an automorphism. But then v is a retraction, a contradiction. This establishes our claim. Therefore, there exists $k' : V \longrightarrow E$ such that $kk' = 1_V$. Then $guk' = vkk' = v$ and g is indeed right almost split. Because right minimality of g follows from Corollary II.2.30(b), this finishes the proof of (c).

Dually, one proves that (c) implies (b). Therefore, (b) and (c) are equivalent. But now the conjunction of (b) and (c) implies (a) because minimal almost split morphisms are always irreducible, owing to Lemma II.2.23. □

As a first consequence, we establish that, if an almost split sequence exists, then it is uniquely determined up to isomorphism by one of its end terms.

Corollary II.2.32. *An almost split sequence* $0 \longrightarrow L \xrightarrow{f} M \xrightarrow{g} N \longrightarrow 0$ *is uniquely determined by* L *(or by* N*) up to isomorphism.*

Proof. Let $0 \longrightarrow L \xrightarrow{f} M \xrightarrow{g} N \longrightarrow 0$ and $0 \longrightarrow L' \xrightarrow{f'} M' \xrightarrow{g'} N' \longrightarrow 0$ be almost split sequences. Assume $L = L'$. Because f and f' are left minimal almost split, it follows from Proposition II.2.19 that there exists an isomorphism $h : M \longrightarrow M'$ such that $hf = f'$. Passing to cokernels, we get an isomorphism $h' : N \longrightarrow N'$ such that $h'g = g'h$. Thus, the sequences are isomorphic. The proof is dual if we assume that $N = N'$. □

We also obtain another characterisation of almost split sequences that will be useful in the next section.

Corollary II.2.33. *Let* $0 \longrightarrow L \xrightarrow{f} M \xrightarrow{g} N \longrightarrow 0$ *be a short exact sequence. The following conditions are equivalent:*

(a) *The sequence is almost split.*
(b) *The module N is indecomposable and the morphism f is left almost split.*
(c) *The module L is indecomposable and the morphism g is right almost split.*

Proof. Because (a) clearly implies (b) and (c), it suffices, because of duality, to prove that (b) implies (a). Because of Theorem II.2.31, we just need to show that f is left minimal, and this follows from Corollary II.2.30(a). □

Exercises for Section II.2

Exercise II.2.1.

(a) Prove that the following statements are equivalent for an epimorphism $f : P \longrightarrow M$ with P projective:

  (i) f is a projective cover.
  (ii) f is superfluous.
  (iii) f is right minimal.

(b) Prove that the following statements are equivalent for a monomorphism $f : M \longrightarrow I$ with I injective:

  (i) f is an injective envelope.
  (ii) f is essential.
  (iii) f is left minimal.

Exercise II.2.2.

(a) Let $f : L \longrightarrow M$, $g : M \longrightarrow N$ be epimorphisms. Prove that:

  (i) If both g and f are superfluous, then so is gf.
  (ii) If gf is superfluous, then so is f.

(b) Let $f : L \longrightarrow M$, $g : M \longrightarrow N$ be monomorphisms. Prove that :

  (i) If both g and f are essential, then so is gf.
  (ii) If gf is essential, then so is g.

Exercise II.2.3. Prove that a monomorphism $f : L \longrightarrow M$ is essential if and only if $\operatorname{Im} f$ has a nonzero intersection with every nonzero submodule of M. Deduce that, if $f : L \longrightarrow M$ is an essential monomorphism with L injective, then f is an isomorphism.

Exercise II.2.4. Consider a morphism in mod A of the form

$$\begin{pmatrix} f & g \\ 0 & h \end{pmatrix} : M \oplus N \longrightarrow M' \oplus N'$$

with $f : M \longrightarrow M'$, $g : N \longrightarrow M'$ and $h : N \longrightarrow N'$. Prove that:

(a) If $\begin{pmatrix} f & g \\ 0 & h \end{pmatrix}$ is an isomorphism, then f is a section and h is a retraction.
(b) If $\begin{pmatrix} f & g \\ 0 & h \end{pmatrix}$ is an isomorphism, and either f or h is an isomorphism, then so is the other.
(c) If both f and h are isomorphisms, then so is $\begin{pmatrix} f & g \\ 0 & h \end{pmatrix}$.

Exercise II.2.5.

(a) Let $f : L \longrightarrow M$ be an irreducible monomorphism and M' a proper submodule of M containing $\operatorname{Im} f$. Prove that $\operatorname{Im} f$ is a direct summand of M'.
(b) Let $g : M \longrightarrow N$ be an irreducible epimorphism and M' a nonzero submodule of M contained in $\operatorname{Ker} g$. Prove that N is a direct summand of M/M'.

Exercise II.2.6. Let $0 \longrightarrow L \xrightarrow{f} M \xrightarrow{g} N \longrightarrow 0$ be a short exact sequence with indecomposable middle term M. Prove that:

(a) If f is irreducible, then each irreducible morphism $h : X \longrightarrow N$ is surjective.
(b) If g is irreducible, then each irreducible morphism $k : L \longrightarrow Y$ is injective.

Exercise II.2.7. Let $f : L \longrightarrow M$ be an irreducible morphism and N an A-module. Prove that:

(a) If $\operatorname{Hom}_A(M, N) = 0$, then $\operatorname{Ext}^1_A(N, f)$ is a monomorphism.
(b) If $\operatorname{Hom}_A(N, L) = 0$, then $\operatorname{Ext}^1_A(f, N)$ is a monomorphism.

Exercise II.2.8. Let $0 \longrightarrow L \xrightarrow{f} M \xrightarrow{g} N \longrightarrow 0$ be a nonsplit short exact sequence. Prove that:

(a) f is irreducible if and only if every subfunctor F of $\operatorname{Hom}_A(-, N)$ either contains or is contained in the image of $\operatorname{Hom}_A(-, g)$.
(b) g is irreducible if and only if every subfunctor F of $\operatorname{Hom}_A(L, -)$ either contains or is contained in the image of $\operatorname{Hom}_A(f, -)$.

Exercise II.2.9.

(a) Prove that a morphism $f : L \longrightarrow M$ is left almost split if and only if it is radical, L is indecomposable, and if $U \not\cong L$ is indecomposable, then every morphism $u : L \longrightarrow U$ factors through f.
(b) Prove that a morphism $g : M \longrightarrow N$ is right almost split if and only if it is radical, N is indecomposable, and if $V \not\cong N$ is indecomposable, then every morphism $v : V \longrightarrow N$ factors through g.

Exercise II.2.10. Let $0 \longrightarrow L \xrightarrow{f} M \xrightarrow{g} N \longrightarrow 0$ be an almost split sequence.

(a) Let N' be a proper submodule of N. Show that the almost split sequence induces a split exact sequence:

$$0 \longrightarrow L \longrightarrow g^{-1}\left(N'\right) \longrightarrow N' \longrightarrow 0$$

(b) Let L' be a proper submodule of L. Show that the almost split sequence induces a split exact sequence:

$$0 \longrightarrow L/L' \longrightarrow M/f\left(L'\right) \longrightarrow N \longrightarrow 0.$$

Exercise II.2.11. Let $\xi : 0 \longrightarrow L \xrightarrow{f} M \xrightarrow{g} N \longrightarrow 0$ be a nonsplit short exact sequence with L, N indecomposable. Show that the following conditions are equivalent:

(a) The sequence ξ is almost split.
(b) For every radical morphism $u : L \longrightarrow U$, we have $\operatorname{Ext}^1_A(N, u)(\xi) = 0$.
(c) For every radical morphism $v : V \longrightarrow N$, we have $\operatorname{Ext}^1_A(v, L)(\xi) = 0$.

Exercise II.2.12. Let $0 \longrightarrow L \xrightarrow{f} M \xrightarrow{g} N \longrightarrow 0$ be an almost split sequence. Prove that:

(a) For every commutative diagram with nonsplit exact rows

$$\begin{array}{ccccccccc} 0 & \longrightarrow & L & \xrightarrow{f} & M & \xrightarrow{g} & N & \longrightarrow & 0, \\ & & \downarrow{\scriptstyle u} & & \downarrow{\scriptstyle v} & & \| & & \\ 0 & \longrightarrow & L' & \xrightarrow{f'} & M' & \xrightarrow{g'} & N & \longrightarrow & 0 \end{array}$$

the morphisms u and v are sections.

(b) For every commutative diagram with nonsplit exact rows

$$\begin{array}{ccccccccc} 0 & \longrightarrow & L & \xrightarrow{f'} & M' & \xrightarrow{g'} & N' & \longrightarrow & 0, \\ & & \| & & \downarrow{\scriptstyle u} & & \downarrow{\scriptstyle v} & & \\ 0 & \longrightarrow & L & \xrightarrow{f} & M & \xrightarrow{g} & N & \longrightarrow & 0 \end{array}$$

the morphisms u and v are retractions.

Exercise II.2.13. Let $0 \longrightarrow L \xrightarrow{f} \bigoplus_{i=1}^{t} M_i \xrightarrow{g} N \longrightarrow 0$ be an almost split sequence with the M_i indecomposable. Prove that, for every i, we have $l(M_i) \neq l(N)$.

Exercise II.2.14. Let $0 \longrightarrow L \xrightarrow{f} M \xrightarrow{g} N \longrightarrow 0$ be an almost split sequence. Prove that:

(a) If M is projective, then $g : M \longrightarrow N$ is a projective cover.
(b) If M is injective, then $f : L \longrightarrow M$ is an injective envelope.

Exercise II.2.15. Let A be given by the quiver:

$$1 \longleftarrow 2 \longleftarrow 3$$

Prove that the sequences:

(a) $0 \longrightarrow S_1 \longrightarrow P_2 \longrightarrow S_2 \longrightarrow 0.$
(b) $0 \longrightarrow P_2 \longrightarrow S_2 \oplus P_3 \longrightarrow I_2 \longrightarrow 0.$
(c) $0 \longrightarrow S_2 \longrightarrow I_2 \longrightarrow S_3 \longrightarrow 0.$

(where all morphisms are either inclusions or projections) are exact and almost split.

Exercise II.2.16. Let A be given by the quiver

$$1 \underset{\beta}{\overset{\alpha}{\leftrightarrows}} 2$$

bound by $\alpha\beta = 0$. Prove that the sequences:

(a) $0 \longrightarrow S_2 \longrightarrow P_1 \longrightarrow S_1 \longrightarrow 0$
(b) $0 \longrightarrow P_1 \longrightarrow S_1 \oplus P_2 \longrightarrow I_1 \longrightarrow 0$
(c) $0 \longrightarrow S_1 \longrightarrow I_1 \longrightarrow S_2 \longrightarrow 0$

(where all morphisms are either inclusions or projections) are exact and almost split.

Exercise II.2.17. Let A be given by the quiver

$$1 \xleftarrow{\gamma} 2 \xleftarrow{\beta} 3 \xleftarrow{\alpha} 4$$

bound by $\alpha\beta\gamma = 0$. Prove that the sequences

(a) $0 \longrightarrow S_1 \longrightarrow P_2 \longrightarrow S_2 \longrightarrow 0$
(b) $0 \longrightarrow P_2 \longrightarrow S_2 \oplus P_3 \longrightarrow P_3/S_1 \longrightarrow 0$
(c) $0 \longrightarrow S_2 \longrightarrow P_3/S_1 \longrightarrow S_3 \longrightarrow 0$
(d) $0 \longrightarrow P_3/S_1 \longrightarrow P_4 \oplus S_3 \longrightarrow I_3 \longrightarrow 0$
(e) $0 \longrightarrow S_3 \longrightarrow I_3 \longrightarrow S_4 \longrightarrow 0$

(where all morphisms are either inclusions or projections) are exact and almost split.

Exercise II.2.18. Let A be given by the quiver:

$$1 \underset{\beta}{\overset{\alpha}{\rightleftarrows}} 2$$

bound by $\alpha\beta\alpha = 0$ and $\beta\alpha\beta = 0$. Prove that the sequences:

(a) $0 \longrightarrow S_1 \longrightarrow P_2/S_2 \longrightarrow S_2 \longrightarrow 0$.
(b) $0 \longrightarrow S_2 \longrightarrow P_1/S_1 \longrightarrow S_1 \longrightarrow 0$.
(c) $0 \longrightarrow P_2/S_2 \longrightarrow P_1 \oplus S_2 \longrightarrow P_1/S_1 \longrightarrow 0$.
(d) $0 \longrightarrow P_1/S_1 \longrightarrow P_2 \oplus S_1 \longrightarrow P_2/S_2 \longrightarrow 0$.

(where all morphisms are either inclusions or projections) are exact and almost split.

II.3 The existence of almost split sequences

II.3.1 The functor category Fun A

In the previous section, we defined and studied properties of almost split sequences. Our objective now is to prove the existence theorem of these sequences, due to Auslander and Reiten. The theorem asserts that, if A is a finite dimensional **k**-algebra, and N an indecomposable nonprojective A-module, or dually L an indecomposable noninjective A-module, then there exists an almost split sequence:

$$0 \longrightarrow L \longrightarrow M \longrightarrow N \longrightarrow 0.$$

A consequence of this theorem is the existence of enough minimal almost split morphisms in the module category: for every indecomposable module L, there exists a left minimal almost split morphism $L \longrightarrow M$, and, dually, for every indecomposable module N, there exists a right minimal almost split morphism $M \longrightarrow N$.

There are many proofs of this existence theorem. The proof we present in this section uses a functorial approach to the theory. There are several reasons for adopting this point of view. Indeed, it is well-known to specialists that the category of **k**-functors from a module category into the category mod **k** of finite dimensional **k**-vector spaces is, in several aspects, better behaved than the module category itself. In addition, historically, it was the functorial approach that supplied both the original inspiration and the original proofs of many results of the Auslander–Reiten theory. Finally, the proof we present is relatively easy and elementary.

Let Fun A be the category whose objects are the contravariant **k**-functors from mod A to mod **k** and whose morphisms are the functorial morphisms. Strictly speaking, given objects F, G in Fun A, the functorial morphisms from F to G do not usually constitute a set. However, the class of objects of every skeleton of mod A is a set. Because we do not distinguish between isomorphic objects, we may identify mod A with one of its skeletons and then the class Hom (F, G) of functorial morphisms from F to G becomes a set; thus, the category Fun A is well-defined. It follows from well-known results of category theory that the category Fun A is **k**-linear and actually abelian.

The most efficient tool for translating statements about modules into statements about functors, and vice versa, is **Yoneda's lemma**, which we now state and prove.

Theorem II.3.1 (Yoneda's lemma). *Let $\mathscr{C}$ be a* **k***-linear category, $F : \mathscr{C} \longrightarrow$ mod* **k** *a contravariant* **k***-functor and X an object in $\mathscr{C}$. There is an isomorphism of vector spaces:*

$$\varepsilon\colon \operatorname{Hom}(\operatorname{Hom}_{\mathscr{C}}(-, X), F) \longrightarrow F(X) \text{ given by } \varphi \mapsto \varphi_X(1_X).$$

Proof. Clearly, ε maps $\operatorname{Hom}(\operatorname{Hom}_{\mathscr{C}}(-, X), F)$ into $F(X)$. To prove that it is bijective, we construct its inverse σ.

Let $x \in F(X)$ and Y an arbitrary object in $\mathscr{C}$. We define the functorial morphism $\sigma(x)_Y : \operatorname{Hom}_{\mathscr{C}}(Y, X) \longrightarrow F(Y)$ as follows: let $f \in \operatorname{Hom}_{\mathscr{C}}(Y, X)$, then set $\sigma(x)_Y(f) = F(f)(x)$. Indeed, if $f : Y \longrightarrow X$, then $F(f) : F(X) \longrightarrow F(Y)$ and thus $F(f)(x) \in F(Y)$.

We first prove that $\sigma(x) : \operatorname{Hom}_{\mathscr{C}}(-, X) \longrightarrow F$ is a functorial morphism. Let $g : Y \longrightarrow Z$ be an arbitrary morphism. We must prove that the square

$$\begin{array}{ccc} \operatorname{Hom}_{\mathscr{C}}(Z,X) & \xrightarrow{\operatorname{Hom}_{\mathscr{C}}(g,X)} & \operatorname{Hom}_{\mathscr{C}}(Y,X) \\ \downarrow{\scriptstyle \sigma(x)_Z} & & \downarrow{\scriptstyle \sigma(x)_Y} \\ F(Z) & \xrightarrow{F(g)} & F(Y) \end{array}$$

commutes. Indeed, let $f \in \operatorname{Hom}_{\mathscr{C}}(Z, X)$. Then we have

$$\begin{aligned} F(g)\sigma(x)_Z(f) &= F(g)F(f)(x) = F(fg)(x) = \sigma(x)_Y(fg) \\ &= \sigma(x)_Y \operatorname{Hom}_{\mathscr{C}}(g, X)(f), \end{aligned}$$

which establishes our claim.

We now prove that ε and σ are mutually inverse. Let $x \in F(X)$, then

$$\varepsilon\sigma(x) = \sigma(x)_X(1_X) = F(1_X)(x) = 1_{F(X)}(x) = x.$$

Let φ be a functorial morphism from $\operatorname{Hom}_{\mathscr{C}}(-, X)$ to F, and Y an object in $\mathscr{C}$. We claim that $\varphi_Y = \sigma\varepsilon(\varphi)_Y$. Let $f \in \operatorname{Hom}_{\mathscr{C}}(Y, X)$, we have a commutative square

$$\begin{array}{ccc} \operatorname{Hom}_{\mathscr{C}}(X,X) & \xrightarrow{\operatorname{Hom}_{\mathscr{C}}(f,X)} & \operatorname{Hom}_{\mathscr{C}}(Y,X) \\ \downarrow{\scriptstyle \varphi_X} & & \downarrow{\scriptstyle \varphi_Y} \\ F(X) & \xrightarrow{F(f)} & F(Y) \end{array}$$

from which we deduce:

$$\sigma\varepsilon(\varphi)_Y(f) = F(f)(\varepsilon(\varphi)) = F(f)\varphi_X(1_X) = \varphi_Y \operatorname{Hom}_{\mathscr{C}}(f, X)(1_X) = \varphi_Y(f).$$

This establishes our claim and hence the bijectivity of ε. Finally, ε is easily seen to be a morphism of **k**-vector spaces. □

Clearly, there also exists a version of Yoneda's lemma using covariant functors instead of contravariant ones. We leave to the reader its easy formulation and proof.

In the sequel, we need not only the existence of Yoneda's bijections ε and σ, but also the explicit formulae expressing these bijections. We now consider the case where $\mathscr{C} = \operatorname{mod} A$.

Corollary II.3.2. *Let M, N be modules and F a subfunctor of* $\operatorname{Hom}_A(-, N)$. *Then there exists an isomorphism of* **k***-vector spaces* $F(M) \cong \operatorname{Hom}(\operatorname{Hom}_{\mathscr{C}}(-, M), F)$ *given by* $f \longmapsto \operatorname{Hom}_A(-, f)$. *If, in particular,* $F = \operatorname{Hom}_A(-, N)$, *then this map yields an isomorphism of vector spaces* $\operatorname{Hom}_A(M, N) \cong \operatorname{Hom}(\operatorname{Hom}_A(-, M), \operatorname{Hom}_A(-, N))$.

Proof. Let $f \in F(M)$. The Yoneda isomorphism σ applied to f gives a functorial morphism $\sigma(f) : \operatorname{Hom}_A(-, M) \longrightarrow F$ defined as follows. For every object X and morphism $g : X \longrightarrow M$, we have, as seen in the proof of Theorem II.3.1,

$$\sigma(f)_X(g) = \operatorname{Hom}_A(g, N)(f) = fg = \operatorname{Hom}_A(X, f)(g).$$

Thus, $\sigma(f)_X = \operatorname{Hom}_A(X, f)$ for every object X, that is, $\sigma(f) = \operatorname{Hom}_A(-, f)$. This completes the proof. □

The best known consequence of Yoneda's lemma is the projectivity of the Hom functor. An object H in $\operatorname{Fun} A$ is called **projective** if, for every functorial epimorphism $\varphi : F \longrightarrow G$ and every functorial morphism $\eta : H \longrightarrow G$, there exists a functorial morphism $\xi : H \longrightarrow F$ such that $\varphi\xi = \eta$, that is, such that the following diagram is commutative:

$$\begin{array}{ccc} & & H \\ & \overset{\xi}{\swarrow} & \downarrow{\scriptstyle \eta} \\ F & \xrightarrow{\varphi} & G \end{array}$$

Corollary II.3.3. *Let M be an A-module. Then,* $\operatorname{Hom}_A(-, M)$ *is a projective object in* $\operatorname{Fun} A$.

Proof. Let $\varphi : F \longrightarrow G$ be a functorial epimorphism and $\eta : \operatorname{Hom}_A(-, M) \longrightarrow G$ a functorial morphism. Then, φ induces a morphism

$$\varphi^* : \operatorname{Hom}(\operatorname{Hom}_A(-, M), F) \longrightarrow \operatorname{Hom}(\operatorname{Hom}_A(-, M), G)$$

by $\psi \longmapsto \varphi\psi$. Now, let $\varepsilon_F\colon \mathrm{Hom}(\mathrm{Hom}_A(-,M), F) \longrightarrow F(M)$ and $\varepsilon_G\colon \mathrm{Hom}(\mathrm{Hom}_A(-,M), G) \longrightarrow G(M)$ be the isomorphisms of Yoneda's lemma corresponding to the functors F and G respectively. Then we have a square:

$$\begin{array}{ccc} \mathrm{Hom}(\mathrm{Hom}_A(-,M),F) & \xrightarrow{\varphi^*} & \mathrm{Hom}(\mathrm{Hom}_A(-,M),G) \\ \downarrow{\scriptstyle \varepsilon_F} & & \downarrow{\scriptstyle \varepsilon_G} \\ F(M) & \xrightarrow{\varphi_M} & G(M) \end{array}$$

Let $\psi : \mathrm{Hom}_A(-,M) \longrightarrow F$ be a functorial morphism. Then we have

$$\varphi_M\varepsilon_F(\psi) = \varphi_M\psi_M(1_M) = (\varphi\psi)_M(1_M) = \varepsilon_G(\varphi\psi) = \varepsilon_G\varphi^*(\psi)$$

that is, the above square commutes. Now, φ_M is surjective, and hence so is φ^*. Therefore, there exists $\xi : \mathrm{Hom}_A(-,M) \longrightarrow F$ such that $\eta = \varphi^*(\xi) = \varphi\xi$. □

Yoneda's lemma and its corollaries show that one can reduce several questions about arbitrary modules to questions about projective functors. Because working with projective objects is always easier than working with arbitrary ones, this partially explains why passing from mod A to Fun$A$ turned out to be a fruitful idea.

II.3.2 Simple objects in Fun A

Because the projective objects of the form $\mathrm{Hom}_A(-,M)$, with M a finitely generated A-module, are particularly interesting, we consider the quotients of such objects. A functor F is called **finitely generated** if there exist a (finitely generated) module M_A and a functorial epimorphism:

$$\mathrm{Hom}_A(-,M) \longrightarrow F \longrightarrow 0.$$

We prove that the only finitely generated projective objects in Fun A are precisely the functors of the form $\mathrm{Hom}_A(-,M)$.

Lemma II.3.4. *An object F in* Fun A *is finitely generated projective if and only if there exists an A-module M such that $F \cong \mathrm{Hom}_A(-,M)$. Also, F is indecomposable if and only if M is indecomposable.*

Proof. Because F is finitely generated, there exist a module M_A and a functorial epimorphism $\varphi : \mathrm{Hom}_A(-,M) \longrightarrow F$. Because F is projective, φ is a retraction and there exists $\psi : F \longrightarrow \mathrm{Hom}_A(-,M)$ such that $\varphi\psi = 1_F$. Then, $\psi\varphi : \mathrm{Hom}_A(-,M) \longrightarrow \mathrm{Hom}_A(-,M)$ is an idempotent functorial endomorphism whose image is F. Because of Corollary II.3.3, there exists $f \in \mathrm{End}_A M$ such that $\psi\varphi = \mathrm{Hom}_A(-,f)$, and f is idempotent because so is $\mathrm{Hom}_A(-,f)$.

Consequently, $M' = \operatorname{Im} f$ is a direct summand of M and is such that $\operatorname{Hom}_A\left(-, M'\right)$ is the image of $\operatorname{Hom}_A(-, f) = \psi\varphi$. This shows that $F \cong \operatorname{Hom}_A\left(-, M'\right)$. The last assertion is easy to prove. □

We next prove that every indecomposable finitely generated projective object in Fun A has a unique maximal subobject, which is the radical, exactly as is the case for indecomposable finitely generated projective A-modules.

Lemma II.3.5. *Let M be an indecomposable module. Then,* $\operatorname{rad}_A(-, M)$ *is the unique maximal subfunctor of* $\operatorname{Hom}_A(-, M)$.

Proof. Let F be any proper subfunctor of $\operatorname{Hom}_A(-, M)$. We claim that F is actually a subfunctor of $\operatorname{rad}_A(-, M)$. We must show that, for every indecomposable A-module L, we have $F(L) \subseteq \operatorname{rad}_A(L, M)$. If L is not isomorphic to M, then $\operatorname{rad}_A(L, M) = \operatorname{Hom}_A(L, M)$ and the statement holds trivially. If $L = M$, let $f : M \longrightarrow M$ belong to $F(M)$. Because of Corollary II.3.2, under the Yoneda bijection, the functorial morphism

$$\operatorname{Hom}_A(-, f) : \operatorname{Hom}_A(-, M) \longrightarrow F$$

corresponds to it. Composing it with the proper inclusion $F \hookrightarrow \operatorname{Hom}_A(-, M)$, we get that

$$\operatorname{Hom}_A(-, f) : \operatorname{Hom}_A(-, M) \longrightarrow \operatorname{Hom}_A(-, M)$$

is not an isomorphism. But then neither is f. Hence, $f \in \operatorname{rad}_A(M, M)$, as required. □

An object in Fun A is called **simple** if it is nonzero and has only two subobjects, namely itself and the zero functor. It follows immediately from Lemma II.3.5 above that, for every indecomposable A-module M, the functor

$$S_M = \operatorname{Hom}_A(-, M) \,/\, \operatorname{rad}_A(-, M)$$

is simple.

Because $\operatorname{End}_A M$ is local, $S_M(M)$ is a skew field. Applying Yoneda's lemma, the space of functorial morphisms $\operatorname{Hom}(\operatorname{Hom}_A(-, M), S_M)$ is also a skew field. Therefore, there exists a (unique up to multiples by elements of the skew field) nonzero functorial morphism

$$\pi_M : \operatorname{Hom}_A(-, M) \longrightarrow S_M,$$

which is an epimorphism because S_M is simple. We now prove that conversely, every simple object in Fun A is of the form S_M for some indecomposable module M. Furthermore, we also prove that the morphism π_M is actually a projective cover. We define projective covers in Fun A exactly as we do in mod A: an epimorphism $\varphi : H \longrightarrow F$ with H projective is a **projective cover** if, whenever

$\varphi' : H' \longrightarrow F$ is another epimorphism with H' projective, there exists an epimorphism $\eta : H' \longrightarrow H$ such that $\varphi' = \varphi\eta$. In particular, η is a retraction so that H is a direct summand of H'.

Lemma II.3.6. *Let S be a simple object in Fun A. Up to isomorphism, there exists a unique indecomposable A-module M such that $S(M) \neq 0$, and then $S \cong S_M$. In addition, the functorial morphism $\pi_M : \mathrm{Hom}_A(-, M) \longrightarrow S_M$ is a projective cover.*

Proof. Because of Yoneda's lemma, for every module X, we have $S(X) \neq 0$ if and only if there exists a functorial morphism $\mathrm{Hom}_A(-, X) \longrightarrow S$ that is necessarily an epimorphism because S is simple. Because $S \neq 0$, there exists at least an indecomposable module M such that $S(M) \neq 0$. Assume that X is a module such that $S(X) \neq 0$. The projectivity of the functors $\mathrm{Hom}_A(-, M)$ and $\mathrm{Hom}_A(-, X)$ and Corollary II.3.2 yield morphisms $u: M \longrightarrow X$ and $v: X \longrightarrow M$ such that we have a commutative diagram with exact rows

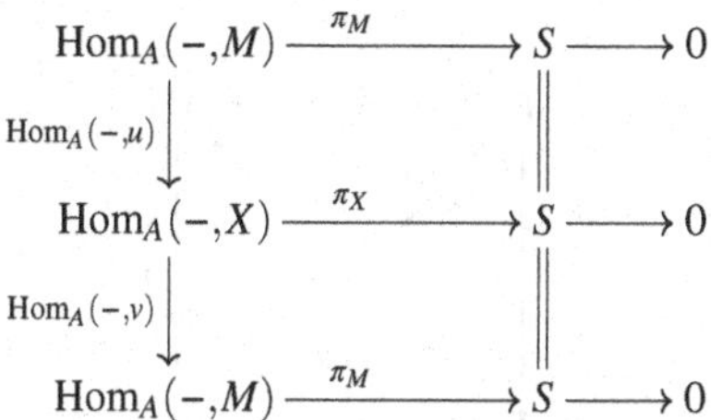

$$\begin{array}{ccccc}
\mathrm{Hom}_A(-,M) & \xrightarrow{\pi_M} & S & \longrightarrow & 0 \\
\downarrow{\scriptstyle \mathrm{Hom}_A(-,u)} & & \| & & \\
\mathrm{Hom}_A(-,X) & \xrightarrow{\pi_X} & S & \longrightarrow & 0 \\
\downarrow{\scriptstyle \mathrm{Hom}_A(-,v)} & & \| & & \\
\mathrm{Hom}_A(-,M) & \xrightarrow{\pi_M} & S & \longrightarrow & 0
\end{array}$$

The indecomposability of M implies that End M is local. Hence, $vu : M \longrightarrow M$ is nilpotent or invertible. If it were nilpotent, and $m > 0$ were such that $(vu)^m = 0$, then we would get the contradiction $\pi_M = \pi_M \mathrm{Hom}_A\left(-, (vu)^m\right) = 0$. Therefore, vu is invertible; thus, $v : X \longrightarrow M$ is a retraction. This shows that $S(X) \neq 0$ if and only if M is a direct summand of X. In particular, the indecomposable module M is unique up to isomorphism.

Replacing, in the proof above, $\mathrm{Hom}_A(-, X)$ by a projective functor F such that there exists a nonzero functorial morphism $F \longrightarrow S$, the same argument gives that $\pi_M : \mathrm{Hom}_A(-, M) \longrightarrow S_M$ is a projective cover morphism.

Finally, because S is simple and $\mathrm{rad}_A(-, M)$ is the unique maximal subfunctor of $\mathrm{Hom}_A(-, M)$, we have $S \cong \mathrm{Hom}_A(-, M) / \mathrm{rad}_A(-, M) = S_M$. □

Corollary II.3.7. *Let M, N be A-modules, with M indecomposable. Then, $S_M(N) \neq 0$ if and only if M is isomorphic to a direct summand of N.*

Proof. This was shown in the course of the proof of Lemma II.3.6 □

II.3.3 Projective resolutions of simple functors

Our first lemma is an easy exercise of homological algebra.

Lemma II.3.8. *Let $\mathscr{C}$ be an abelian category and (P, p_1, p_2) the fibered product of $f_1 : M_1 \longrightarrow M$, $f_2 : M_2 \longrightarrow M$ in $\mathscr{C}$. Let $f_1 = j_1 q_1$, $p_2 = j_2 q_2$ be the canonical factorisations through $K_1 = \operatorname{Im} f_1$ and $K_2 = \operatorname{Im} p_2$ respectively. Then*

(a) *There exists a unique $f : K_2 \longrightarrow K_1$ such that $j_1 f = f_2 j_2$.*
(b) *(P, p_1, q_2) is a fibered product of $q_1 : M_1 \longrightarrow K_1$ and $f : K_2 \longrightarrow K_1$.*
(c) *$\operatorname{Ker} f_1 \cong \operatorname{Ker} p_2$.*

Proof.

(a) Let K denote the cokernel of f_1, so that we have a short exact sequence

$$0 \longrightarrow K_1 \xrightarrow{j_1} M \xrightarrow{h} K \longrightarrow 0.$$

Then, $hf_2 j_2 q_2 = hf_2 p_2 = hf_1 p_1 = 0$. Hence, $hf_2 j_2 = 0$, because q_2 is an epimorphism. Therefore, $f_2 j_2$ factors through $\operatorname{Ker} h = K$, that is, there exists $f : K_2 \longrightarrow K_1$ such that $j_1 f = f_2 j_2$.

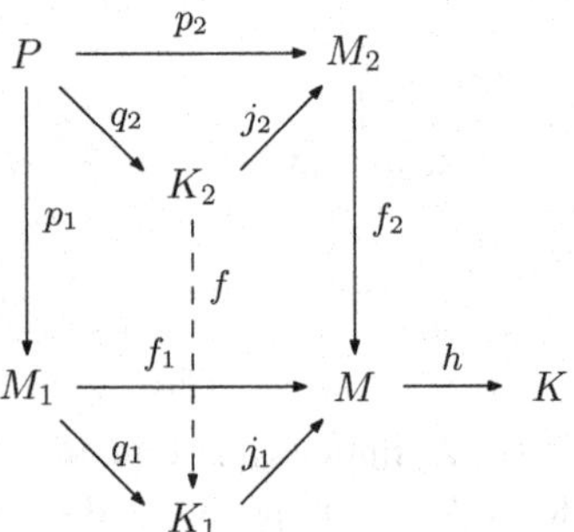

The uniqueness of f follows from the fact that j_1 is a monomorphism.

(b) First, we have $j_1 f q_2 = f_2 j_2 q_2 = f_2 p_2 = f_1 p_1 = j_1 q_1 p_1$. Hence, $f q_2 = q_1 p_1$ because j_1 is a monomorphism.

Let (U, u_1, u_2) be such that $f u_2 = q_1 u_1$. Then, $f_2 j_2 u_2 = j_1 f u_2 = j_1 q_1 u_1 = f_1 u_1$. The universal property of P yields a unique $u : U \longrightarrow P$ such that $u_1 = p_1 u$ and $j_2 u_2 = p_2 u = j_2 q_2 u$. Because j_2 is a monomorphism, the latter equality is equivalent to $u_2 = q_2 u$. This completes the proof of (b).

(c) We know that $\operatorname{Ker} f_1 \cong \operatorname{Ker} q_1$ and $\operatorname{Ker} p_2 \cong \operatorname{Ker} q_2$. The result then follows from (a), (b) and the commutative diagram with exact rows

$$\begin{array}{ccccccccc}
0 & \longrightarrow & \operatorname{Ker} p_2 & \longrightarrow & P & \xrightarrow{q_2} & K_2 & \longrightarrow & 0 \\
 & & & & \downarrow{\scriptstyle p_1} & & \downarrow{\scriptstyle f} & & \\
0 & \longrightarrow & \operatorname{Ker} f_1 & \longrightarrow & M_1 & \xrightarrow{q_1} & K_1 & \longrightarrow & 0
\end{array}$$

□

We now let N be an indecomposable module and examine the projective resolution of the simple functor S_N in the category Fun A. We start by considering the case where N is projective.

Lemma II.3.9. *Let N be an indecomposable A-module. Then, N is projective if and only if the simple functor S_N admits a projective resolution of the form*

$$0 \longrightarrow \mathrm{Hom}_A(-, M) \longrightarrow \mathrm{Hom}_A(-, N) \longrightarrow S_N \longrightarrow 0.$$

Proof. Assume first that N is projective. Because of Lemma II.3.5, we have a short exact sequence of functors:

$$0 \longrightarrow \mathrm{rad}_A(-, N) \longrightarrow \mathrm{Hom}_A(-, N) \longrightarrow S_N \longrightarrow 0.$$

Because of Corollary II.1.10, for every module X, the vector space $\mathrm{rad}_A(X, N)$ consists of the nonretractions from X to N. But N is projective; therefore, this space coincides with the set of nonsurjections from X to N, that is, the morphisms from X to N whose image lies in the unique maximal submodule $\mathrm{rad}\, N$ of N. Therefore, $\mathrm{rad}_A(X, N) = \mathrm{Hom}_A(X, \mathrm{rad}\, N)$ and the previous sequence becomes:

$$0 \longrightarrow \mathrm{Hom}_A(-, \mathrm{rad}\, N) \longrightarrow \mathrm{Hom}_A(-, N) \longrightarrow S_N \longrightarrow 0.$$

This completes the proof of this implication.

Conversely, assume that N is not projective and we have a short exact sequence of functors of the form:

$$0 \longrightarrow \mathrm{Hom}_A(-, M) \longrightarrow \mathrm{Hom}_A(-, N) \longrightarrow S_N \longrightarrow 0.$$

Evaluating this sequence on the module A_A yields a short exact sequence:

$$0 \longrightarrow M \longrightarrow N \longrightarrow S_N(A) \longrightarrow 0.$$

Because N is not projective, $S_N(A) = 0$ where we used Corollary II.3.7. Therefore, $M \cong N$; hence, $\mathrm{Hom}_A(-, M) \cong \mathrm{Hom}_A(-, N)$ and $S_N = 0$, an absurdity that completes the proof. □

The main result of this section, when translated into module language, will imply the existence theorem for almost split sequences. It asserts that simple objects in Fun A have projective resolutions of length at most two and exhibits a minimal projective resolution for such an object.

Theorem II.3.10. *Let N be an indecomposable A-module. The simple functor S_N admits a minimal projective resolution of the form:*

$$0 \longrightarrow \mathrm{Hom}_A(-, L) \longrightarrow \mathrm{Hom}_A(-, M) \longrightarrow \mathrm{Hom}_A(-, N) \longrightarrow S_N \longrightarrow 0.$$

If N is projective, then $L = 0$. Otherwise, L is indecomposable.

Proof. The case where N is projective follows from Lemma II.3.9; thus, we may assume that N is not projective. Let

$$P_1 \xrightarrow{p_1} P_0 \xrightarrow{p_0} N \longrightarrow 0$$

be a minimal projective presentation of N. It induces an exact sequence in Fun A of the form

$$\mathrm{D}\,\mathrm{Hom}_A(P_1,-) \xrightarrow{\mathrm{D}\,\mathrm{Hom}_A(p_1,-)} \mathrm{D}\,\mathrm{Hom}_A(P_0,-) \xrightarrow{\mathrm{D}\,\mathrm{Hom}_A(p_0,-)} \mathrm{D}\,\mathrm{Hom}_A(N,-) \longrightarrow 0.$$

Now, we recall from Lemma I.1.18 that the Nakayama functor $\nu = \mathrm{D}\,\mathrm{Hom}_A(-, A)$ induces, for each projective module P, a functorial isomorphism $\mathrm{D}\,\mathrm{Hom}_A(P, -) \cong \mathrm{Hom}_A(-, \nu P)$. Therefore, the previous exact sequence may be rewritten as

$$\mathrm{Hom}_A(-, \nu P_1) \xrightarrow{\mathrm{Hom}_A(-,\nu p_1)} \mathrm{Hom}_A(-, \nu P_0) \xrightarrow{\theta} \mathrm{D}\,\mathrm{Hom}_A(N, -) \longrightarrow 0$$

where θ denotes the composition of the isomorphism $\mathrm{Hom}_A(-, \nu P_0) \cong \mathrm{D}\,\mathrm{Hom}_A(P_0, -)$ with the morphism $\mathrm{D}\,\mathrm{Hom}_A(p_0, -)$. We claim that there exists a functorial morphism from $\mathrm{Hom}_A(-, N)$ to $\mathrm{D}\,\mathrm{Hom}_A(N, -)$ having the simple functor S_N as its image. Indeed, define a functorial morphism $\eta_N : S_N \longrightarrow \mathrm{D}\,\mathrm{Hom}_A(N, -)$ by sending the residual class $\overline{g}$ of a morphism $g \in \mathrm{End}_A N$ modulo $\mathrm{rad}\,\mathrm{End}_A N$ to the linear form mapping $f \in \mathrm{End}_A N$ to the residual class $\overline{gf}$ of the composition gf in $\mathbf{k}$. It is easily seen that η_N is well-defined and nonzero. In addition, η_N is a monomorphism, because S_N is simple. Because the projective cover morphism $\pi_N : \mathrm{Hom}_A(-, N) \longrightarrow S_N$ is an epimorphism, the composition $\eta_N\pi_N : \mathrm{Hom}_A(-, N) \longrightarrow \mathrm{D}\,\mathrm{Hom}_A(N, -)$ is nonzero and admits S_N as its image. This establishes our claim.

The projectivity of $\mathrm{Hom}_A(-, N)$ and Corollary II.3.2 yield a morphism $u : N \longrightarrow \nu P_0$ such that $\theta\,\mathrm{Hom}_A(-, u) = \eta_N\pi_N$, that is, the following diagram is commutative.

$$\begin{array}{ccccc} & & \mathrm{Hom}_A(-, N) & & \\ & \swarrow{\scriptstyle \mathrm{Hom}_A(-,u)} & \downarrow{\scriptstyle \eta_N\pi_N} & & \\ \mathrm{Hom}_A(-, \nu P_0) & \xrightarrow{\theta} & \mathrm{D}\,\mathrm{Hom}_A(N, -) & \longrightarrow & 0 \end{array}$$

Let M be the fibered product of u and νp_1. Setting $L = \mathrm{Ker}\,\nu p_1$, we get a commutative diagram with exact rows

$$\begin{array}{ccccccc} 0 & \longrightarrow & L & \xrightarrow{f} & M & \xrightarrow{g} & N \\ & & \| & & \downarrow{\scriptstyle v} & & \downarrow{\scriptstyle u} \\ 0 & \longrightarrow & L & \longrightarrow & \nu P_1 & \xrightarrow{\nu p_1} & \nu P_0 \end{array}$$

Indeed, the exactness of the lower row is obvious and the exactness of the upper row follows from Lemma II.3.8. We deduce a commutative diagram with an exact lower row

$$\begin{array}{ccccccccc}
0 \to \operatorname{Hom}_A(-,L) & \xrightarrow{\operatorname{Hom}_A(-,f)} & \operatorname{Hom}_A(-,M) & \xrightarrow{\operatorname{Hom}_A(-,g)} & \operatorname{Hom}_A(-,N) & \longrightarrow & S_N & \longrightarrow & 0 \\
\| & & \downarrow{\scriptstyle \operatorname{Hom}_A(-,\nu)} & & \downarrow{\scriptstyle \operatorname{Hom}_A(-,u)} & & \downarrow{\scriptstyle \eta_N} & & \\
0 \to \operatorname{Hom}_A(-,L) & \longrightarrow & \operatorname{Hom}_A(-,\nu P_1) & \xrightarrow{\operatorname{Hom}_A(-,\nu p_1)} & \operatorname{Hom}_A(-,\nu P_0) & \longrightarrow & D\operatorname{Hom}_A(N,-) & \longrightarrow & 0
\end{array}$$

We claim that the upper row is also exact. It suffices to prove its exactness at $\operatorname{Hom}_A(-, N)$. Because of Lemma II.3.5, this amounts to showing that, for every module X, we have $\operatorname{Im}\operatorname{Hom}_A(X, g) = \operatorname{rad}_A(X, N)$.

Because of commutativity, we have

$$\eta_N \pi_N \operatorname{Hom}_A(-, g) = \theta \operatorname{Hom}_A(-, \nu p_1) \operatorname{Hom}_A(-, v) = 0,$$

using the exactness of the lower row. Because η_N is a monomorphism, this implies that $\pi_N \operatorname{Hom}_A(-, g) = 0$ and so $\operatorname{Im}\operatorname{Hom}_A(-, g) \subseteq \operatorname{Ker}\pi_N = \operatorname{rad}_A(-, N)$. Therefore, for every module X, we have $\operatorname{Im}\operatorname{Hom}_A(X, g) \subseteq \operatorname{rad}_A(X, N)$. Conversely, assume that $h \in \operatorname{rad}_A(X, N)$, that is, $h : X \longrightarrow N$ is not a retraction. Then $\pi_{N,X}(h) = 0$ and commutativity yield $\theta \operatorname{Hom}_A(X, u)(h) = 0$, that is, $\theta(uh) = 0$. Because the lower row is exact, there exists $h' : X \longrightarrow \nu P_1$ such that $uh = (\nu p_1)h'$. The universal property of M yields $k : X \longrightarrow M$ such that $gk = h$ and $uk = h'$. In particular, $h \in \operatorname{Im}\operatorname{Hom}_A(X, g)$. This establishes our claim.

We have thus finished proving that the upper sequence is a projective resolution of S_N.

We next prove that L is indecomposable. The exact sequence

$$0 \longrightarrow L \longrightarrow \nu P_1 \xrightarrow{\nu p_1} \nu P_0$$

is the start of an injective coresolution of L. If L were decomposable, then every direct sum decomposition of L induces a direct sum decomposition of the morphism νp_1, and thus of p_1, contradicting the minimality of the given projective presentation $P_1 \longrightarrow P_0 \longrightarrow N \longrightarrow 0$.

It only remains to prove that the constructed projective resolution of S_N is minimal. Now, if this is not the case, then the indecomposability of L implies that there exists a direct sum decomposition $M \cong M' \oplus L$ such that we have a short exact sequence of functors

$$0 \longrightarrow \operatorname{Hom}_A\left(-, M'\right) \longrightarrow \operatorname{Hom}_A(-, N) \longrightarrow S_N \longrightarrow 0.$$

But then, because of Lemma II.3.9, N is projective, a contradiction. The proof is now complete. □

It turns out that the minimal projective resolution of S_N we just constructed induces an almost split sequence.

Proposition II.3.11. *Let N be an indecomposable nonprojective A-module and*

$$0 \longrightarrow \mathrm{Hom}_A(-, L) \xrightarrow{\mathrm{Hom}_A(-,f)} \mathrm{Hom}_A(-, M) \xrightarrow{\mathrm{Hom}_A(-,g)} \mathrm{Hom}_A(-, N) \longrightarrow S_N \longrightarrow 0$$

a minimal projective resolution of S_N. Then the sequence

$$0 \longrightarrow L \xrightarrow{f} M \xrightarrow{g} N \longrightarrow 0$$

is exact and almost split.

Proof. We evaluate the sequence of functors on the module A_A. Because N is nonprojective, it follows from Corollary II.3.7 that $S_N(A) = 0$. We thus get a short exact sequence

$$0 \longrightarrow L \xrightarrow{f} M \xrightarrow{g} N \longrightarrow 0$$

with L indecomposable. Because of Corollary II.2.33, it suffices to show that the morphism g is right almost split.

Indeed, assume first that g is not a radical morphism. Then, g is a retraction and there exists $g' : N \longrightarrow M$ such that $gg' = 1_N$. But then, for every $h \in \mathrm{End}_A N$, we have:

$$h = gg'h = \mathrm{Hom}_A(N, g)\left(g'h\right) \in \mathrm{Im}\,\mathrm{Hom}_A(N, g) = \mathrm{Ker}\,\pi_{N,N}.$$

This implies $S_N(N) = 0$, a contradiction. Hence, g is a radical morphism.

Now, let V be an A-module and $v \in \mathrm{rad}_A(V, N)$. Because $\mathrm{rad}_A(V, N) = \mathrm{Im}\,\mathrm{Hom}_A(V, g)$, as we saw in the proof of Theorem II.3.10, there exists $v' : V \longrightarrow M$ such that $v = \mathrm{Hom}_A(V, g)(v') = gv'$. This completes the proof. □

The proof of Theorem II.3.10 contains a construction procedure for the almost split sequence of Proposition II.3.11: indeed, the morphism $g : M \longrightarrow N$ is obtained by taking the fibered product of $u : N \longrightarrow \nu P_0$ and $\nu p_1 : \nu P_1 \longrightarrow \nu P_0$; and the morphism $f : L \longrightarrow M$ is just the kernel of g. We use this remark in the following chapter.

We deduce the main existence theorem of Auslander and Reiten.

Theorem II.3.12. *Let N be an indecomposable nonprojective A-module, or L an indecomposable noninjective A-module. Then, there exists an almost split sequence*

$$0 \longrightarrow L \xrightarrow{f} M \xrightarrow{g} N \longrightarrow 0.$$

Additionally, this sequence is uniquely determined by N, or by L, up to isomorphism.

Proof. Assume that N is an indecomposable nonprojective module. Then the existence statement follows directly from Theorem II.3.10 and Proposition II.3.11. If L is an indecomposable noninjective module, then $\mathrm{D}L$ is an indecomposable nonprojective left A-module; thus, there exists an almost split sequence

$$0 \longrightarrow N' \longrightarrow M' \longrightarrow \mathrm{D}L \longrightarrow 0$$

in mod A^{op}. Applying the duality functor D yields the required almost split sequence in mod A. Finally, the uniqueness assertion follows from Corollary II.2.32. □

As an easy consequence, we get that the module category contains enough minimal almost split morphisms.

Corollary II.3.13.

(a) *If N is an indecomposable A-module, then there exists a right minimal almost split morphism $g : M \longrightarrow N$.*
(b) *If L is an indecomposable A-module, then there exists a left minimal almost split morphism $f : L \longrightarrow M$.*

Proof. We only prove (a), because the proof of (b) is dual.

If N is projective, then the inclusion $\operatorname{rad} N \hookrightarrow N$ is right minimal almost split. Otherwise, there exists an almost split sequence

$$0 \longrightarrow L \xrightarrow{f} M \xrightarrow{g} N \longrightarrow 0$$

in which the morphism g is right minimal almost split. □

The reader may wonder why, in this section, we decided to work with contravariant functors instead of the perhaps more familiar covariant ones. This is because the almost split sequence $0 \longrightarrow L \longrightarrow M \longrightarrow N \longrightarrow 0$ is very easy to read from the minimal projective resolution of Theorem II.3.10. In fact, one could work equally well with the category Fun A^{op} of the covariant functors from mod A to mod $\mathbf{k}$. But this is left to the exercises.

Exercises for Section II.3

Exercise II.3.1. Prove that the category Fun A is an abelian $\mathbf{k}$-linear category.

Exercise II.3.2. Let M, N be A-modules. Prove that the following conditions are equivalent:

(a) $M \cong N$ in mod A.
(b) $\operatorname{Hom}_A(-, M) \cong \operatorname{Hom}_A(-, N)$ in Fun A.
(c) $\operatorname{Hom}_A(M, -) \cong \operatorname{Hom}_A(N, -)$ in Fun A^{op}.

Exercise II.3.3. Let M, N be A-modules.

(a) Assume that there exists a monomorphism (or an epimorphism) $\operatorname{Hom}_A(-, M) \longrightarrow \operatorname{Hom}_A(-, N)$. Prove that there exists a monomorphism (or an epimorphism respectively) $M \longrightarrow N$.
(b) Assume that there exists a monomorphism (or an epimorphism) $\operatorname{Hom}_A(M, -) \longrightarrow \operatorname{Hom}_A(N, -)$. Prove that there exists a monomorphism (or an epimorphism respectively) $N \longrightarrow M$.

Exercise II.3.4. Prove that, for every A-module M

(a) The functor $\operatorname{D}\operatorname{Hom}_A(M, -)$ is an injective object in $\operatorname{Fun} A$.
(b) The functor $\operatorname{D}\operatorname{Hom}_A(-, M)$ is an injective object in $\operatorname{Fun} A^{op}$.

Exercise II.3.5. Let M be an indecomposable nonprojective A-module. Prove that the composition of the projective cover $\pi_M : \operatorname{Hom}_A(-, M) \longrightarrow S_M$ with the morphism $\eta_M : S_M \longrightarrow \operatorname{D}\operatorname{Hom}_A(M, -)$ of the proof of Theorem II.3.10 is the morphism that assigns to $f \in \operatorname{Hom}_A(X, M)$ the linear form $g \mapsto \overline{fg}$ on $\operatorname{Hom}_A(M, X)$, where $\overline{fg} \in \operatorname{End} M / \operatorname{rad} \operatorname{End} M$ is the residual class of $fg \in \operatorname{End} M$ modulo the radical.

Exercise II.3.6.

(a) Let N be an indecomposable A-module. Prove that a morphism $g\colon M \longrightarrow N$ is right almost split if and only if the corresponding sequence

$$\operatorname{Hom}_A(-, M) \xrightarrow{\operatorname{Hom}_A(-,g)} \operatorname{Hom}_A(-, N) \xrightarrow{\pi_N} S_N \longrightarrow 0$$

is a projective presentation. In addition, this is a minimal projective presentation if and only if g is right minimal almost split.
(b) Let L be an indecomposable A-module. Prove that a morphism $f\colon L \longrightarrow M$ is left almost split if and only if the corresponding sequence

$$\operatorname{Hom}_A(M, -) \xrightarrow{\operatorname{Hom}_A(f,-)} \operatorname{Hom}_A(L, -) \xrightarrow{\eta^L} S^L \longrightarrow 0$$

is a projective presentation. In addition, this is a minimal projective presentation if and only if f is left minimal almost split.

Exercise II.3.7. Let $0 \longrightarrow F_1 \longrightarrow F_2 \longrightarrow F_3 \longrightarrow 0$ be a short exact sequence in $\operatorname{Fun} A$. Prove that:

(a) If F_1 and F_3 are finitely generated, then so is F_2.
(b) If F_2 is finitely generated, then so is F_3.

Exercise II.3.8. Let F be a finitely generated object in Fun A. Prove that:

(a) If M is an A-module of least dimension such that there exists an epimorphism $\varphi\colon \operatorname{Hom}_A(-, M) \longrightarrow F$, then φ is a projective cover.
(b) If M_1 and M_2 are such that $\varphi_1\colon \operatorname{Hom}_A(-, M_1) \longrightarrow F$ and $\varphi_2\colon \operatorname{Hom}_A(-, M_2) \longrightarrow F$ are projective covers, then there exists an isomorphism $f\colon M_1 \longrightarrow M_2$ such that $\varphi_2 \operatorname{Hom}_A(-, f) = \varphi_1$.

II.4 Factorising radical morphisms

II.4.1 Higher powers of the radical

The radical being an ideal in the module category, it is possible to form its powers in the usual way. We have already defined at the beginning of Subsection II.2.1 the radical square $\operatorname{rad}^2_A(L, N)$, for modules L, N. Following the same idea, we can define, inductively, for all $m > 1$,

$$\operatorname{rad}^m_A = \operatorname{rad}^{m-1}_A \cdot \operatorname{rad}_A$$

that is, for the modules L, N, we define $\operatorname{rad}^m_A(L, N)$ to consist of all compositions gf with $g \in \operatorname{rad}^{m-1}_A(M, N)$ and $f \in \operatorname{rad}_A(L, M)$ for some module M (where we agree that $\operatorname{rad}^0_A = \operatorname{Hom}_A$).

It is easily seen that, for every $m > 1$, the set $\operatorname{rad}^m_A(L, N)$ is a **k**-subspace of $\operatorname{rad}^{m-1}_A(L, N)$; thus, we have an infinite chain of inclusions:

$$\operatorname{Hom}_A(L, N) \supseteq \operatorname{rad}_A(L, N) \supseteq \operatorname{rad}^2_A(L, N) \supseteq \dots$$

We also set:

$$\operatorname{rad}^\infty_A = \bigcap_{m \geq 1} \operatorname{rad}^m_A .$$

This is the **infinite radical** of the module category.

The following easy lemma is particularly important.

Lemma II.4.1. *Given the modules M, N there exists a least integer $m \geq 1$ (depending on M and N) such that* $\operatorname{rad}^\infty_A(M, N) = \operatorname{rad}^m_A(M, N)$.

Proof. Indeed, $\operatorname{Hom}_A(M, N)$ is a finite dimensional vector space; hence, the sequence of subspaces:

$$\operatorname{Hom}_A(M, N) \supseteq \operatorname{rad}_A(M, N) \supseteq \operatorname{rad}^2_A(M, N) \supseteq \dots \supseteq \operatorname{rad}^\infty_A(M, N)$$

must eventually stabilise. □

Example II.4.2. Let A be given by the quiver

$$1 \bullet \underset{\beta}{\overset{\alpha}{\rightleftarrows}} \bullet 2$$

bound by $\alpha\beta\alpha\beta\alpha = 0$. Then, the projection morphism f from the indecomposable projective module

$$P_1 = \begin{smallmatrix} 1 \\ 2 \\ 1 \\ 2 \\ 1 \end{smallmatrix}$$

to its socle $S_1 = 1$ belongs to $\mathrm{rad}^2_A(P_1, P_1)$. Indeed, letting

$$M = \begin{smallmatrix} 1 \\ 2 \\ 1 \end{smallmatrix}$$

and $g : P_1 \longrightarrow M$, $h : M \longrightarrow S_1$ be the projection morphisms, both are radical morphisms, because they are nonisomorphisms and we clearly have $f = hg$.

Example II.4.3. As a second example, we show a morphism lying in the infinite radical of the module category. Let A be the Kronecker algebra, given by the quiver

$$1 \bullet \underset{\beta}{\overset{\alpha}{\leftleftarrows}} \bullet 2$$

Consider the indecomposable modules $L = S_1$ and $M = \begin{smallmatrix} 2 \\ 1 \end{smallmatrix}$. There is an obvious nonzero morphism f embedding L as the socle of M, given by the left multiplication by α. Clearly, f is radical. We prove that $f \in \mathrm{rad}^\infty_A(L, M)$. For this purpose, we construct an infinite family of nonisomorphic indecomposable modules $(L_i)_{i \geq 0}$ starting with $L_0 = L$ and radical morphisms $g_i : L_{i-1} \longrightarrow L_i$ such that, for every i, f factors through $g_i \dots g_1$. This implies the statement. Let L_1 be the indecomposable projective module:

P_2

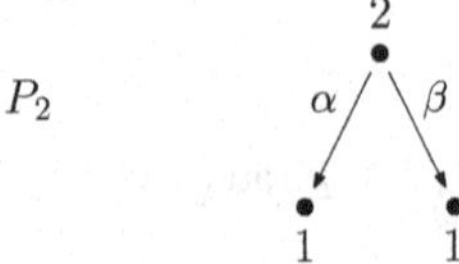

There exist two linearly independent embeddings of $L = S_1$ as a socle factor of P_2, given by left multiplication by the arrows α and β respectively. We call these embeddings $j_\alpha^{(1)}$ and $j_\beta^{(1)}$ respectively. Because $\operatorname{rad} P_2 = S_1^2$, the morphism $j^{(1)} = \begin{pmatrix} j_\alpha^{(1)} \\ j_\beta^{(1)} \end{pmatrix} : S_1^2 \longrightarrow P_2$ is right minimal almost split and the morphisms $j_\alpha^{(1)}, j_\beta^{(1)}$ are irreducible. Also, $M = \operatorname{Coker} j_\beta^{(1)}$. Set $g_1 = j_\alpha^{(1)} : S_1 \longrightarrow P_2$. We now construct L_2. Let $L_2^1 = L_1 \oplus P_2 = P_2^2$. Then we have four linearly independent embeddings of S_1 as a direct summand of the socle of L_2^1, as shown in the picture below

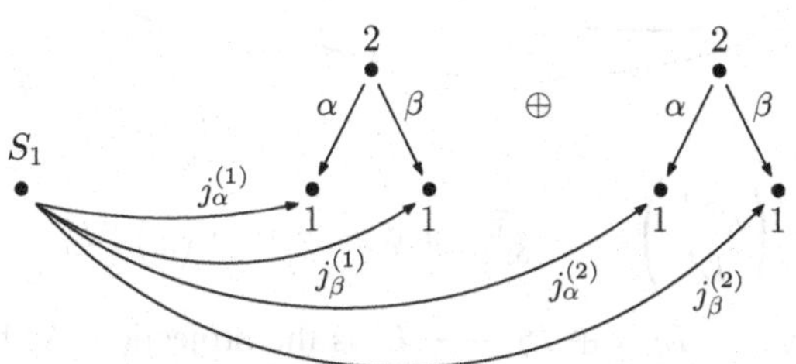

Let $L_2 = \operatorname{Coker} \begin{pmatrix} j_\beta^{(1)} \\ j_\alpha^{(2)} \end{pmatrix} \cong P_2^2/S_1$. Because the morphism $\begin{pmatrix} j_\beta^{(1)} \\ j_\alpha^{(2)} \end{pmatrix}$ identifies S_1 to the codomains of $j_\beta^{(1)}, j_\alpha^{(2)}$, we get that L_2 is a module of the form

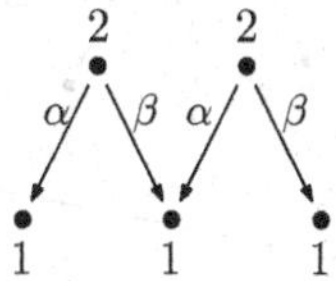

that is, denoting by $\{e_2', \alpha', \beta'\}$ and $\{e_2'', \alpha'', \beta''\}$ the basis vectors of the two copies of P_2 in the direct sum L_2', we get that a $\mathbf{k}$-basis of the five-dimensional module L_2 is given by $\{e_2', e_2'', e_2'\alpha' = \alpha', e_2'\beta' = e_2''\alpha'', e_2''\beta'' = \beta''\}$. Its top is the two-dimensional space with the basis $\{e_2', e_2''\}$.

This implies that L_2 is indecomposable. Indeed, assume that $L_2 = L_2' \oplus L_2''$. Then, $\operatorname{top} L_2 = \operatorname{top} L_2' \oplus \operatorname{top} L_2''$. If $\operatorname{top} L_2'$ contains both basis vectors e_2', e_2'', then $L_2' = L_2$ because L_2' is a submodule. The situation is similar if $\operatorname{top} L_2''$ contains both e_2', e_2''. Suppose, thus, that $e_2' \in L_2'$ and $e_2'' \in L_2''$. Then $e_2'\beta' = e_2''\alpha'' \in L_2' \cap L_2''$. But the latter should be zero, a contradiction.

Denoting the projection by $p_2 : L_2^1 \longrightarrow L_2$, we see that the composition $g_2 = p_2 \binom{1}{0} : P_2 = L_1 \xrightarrow{\binom{1}{0}} L_2' = L_1 \oplus P_2 \xrightarrow{p} L_2$ is an embedding such that $g_2 g_1 \neq 0$. We continue inductively. Assume that we have constructed indecomposable modules $L_1, L_2, \ldots, L_{i-1}$ and monomorphisms $g_1, g_2, \ldots, g_{i-1}$ such that $g_{i-1} \ldots g_2 g_1 \neq 0$. We set $L'_i = L_{i-1} \oplus P_2$ and consider the two embeddings of S_1 as a direct summand of the socle of L'_i, which are shown in the picture below

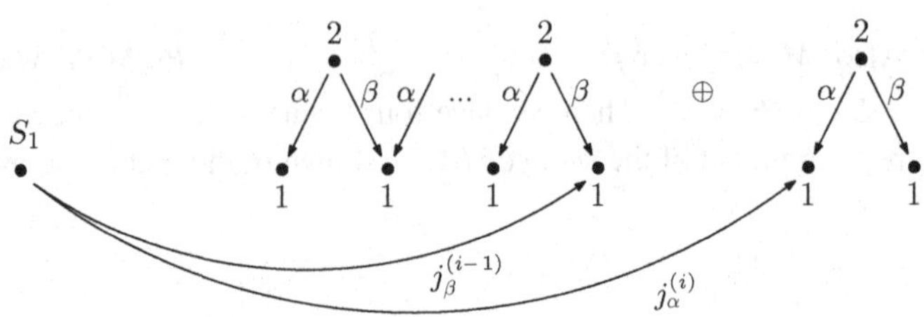

We let $L_i = \operatorname{Coker}\begin{pmatrix} j_\beta^{(i-1)} \\ j_\alpha^{(i)} \end{pmatrix} \cong \frac{L_{i-1} \oplus P_2}{S_1}$ and $g_i = p_{i-1}\begin{pmatrix} 1 \\ 0 \end{pmatrix} : L_{i-1} \xrightarrow{\binom{1}{0}} L_{i-1} \oplus P_2 \xrightarrow{p_{i-1}} L_i$, where $p_{i-1} : L_{i-1} \oplus P_2 \longrightarrow L_i$ is the projection. As before, L_i is indecomposable and g_i is a monomorphism such that $g_i g_{i-1} \ldots g_1 \neq 0$. For each $i \geq 1$, we have an epimorphism $h_i : L_i \longrightarrow M$ defined by sending the "first" top summand S_2 of L_i to top M, and the other summands to zero. Then we easily see that $h_i g_i \ldots g_1 = f$, as required.

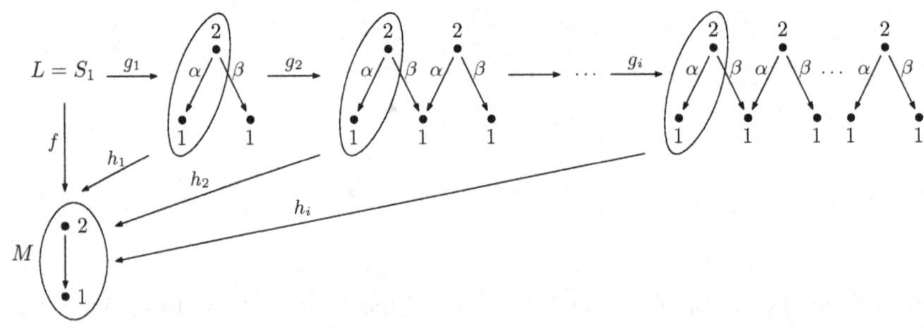

We have $\dim_{\mathbf{k}} \operatorname{Hom}_A(L, M) = 1$, and also $\dim_{\mathbf{k}} \operatorname{rad}_A^\infty(L, M) \geq 1$ because $f \in \operatorname{rad}_A^\infty(L, M)$ is nonzero. Hence, we have

$$\operatorname{Hom}_A(L, M) = \operatorname{rad}_A(L, M) = \ldots = \operatorname{rad}_A^\infty(L, M)$$

in this case. That is, the integer m of Lemma II.4.1 is here equal to 1. Similarly, the nonzero morphism $M \longrightarrow S_2$ given by right multiplication by e_2 also belongs to the infinite radical.

We have proved in passing that the Kronecker algebra is representation-infinite: indeed, we have exhibited an infinite family of nonisomorphic indecomposable modules. As a consequence, every path algebra having multiple arrows is representation-infinite, see Exercise II.4.9.

II.4.2 Factorising radical morphisms

Our first proposition asserts that every morphism lying in a finite power of the radical may be written as a sum of compositions of irreducible morphisms.

Proposition II.4.4. *Let M, N be indecomposable modules and $f \in \operatorname{rad}_A^n(M, N)$ for some $n \geq 2$. Then:*

(a) *There exist $s \geq 1$, indecomposable modules $X_1, \ldots, X_s$ and morphisms $M \xrightarrow{h_i} X_i \xrightarrow{g_i} N$ with $h_i \in \operatorname{rad}_A(M, X_i)$ and g_i a sum of compositions of $n-1$ irreducible morphisms between indecomposables and $f = \sum_{i=1}^{s} g_i h_i$. In addition, if $f \notin \operatorname{rad}_A^{n+1}(M, N)$, then at least one of the h_i is irreducible and f can be written as $f = u + v$, where $u \neq 0$ is a sum of compositions of irreducible morphisms between indecomposables and $v \in \operatorname{rad}_A^{n+1}(M, N)$.*

(b) *There exist $s \geq 1$, indecomposable modules $X_1, \ldots, X_s$ and morphisms $M \xrightarrow{h_i} X_i \xrightarrow{g_i} N$ with $g_i \in \operatorname{rad}_A(X_i, N)$ and h_i a sum of compositions of $n-1$ irreducible morphisms between indecomposables and $f = \sum_{i=1}^{s} g_i h_i$. In addition, if $f \notin \operatorname{rad}_A^{n+1}(M, N)$, then at least one of the g_i is irreducible and f can be written as $f = u + v$, where $u \neq 0$ is a sum of compositions of irreducible morphisms between indecomposables and $v \in \operatorname{rad}_A^{n+1}(M, N)$.*

Proof. We prove only (a) because the proof of (b) is similar.

Both statements in (a) are proven by induction on n.

Assume first that $n = 2$. Because of Corollary II.3.13, there exists a right minimal almost split morphism $g: E \longrightarrow N$. Let $E = \oplus_{i=1}^{s} E_i$ be a decomposition of E into indecomposable summands and $g_i = g|_{E_i}$. Then, there exists a morphism

$$h = \begin{pmatrix} h_1 \\ \vdots \\ h_s \end{pmatrix} : M \longrightarrow E \text{ such that } f = \textstyle\sum_{i=1}^{s} g_i h_i:$$

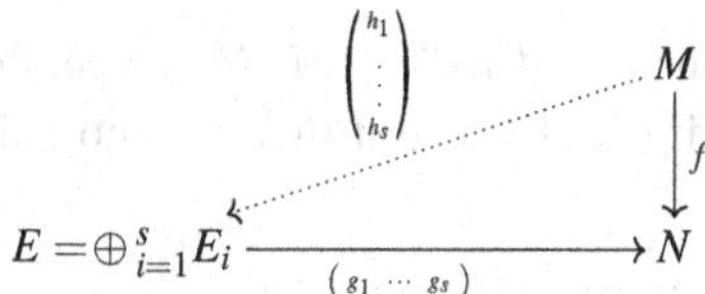

Because $f \in \operatorname{rad}_A^2(M, N)$, it is not irreducible, that is, none of the h_i is an isomorphism. Consequently, $h_i \in \operatorname{rad}(M, E_i)$ for all i. This proves (a).

If, on the other hand, $f \notin \mathrm{rad}_A^3(M, N)$, then there exists i such that $h_i \notin \mathrm{rad}_A^2(M, E_i)$. But then such an h_i is irreducible and we have proven (b).

Suppose now that $n \geq 3$ and $f \in \mathrm{rad}_A^n(M, N)$. By definition, there exist an A-module Y and morphisms $f' \in \mathrm{rad}_A(M, Y)$ and $f'' \in \mathrm{rad}_A^{n-1}(Y, N)$ such that $f = f''f'$. Let $Y = \oplus_{i=1}^t Y_i$ be a decomposition of Y into indecomposable summands so that the morphisms f' and f'' can be written as

$$M \xrightarrow{f' = \begin{pmatrix} f_1' \\ \vdots \\ f_t' \end{pmatrix}} \oplus_{i=1}^t Y_i \xrightarrow{f'' = (f_1'' \cdots f_t'')} N;$$

thus, $f = \sum_{i=1}^t f_i'' f_i'$. Then we have $f_i'' \in \mathrm{rad}_A^{n-1}(Y_i, N)$ for each i. Applying the induction hypothesis yields, for each i, a finite set of indecomposables $Z_{i1}, \ldots, Z_{is_i}$ and morphisms

$$Y_i \xrightarrow{g_{ij}'} Z_{ij} \xrightarrow{g_{ij}} N$$

such that $g_{ij}' \in \mathrm{rad}_A(Y_i, Z_{ij})$ for each j. In addition, each g_{ij} is a sum of compositions of $n-2$ irreducible morphisms between indecomposable modules such that $f_i'' = \sum_{j=1}^{s_i} g_{ij} g_{ij}'$. Because each $g_{ij}' f_i' \colon M \longrightarrow Z_{ij}$ belongs to $\mathrm{rad}_A^2(M, Z_{ij})$, the case $n = 2$ above yields $g_{ij}' f_i' = \sum_{l=1}^{m_{ij}} h_{ijl}' h_{ijl}$ where, for each l, $h_{ijl}' \colon E_{ijl} \longrightarrow Z_{ijl}$ is an irreducible morphism between indecomposables and $h_{ijl} \in \mathrm{rad}_A(M, E_{ijl})$. Substituting, we get

$$f = f''f' = \sum_{i=1}^{s} f_i'' f_i' = \sum_{i,j} g_{ij} g_{ij}' f_i' = \sum_{i,j,l} g_{ij} h_{ijl}' h_{ijl}.$$

Because $h_{ijl} \in \mathrm{rad}_A(M, E_{ijl})$ and each $g_{ij} h_{ijl}' \colon M \longrightarrow E_{ijl}$ is a sum of compositions of $n-1$ irreducible morphisms, this finishes the proof of the first part of (a).

For the second part, assume $f \notin \mathrm{rad}_A^{n+1}(M, N)$. Then, there exist indices i, j, l above such that $h_{ijl} \notin \mathrm{rad}_A^2(M, E_{ijl})$. But then the morphism h_{ijl} is irreducible. The statement follows. □

Corollary II.4.5. *Let M, N be indecomposable modules. Then, every radical morphism $f \in \mathrm{rad}_A(M, N)$ can be written as $f = u + v$, where u is a sum of compositions of irreducible morphisms, and $v \in \mathrm{rad}_A^\infty(M, N)$.*

Proof. If $f \in \mathrm{rad}_A^\infty(M, N)$, then there is nothing to prove. Assume thus that $f \notin \mathrm{rad}_A^\infty(M, N)$. Then, there exists $n > 0$ such that $f \in \mathrm{rad}_A^n(M, N) \setminus$

$\mathrm{rad}_A^{n+1}(M, N)$. Applying Proposition II.4.4(b), we get $f = u_0 + v_1$, where u_0 is a sum of compositions of irreducible morphisms between indecomposable modules and $v_1 \in \mathrm{rad}_A^{n+1}(M, N)$. Repeating the same procedure with v_1 we get that $v_1 = u_1 + v_2$, where u_1 is a sum of compositions of irreducible morphisms between indecomposable modules and $v_2 \in \mathrm{rad}_A^m(M, N)$, with $m > n + 1$. Then, $f = (u_0 + u_1) + v_2$, and we repeat this procedure again. It stops after finitely many steps because there exists $l > 0$ such that $\mathrm{rad}_A^l(M, N) = \mathrm{rad}_A^\infty(M, N)$. □

Corollary II.4.6. *Let M, N be indecomposable modules, and $f \in \mathrm{rad}_A(M, N)$. If $\mathrm{rad}_A^\infty(M, N) = 0$. Then, f is a sum of compositions of irreducible morphisms.*

Proof. This follows from Corollary II.4.5 above. □

II.4.3 Paths

The results of Subsection II.4.2 may be reformulated using the notion of path, which we now define. Paths are used to visualise statements about compositions of morphisms that may otherwise look technical.

Definition II.4.7. Let M, N be indecomposable modules. A **path** from M to N in $\mathrm{ind}\, A$ (denoted as $M \rightsquigarrow N$) of **length** t is a sequence

$$M = M_0 \xrightarrow{f_1} M_1 \xrightarrow{f_2} M_2 \longrightarrow \cdots \longrightarrow M_{t-1} \xrightarrow{f_t} M_t = N$$

where all M_i are indecomposable modules and all f_i are nonzero morphisms. We then say that M is a **predecessor** of N, or that N is a **successor** of M. This path is called a **radical path** if all f_i are radical morphisms. It is a **path of irreducible morphisms** if all f_i are irreducible.

For instance, if $0 \longrightarrow L \longrightarrow M \longrightarrow N \longrightarrow 0$ is an almost split sequence, and M' is an indecomposable summand of M, then we have a path of irreducible morphisms $L \longrightarrow M' \longrightarrow N$ of length two.

As we do in the case of quivers (see Subsection I.2.1), we agree to associate with each module M a path of length zero, called the trivial, or the stationary path at M.

From Subsection II.4.2, we can already derive existence results for paths of irreducible morphisms.

Corollary II.4.8. *Let M, N be indecomposable modules and $f: M \longrightarrow N$ a nonzero radical morphism.*

(a) *If $f \in \mathrm{rad}_A^n(M, N) \setminus \mathrm{rad}_A^{n+1}(M, N)$ for some $n \geq 1$, then there exists a path of irreducible morphisms $M \rightsquigarrow N$ of length n.*
(b) *If $\mathrm{rad}_A^\infty(M, N) = 0$, then there exists a path of irreducible morphisms $M \rightsquigarrow N$.*

Proof.

(a) This follows from Proposition II.4.4(a).
(b) This follows from Corollary II.4.6.

□

Suppose now $\mathrm{rad}_A^\infty(M,N) \neq 0$ with M, N indecomposable. We prove that, in this case, there exist paths of irreducible morphisms of arbitrary length, starting with M or ending with N.

Proposition II.4.9. *Let M, N be indecomposable modules such that $\mathrm{rad}_A^\infty(M,N) \neq 0$. Then, for every $i \geq 0$, there exist:*

(a) *a path of irreducible morphisms*

$$M = M_0 \xrightarrow{f_1} M_1 \longrightarrow \dots \longrightarrow M_{t-1} \xrightarrow{f_i} M_i$$

and a morphism $g_i \in \mathrm{rad}_A^\infty(M_i, N)$ such that $g_i f_i \dots f_1 \neq 0$, and

(b) *a path of irreducible morphisms*

$$N_i \xrightarrow{g_i} N_{i-1} \longrightarrow \dots \longrightarrow N_1 \xrightarrow{g_1} N_0 = N$$

and a morphism $f_i \in \mathrm{rad}_A^\infty(M, N_i)$ such that $g_1 \dots g_i f_i \neq 0$.

Proof. We only prove (a), because (b) follows by duality.

Let $h = \begin{pmatrix} h_1 \\ \vdots \\ h_t \end{pmatrix} : M \longrightarrow \oplus_{j=1}^t E_j$ be left minimal almost split with the E_j indecomposable. We know that, for each j, there exists a least m_j such that $\mathrm{rad}_A^{m_j}(E_j, N) = \mathrm{rad}_A^\infty(E_j, N)$, see Lemma II.4.1. Let $m = \max\{m_j : 1 \leq j \leq t\}$. Then, $\mathrm{rad}_A^m(E_j, N) = \mathrm{rad}_A^\infty(E_j, N)$ for all j.
Now, let $f \in \mathrm{rad}^\infty(M,N)$ be a nonzero morphism. Because the infinite radical is the intersection of all powers of the radical, we have, in particular, $f \in \mathrm{rad}_A^{m+1}(M,N)$. Then, we can write $f = \sum_{i=1}^{s} g_i f_i$ with the $g_i \in \mathrm{rad}_A^m(X_i, N)$, $f_i \in \mathrm{rad}_A(M, X_i)$ and the X_i indecomposable. We can assume $g_i f_i \neq 0$ for all i.
Because $f_1 \in \mathrm{rad}_A(M, X_1)$ is not an isomorphism, it factors through h, that is, there exists $l = (l_1 \cdots l_t) \colon \oplus_{j=1}^t E_j \longrightarrow X_1$ such that $f_1 = lh = \sum_{j=1}^t l_j h_j$. Because $g_1 f_1 \neq 0$, there exists j such that $g_1 l_j h_j \neq 0$. Now we have $g_1 l_j \in \mathrm{rad}_A^m(E_j, N) = \mathrm{rad}_A^\infty(E_j, N)$ and h_j is irreducible. Repeating the same procedure with $g_1 l_j$, the result follows from an easy induction. □

In Example II.4.3 above, we constructed precisely a path and a morphism as in part (a) of the proposition.

Exercises for Section II.4

Exercise II.4.1. Let L, M be indecomposable modules and $f: L \longrightarrow M$ a morphism. Assume that f is neither a monomorphism nor an epimorphism. Prove that $f \in \operatorname{rad}^2(L, M)$.

Exercise II.4.2. Let M, N be indecomposable A-modules and $f: M \longrightarrow N$ a nonzero radical morphism. Prove that

(a) If $\operatorname{rad}_A^m(-, N) = 0$ for some $m \geq 0$, then f is a sum of compositions of irreducible morphisms between indecomposable modules.
(b) If $\operatorname{rad}_A^m(M, -) = 0$ for some $m \geq 0$, then f is a sum of compositions of irreducible morphisms between indecomposable modules.

Conclude that, if A is such that $\operatorname{rad}_A^m = 0$ for some $m \geq 0$, then every nonzero radical morphism is a sum of compositions of irreducible morphisms between indecomposable modules.

Exercise II.4.3. Let A be a finite dimensional algebra and M, N modules. Prove that the standard duality $\mathrm{D} = \operatorname{Hom}_k(-, \mathbf{k})$ induces isomorphisms:

(a) $\operatorname{rad}_A^m(M, N) \cong \operatorname{rad}_{A^{op}}^m(\mathrm{D}N, \mathrm{D}M)$, for each $m \geq 1$.
(b) $\operatorname{rad}_A^\infty(M, N) \cong \operatorname{rad}_{A^{op}}^\infty(\mathrm{D}N, \mathrm{D}M)$.

Exercise II.4.4. A functor $F: \operatorname{mod} A \longrightarrow \operatorname{mod} \mathbf{k}$ is called **support-finite** if there are only finitely many isoclasses of indecomposable A-modules M such that $F(M) \neq 0$. Let M be an A-module. Prove that:

(a) $\operatorname{Hom}_A(-, M)$ is support-finite if and only if there exists $m \geq 0$ such that $\operatorname{rad}_A^m(-, M) = 0$.
(b) $\operatorname{Hom}_A(M, -)$ is support-finite if and only if there exists $m \geq 0$ such that $\operatorname{rad}_A^m(M, -) = 0$.

Exercise II.4.5. Let $M = M_0 \xrightarrow{f_1} M_1 \longrightarrow \cdots \xrightarrow{f_t} M_t = N$ be a path in $\operatorname{mod} A$, and assume that there exists i such that $f_{i+1} \in \operatorname{rad}_A^\infty(M_i, M_{i+1})$ and $0 \leq i < t$. Prove that, for every pair of integers $s, t \geq 0$, there exist paths of irreducible morphisms $M_i = M'_0 \longrightarrow M'_1 \longrightarrow \ldots M'_s$ and $N'_t \longrightarrow \cdots \longrightarrow N'_1 \longrightarrow N'_0 = M_{i+1}$, along with a path $M'_s = X_0 \longrightarrow X_1 \longrightarrow \ldots \longrightarrow X_p = N'_t$.

Exercise II.4.6. Prove the following weaker version of Proposition II.4.9. Let M, N be indecomposable A-modules such that $\operatorname{Hom}_A(M, N) \neq 0$ and assume that there exists no path of irreducible morphisms from M to N of length less than t. Then there exist:

(a) A path of irreducible morphisms

$$M = M_0 \xrightarrow{f_1} M_1 \xrightarrow{f_2} \cdots \xrightarrow{f_{t-1}} M_{t-1} \xrightarrow{f_t} M_t$$

and a morphism $g: M_t \longrightarrow N$ such that $g f_t \cdots f_1 \neq 0$.

(b) A path of irreducible morphisms

$$N_t \xrightarrow{g_t} N_{t-1} \xrightarrow{g_{t-1}} \cdots \xrightarrow{g_2} N_1 \xrightarrow{g_1} N_0 = N$$

and a morphism $f\colon M \longrightarrow N_t$ such that $g_1 \cdots g_t f \neq 0$.

Exercise II.4.7. Let A be given by the quiver

$$1 \longleftarrow 2 \longleftarrow 3$$

Prove that every indecomposable A-module lies on a path in ind A from S_1 to S_3.

Exercise II.4.8. Let A be given by the quiver

$$1 \underset{\beta}{\overset{\alpha}{\rightleftarrows}} 2$$

bound by $\alpha\beta = 0$. Prove that every indecomposable A-module lies on a cycle in ind A from S_1 to itself.

Exercise II.4.9.

(a) Let Q be an acyclic quiver having $t \geq 2$ arrows $\alpha_1, \ldots, \alpha_t$ from a point a to a point b. Prove that the path algebra $\mathbf{k}Q$ is representation-infinite.
(b) Let $A = \mathbf{k}Q/I$ be a bound quiver algebra, with Q acyclic. Prove that, if Q has multiple arrows (as in (a)), then A is representation-infinite.

Chapter III
Constructing almost split sequences

The previous chapter was mainly of a theoretical nature: we defined irreducible morphisms and almost split sequences and started to explore their use for the understanding of the radical of a module category. However, we did not say much about the explicit construction of almost split sequences, even though we pointed out that the proof of Theorem II.3.10 suggests the idea of a construction. Carrying out this construction in practice is quite difficult, and our objective in the present chapter is to explain how it can be done, at least in the easiest cases. In the first section, we prove that the indecomposable end terms of an almost split sequence are related by functors, which are called the Auslander–Reiten translations. In Section III.2, we derive the so-called Auslander–Reiten formulae, which lead us to a second existence proof for almost split sequences. Next, in Section III.3, we show how to apply these results to construct examples of almost split sequences. In the final Section III.4, we relate the Auslander–Reiten translates of a given module over an algebra to that over a quotient algebra.

III.1 The Auslander–Reiten translations

III.1.1 The stable categories

As seen in Corollary II.2.32, an almost split sequence

$$0 \longrightarrow L \longrightarrow M \longrightarrow N \longrightarrow 0$$

is uniquely determined by the indecomposable nonprojective module N, or by the indecomposable noninjective module L. This uniqueness suggests that the relation between N and L might conceal some functoriality.

I. Assem, F. U. Coelho, *Basic Representation Theory of Algebras*, Graduate Texts in Mathematics 283, https://doi.org/10.1007/978-3-030-35118-2_3

Now, if the correspondence between N and L extends to functors, then these functors cannot be defined on the whole module category mod A because an almost split sequence is not split; therefore, projective modules are excluded for the last term N, and injective modules are excluded for the first term L. We thus need functors defined on the quotients of mod A obtained by annihilating the projectives or the injectives respectively.

Given modules M, N, we denote by $\mathscr{P}(M, N)$ the set of all morphisms from M to N in mod A that factor through a projective A-module. Dually, we denote by $\mathscr{I}(M, N)$ the set of all morphisms from M to N that factor through an injective A-module. We show that these data define ideals in mod A.

Lemma III.1.1.

(a) *The assignment* $(M, N) \mapsto \mathscr{P}(M, N)$ *defines an ideal* $\mathscr{P}$ *in* mod A.
(b) *The assignment* $(M, N) \mapsto \mathscr{I}(M, N)$ *defines an ideal* $\mathscr{I}$ *in* mod A.

Proof. We only prove (a), because the proof of (b) is dual.

We first show that, for every M, N in mod A, the set $\mathscr{P}(M, N)$ is a subspace of $\mathrm{Hom}_A(M, N)$. Let $f_1, f_2 \in \mathscr{P}(M, N)$. There exist projective modules P_1, P_2 and morphisms $h_1: M \longrightarrow P_1, h_2: M \longrightarrow P_2, g_1: P_1 \longrightarrow N$ and $g_2: P_2 \longrightarrow N$ such that $f_1 = g_1 h_1$ and $f_2 = g_2 h_2$. But then

$$f_1 + f_2 = g_1 h_1 + g_2 h_2 = (g_1 \; g_2) \begin{pmatrix} h_1 \\ h_2 \end{pmatrix},$$

that is, $f_1 + f_2$ factors through $P_1 \oplus P_2$. Clearly, if $\lambda \in \mathbf{k}$ and $f \in \mathscr{P}(M, N)$, then $\lambda f \in \mathscr{P}(M, N)$.

Let $f \in \mathscr{P}(M, N)$ and $u: L \longrightarrow M$ be any morphism. There exist a projective module P and morphisms $h: M \longrightarrow P, g: P \longrightarrow N$ such that $f = gh$. But then $fu = g(hu)$ also factors through P, and so lies in $\mathscr{P}(L, N)$. Thus, $\mathscr{P}$ is right stable under composition by arbitrary morphisms. Similarly, it is left stable. □

In fact, the construction of the ideals $\mathscr{P}$ and $\mathscr{I}$ are particular cases of a more general construction. Given an A-module X, let add X denote the $\mathbf{k}$-linear full subcategory of mod A consisting of all direct sums of indecomposable summands of X, see Example II.1.3. Then, add X generates an ideal $\mathfrak{X} = \langle \mathrm{add}\, X \rangle$ as follows. For modules M, N, let $\mathfrak{X}(M, N)$ be the set of all morphisms from M to N that factor through an object of add X. It is easy to prove, as above, that this defines an ideal in mod A, see Exercise II.1.10. In this notation, a module is projective, or injective, if and only if it belongs to add A_A, or to add$(DA)_A$ respectively, so that $\mathscr{P} = \langle \mathrm{add}\, A \rangle$ whereas $\mathscr{I} = \langle \mathrm{add}\, DA \rangle$.

This brings us to the following definition.

Definition III.1.2.

(a) The **projectively stable category** of mod A is the quotient category $\underline{\text{mod}}\,A = (\text{mod}\,A)/\mathscr{P}$;
(b) The **injectively stable category** of mod A is the quotient category $\overline{\text{mod}}\,A = (\text{mod}\,A)/\mathscr{I}$.

As seen in Subsection II.1.1, the objects of $\underline{\text{mod}}\,A$ coincide with those of mod A, and the space of morphisms from M to N is the quotient space

$$\underline{\text{Hom}}_A(M, N) = \frac{\text{Hom}_A(M, N)}{\mathscr{P}(M, N)}.$$

There is a projection functor mod $A \longrightarrow \underline{\text{mod}}\,A$ mapping each module to itself and each morphism $f: M \longrightarrow N$ to its residual class $\underline{f} = f + \mathscr{P}(M, N)$.

Similarly, the objects of $\overline{\text{mod}}\,A$ coincide with those of mod A, and the space of morphisms from M to N is the quotient space

$$\overline{\text{Hom}}_A(M, N) = \frac{\text{Hom}_A(M, N)}{\mathscr{I}(M, N)}.$$

There is a projection functor mod $A \longrightarrow \overline{\text{mod}}\,A$ mapping each module to itself and each morphism $f: M \longrightarrow N$ to its residual class $\overline{f} = f + \mathscr{I}(M, N)$.

We now prove that the projective modules are (the only modules) isomorphic to the zero object in $\underline{\text{mod}}\,A$. Dually, the injective modules are (the only modules) isomorphic to the zero object in $\overline{\text{mod}}\,A$. We recall here that an object X in a **k**-linear category $\mathscr{C}$ is isomorphic to the zero object if and only if the identity 1_X on X is equal to the zero morphism from X to itself, and this is the case if and only if $\text{End}_{\mathscr{C}}\, X = 0$, see Exercise III.1.1.

Lemma III.1.3. *Let M be an A-module.*

(a) *M is projective if and only if* $\underline{\text{End}}\, M = 0$;
(b) *M is injective if and only if* $\overline{\text{End}} M = 0$.

Proof. We only prove (a), because the proof of (b) is dual.

Certainly, if M is projective, then $\text{End}\, M = \mathscr{P}(M, M)$ and so $\underline{\text{End}}\, M = 0$. If M is not projective, then we claim that the identity 1_M does not belong to $\mathscr{P}(M, M)$. Indeed, if P is projective and $g: P \longrightarrow M$, $h: M \longrightarrow P$ are such that $1_M = gh$, then h would be a section and M would be projective, a contradiction. This establishes our claim and thus $\underline{\text{End}}\, M \neq 0$. □

III.1.2 Morphisms between projectives and injectives

The main result of this subsection is an alternative description of the stable categories as quotients of the categories of morphisms between projectives or injectives respectively.

Let $\operatorname{mp} A$ be the category whose objects are the morphisms $P_1 \xrightarrow{p} P_0$, where P_0, P_1 are projective A-modules (the letters *mp* stand for *morphism* between *projectives*). A morphism in $\operatorname{mp} A$ from $p\colon P_1 \longrightarrow P_0$ to $p'\colon P_1' \longrightarrow P_0'$ is a pair (u_1, u_0) of morphisms in $\operatorname{mod} A$ such that $u_1\colon P_1 \longrightarrow P_1'$ and $u_0\colon P_0 \longrightarrow P_0'$ satisfy $p'u_1 = u_0 p$, that is, we have a commutative square

$$\begin{array}{ccc} P_1 & \xrightarrow{p} & P_0 \\ \downarrow{\scriptstyle u_1} & & \downarrow{\scriptstyle u_0} \\ P_1' & \xrightarrow{p'} & P_0' \end{array}$$

The composition in $\operatorname{mp} A$ is induced from that in $\operatorname{mod} A$: if $(u_1, u_0)\colon p \longrightarrow p'$ and $(v_1, v_0)\colon p' \longrightarrow p''$, then their composition is

$$(v_1, v_0)(u_1, u_0) = (v_1 u_1, v_0 u_0).$$

A morphism (u_1, u_0) as above is called **negligible** if there exists a morphism $s\colon P_0 \longrightarrow P_1'$ such that

$$p'sp = u_0 p = p'u_1.$$

We denote by $\mathscr{N}_p(p, p')$ the set of negligible morphisms from p to p'. As we shall see, the assignment $(p, p') \mapsto \mathscr{N}_p(p, p')$ defines an ideal $\mathscr{N}_p$ in $\operatorname{mp} A$.

There is an obvious functor $C\colon \operatorname{mp} A \longrightarrow \operatorname{mod} A$ defined by taking cokernels. Namely, if $p\colon P_1 \longrightarrow P_0$ is an object in $\operatorname{mp} A$, then we define $C(p) = \operatorname{Coker} p$ and if (u_1, u_0) is a morphism from $p\colon P_1 \longrightarrow P_0$ to $p'\colon P_1' \longrightarrow P_0'$, then we let $C(u_1, u_0)$ be the unique morphism $u\colon \operatorname{Coker} p \longrightarrow \operatorname{Coker} p'$ obtained by passing to cokernels, that is, such that the following diagram with exact rows is commutative

$$\begin{array}{ccccccc} P_1 & \xrightarrow{p} & P_0 & \longrightarrow & \operatorname{Coker} p & \longrightarrow & 0 \\ \downarrow{\scriptstyle u_1} & & \downarrow{\scriptstyle u_0} & & \vdots{\scriptstyle u} & & \\ P_1' & \xrightarrow{p'} & P_0' & \longrightarrow & \operatorname{Coker} p' & \longrightarrow & 0 \end{array}$$

The composition of C with the projection functor $\operatorname{mod} A \longrightarrow \underline{\operatorname{mod}}\, A$ is denoted by $\underline{C}\colon \operatorname{mp} A \longrightarrow \underline{\operatorname{mod}}\, A$.

Theorem III.1.4. *The functor* $\underline{C}\colon \operatorname{mp} A \longrightarrow \underline{\operatorname{mod}}\, A$ *is full, dense and admits as kernel the set* $\mathscr{N}_p$ *of negligible morphisms in* $\operatorname{mp} A$*. In particular,* $\underline{C}$ *induces an equivalence* $\operatorname{mp} A/\mathscr{N}_p \cong \underline{\operatorname{mod}}\, A$.

Proof. First, $\underline{C}$ is full and dense: every module is the cokernel of a projective presentation and every morphism between modules lifts to a morphism between their projective presentations.

We now claim that, if (u_1, u_0) is a morphism in mp$A$ from $p: P_1 \longrightarrow P_0$ to $p': P_1' \longrightarrow P_0'$, then $\underline{C}(u_1, u_0) = 0$ if and only if (u_1, u_0) is negligible.

Assume first that $\underline{C}(u_1, u_0) = 0$. Then the morphism obtained by passing to the cokernels M, M' of p, p' respectively factors through a projective module P. That is, there exist morphisms $u': P \longrightarrow M'$, $u'': M \longrightarrow P$ such that $u = u'u''$. Thus, we have a commutative diagram with exact rows:

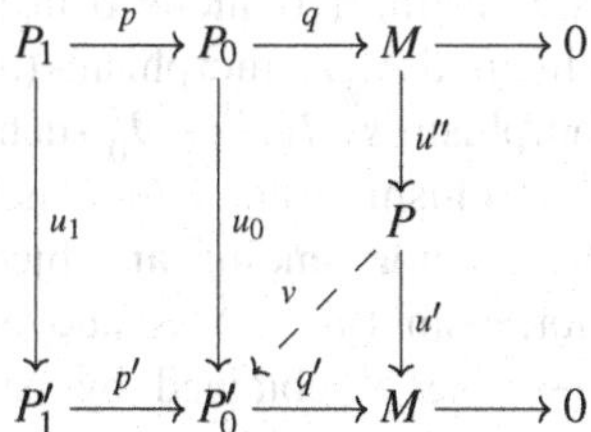

Because P is projective, there exists $v: P \longrightarrow P_0'$ such that $q'v = u'$. But then we have:

$$q'(u_0 - vu''q) = q'u_0 - u'u''q = 0.$$

Therefore, $u_0 - vu''q$ factors through p', which is the kernel of q', that is, there exists $s: P_0 \longrightarrow P_1'$ such that $u_0 - vu''q = p's$. We get $u_0 = vu''q + p's$; hence,

$$u_0 p = (vu''q + p's)p = p'sp.$$

Thus, (u_1, u_0) is negligible.

Conversely, assume that (u_1, u_0) is negligible. There exists $s: P_0 \longrightarrow P_1'$ such that $u_0 p = p'sp$. Consider the commutative diagram with exact rows

$$\begin{array}{ccccccc}
P_1 & \xrightarrow{p} & P_0 & \xrightarrow{q} & M & \longrightarrow & 0 \\
{\scriptstyle u_1}\downarrow & \swarrow{\scriptstyle s} & \downarrow{\scriptstyle u_0} & & \downarrow{\scriptstyle u} & & \\
P_1' & \xrightarrow{p'} & P_0' & \xrightarrow{q'} & M' & \longrightarrow & 0.
\end{array}$$

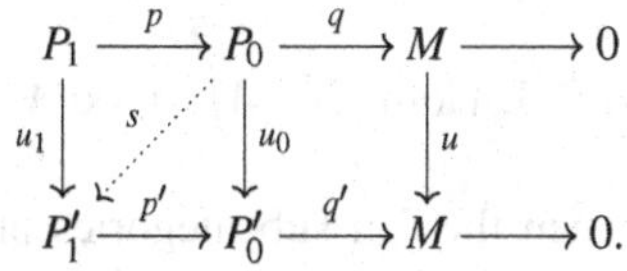

Because $(u_0 - p's)p = 0$, there exists $w: M \longrightarrow P_0'$ such that $u_0 - p's = wq$. But then $uq = q'u_0 = q'(p's + wq) = q'wq$. Now, q is an epimorphism; therefore, $u = q'w$ factors through the projective module P_0'. Hence, $\underline{C}(u_1, u_0) = 0$.

This establishes our claim. The statement of the theorem follows immediately. □

In particular, because $\mathcal{N}_p$ is the kernel of the functor $\underline{C}$, it is an ideal of mp A.

As usual, dual considerations lead to dual results. Let mi A be the category whose objects are the morphisms $j: I_0 \longrightarrow I_1$ between injective A-modules. A morphism in mi A from $j\colon I_0 \longrightarrow I_1$ to $j'\colon I_0' \longrightarrow I_1'$ is a pair of morphisms (u_0, u_1) such that $u_0\colon I_0 \longrightarrow I_0'$ and $u_1\colon I_1 \longrightarrow I_1'$ satisfy $j'u_0 = u_1 j$, that is, the following square is commutative.

$$\begin{array}{ccc} I_0 & \xrightarrow{j} & I_1 \\ \big\downarrow{\scriptstyle u_0} & & \big\downarrow{\scriptstyle u_1} \\ I_0' & \xrightarrow{j'} & I_1' \end{array}$$

The composition of morphisms in mi A is induced in the obvious way from the composition of morphisms in mod A. A morphism (u_0, u_1) as above is called **negligible** if there exists a morphism $s\colon I_1 \longrightarrow I_0'$ such that $j'sj = j'u_0 = u_1 j$, and the set of all negligible morphisms from j to j' is denoted by $\mathcal{N}_i(j, j')$. Let $K\colon \text{mi}\,A \longrightarrow \text{mod}\,A$ be the functor sending an object $j\colon I_0 \longrightarrow I_1$ of mi A to $K(j) = \text{Ker}\, j$ and a morphism (u_0, u_1) as above to the unique morphism $K(u_0, u_1) = u\colon \text{Ker}\, j \longrightarrow \text{Ker}\, j'$ obtained by passing to kernels. Finally, let $\overline{K}\colon \text{mi}\,A \longrightarrow \overline{\text{mod}}\,A$ be the composition of K with the projection functor $\text{mod}\,A \longrightarrow \overline{\text{mod}}\,A$.

Theorem III.1.5. *The functor* $\overline{K}\colon \text{mi}\,A \longrightarrow \overline{\text{mod}}\,A$ *is full, dense and admits as kernel the class* $\mathcal{N}_i$ *of negligible morphisms in* mi A. *In particular,* $\overline{K}$ *induces an equivalence* $\text{mi}\,A/\mathcal{N}_i \cong \overline{\text{mod}}\,A$. □

In particular, $\mathcal{N}_i$ is an ideal in mi A.

III.1.3 The Auslander–Reiten translations

A remarkable consequence of Theorems III.1.4 and III.1.5 is that the categories $\underline{\text{mod}}\,A$ and $\overline{\text{mod}}\,A$ are equivalent. Indeed, recall from Chapter I that the Nakayama functor ν_A, which we denote here for brevity by

$$\nu = - \otimes_A \mathrm{D}A \cong \mathrm{D}\,\mathrm{Hom}_A(-, A)\colon \text{mod}\,A \longrightarrow \text{mod}\,A$$

induces an equivalence between the full subcategories proj A and inj A of mod A consisting respectively of the projective and the injective A-modules, with quasi-inverse given by the functor $\nu^{-1} = \mathrm{Hom}_A(\mathrm{D}A, -)$. In addition, if e is a primitive idempotent in A, then ν maps the indecomposable projective module eA to the indecomposable injective module $\mathrm{D}(Ae)$, corresponding to the same idempotent, see Lemma I.1.18. This leads to the following corollary.

Corollary III.1.6. *There exist equivalences* $\tau : \underline{\text{mod}}\,A \longrightarrow \overline{\text{mod}}\,A$ *and* $\tau^{-1} : \overline{\text{mod}}\,A \longrightarrow \underline{\text{mod}}\,A$.

Proof. Clearly, the Nakayama functor $\nu\colon \operatorname{proj} A \longrightarrow \operatorname{inj} A$ and its quasi-inverse $\nu^{-1}\colon \operatorname{inj} A \longrightarrow \operatorname{proj} A$ induce functors $\operatorname{mp} A \longrightarrow \operatorname{mi} A$ and $\operatorname{mi} A \longrightarrow \operatorname{mp} A$, which are also quasi-inverse. In addition, the image under one of these functors of a negligible morphism in either category is clearly negligible in the other. This shows that ν and ν^{-1} induce an equivalence $\operatorname{mp} A/\mathcal{N}_p \cong \operatorname{mi} A/\mathcal{N}_i$. We take τ as the composition of the equivalences $\underline{\operatorname{mod}}\, A \cong \operatorname{mp} A/\mathcal{N}_p \cong \operatorname{mi} A/\mathcal{N}_i \cong \overline{\operatorname{mod}}\, A$. The functor τ^{-1} is constructed in the same way. □

Definition III.1.7. The equivalences $\tau\colon \underline{\operatorname{mod}}\, A \longrightarrow \overline{\operatorname{mod}}\, A$ and $\tau^{-1}\colon \overline{\operatorname{mod}}\, A \longrightarrow \underline{\operatorname{mod}}\, A$ are called the **Auslander–Reiten translations**. For a module M, the modules τM and $\tau^{-1}M$ are called its **translates**.

It is useful to present in detail the construction of τ and τ^{-1}. Let M be an A-module, considered as an object in $\underline{\operatorname{mod}}\, A$. To view M as an object in $\operatorname{mp} A$, we must find a morphism between projectives of which M is the cokernel, that is, a projective presentation of M. There exist several such presentations, but we always assume implicitly that we are dealing with a minimal projective presentation

$$P_1 \xrightarrow{p} P_0 \longrightarrow M \longrightarrow 0.$$

Following the recipe above, the (right exact) Nakayama functor $\nu = - \otimes_A DA$ yields an exact sequence

$$\nu P_1 \xrightarrow{\nu p} \nu P_0 \longrightarrow \nu M \longrightarrow 0$$

and here νp is an object in $\operatorname{mi} A$ (or, as well, in $\operatorname{mi} A/\mathcal{N}_i$). To pass to $\overline{\operatorname{mod}}\, A$, it suffices to apply the kernel functor, thus obtaining τM. This is summarised in the following lemma.

Lemma III.1.8.

(a) *Let $P_1 \xrightarrow{p} P_0 \longrightarrow M \longrightarrow 0$ be a minimal projective presentation. Then there exists an exact sequence*

$$0 \longrightarrow \tau M \longrightarrow \nu P_1 \xrightarrow{\nu p} \nu P_0 \longrightarrow \nu M \longrightarrow 0.$$

(b) *Let $0 \longrightarrow M \longrightarrow I_0 \xrightarrow{j} I_1$ be a minimal injective copresentation. Then there exists an exact sequence*

$$0 \longrightarrow \nu^{-1}M \longrightarrow \nu^{-1}I_0 \xrightarrow{\nu^{-1}j} \nu^{-1}I_1 \longrightarrow \tau^{-1}M \longrightarrow 0.$$

Proof. We have already proved (a), and the proof of (b) is dual. □

The original approach of Auslander and Reiten is slightly different: it passes through $\operatorname{mod} A^{op}$ and presents each of τ and τ^{-1} as the composition of two dualities. This is natural if one recalls that the Nakayama functor $\nu = D\operatorname{Hom}_A(-, A)$ itself is the composition of two dualities. We outline this approach below.

Let M be an A-module, and consider a minimal projective presentation

$$P_1 \xrightarrow{p} P_0 \longrightarrow M \longrightarrow 0.$$

We apply the left exact functor $(\,-\,)^t = \mathrm{Hom}_A(-, A)\colon\ \mathrm{mod}\,A \longrightarrow \mathrm{mod}\,A^{op}$, thus obtaining an exact sequence of left A-modules:

$$0 \longrightarrow M^t \longrightarrow P_0^t \xrightarrow{p^t} P_1^t \longrightarrow \mathrm{Coker}\,p^t \longrightarrow 0.$$

We denote $\mathrm{Coker}\,p^t$ by $\mathrm{Tr}\,M$ and call it the **transpose** of M. Let $f\colon M \longrightarrow M'$ be a morphism of A-modules, and consider minimal projective presentations $P_1 \xrightarrow{p} P_0 \longrightarrow M \longrightarrow 0$ and $P_1' \xrightarrow{p'} P_0' \longrightarrow M' \longrightarrow 0$ of M and M', respectively. Then, f lifts to morphisms $f_0\colon P_0 \longrightarrow P_0'$ and $f_1\colon P_1 \longrightarrow P_1'$ such that the following diagram with exact rows is commutative

$$\begin{array}{ccccccc}
P_1 & \xrightarrow{p} & P_0 & \longrightarrow & M & \longrightarrow & 0 \\
{\scriptstyle f_1}\downarrow & & {\scriptstyle f_0}\downarrow & & \downarrow{\scriptstyle f} & & \\
P_1' & \xrightarrow{p'} & P_0' & \longrightarrow & M & \longrightarrow & 0
\end{array}$$

Applying the left exact contravariant functor $(\,-\,)^t$, we deduce by passing to cokernels a unique morphism $\mathrm{Tr}\,f\colon\ \mathrm{Tr}\,M' \longrightarrow \mathrm{Tr}\,M$ such that the following diagram with exact rows is commutative

$$\begin{array}{ccccccccc}
0 & \longrightarrow & M^t & \longrightarrow & P_0^t & \longrightarrow & P_1^t & \longrightarrow & \mathrm{Tr}M & \longrightarrow & 0 \\
 & & {\scriptstyle f^t}\uparrow & & {\scriptstyle f_0^t}\uparrow & & {\scriptstyle f_1^t}\uparrow & & \vdots{\scriptstyle \mathrm{Tr}\,f} & & \\
0 & \longrightarrow & M'^t & \longrightarrow & P_0'^t & \longrightarrow & P_1'^t & \longrightarrow & \mathrm{Tr}M' & \longrightarrow & 0.
\end{array}$$

We now prove that Tr defines a functor and actually a duality, called the **transposition**.

Corollary III.1.9. *The above procedure induces dualities* $\underline{\mathrm{mod}}\,A \longrightarrow \underline{\mathrm{mod}}\,A^{op}$ *and* $\overline{\mathrm{mod}}\,A \longrightarrow \overline{\mathrm{mod}}\,A^{op}$. *Actually, we have* $\tau = \mathrm{D\,Tr}$, *and* $\tau^{-1} = \mathrm{Tr\,D}$.

Proof. Let M be an A-module. A minimal projective presentation $P_1 \xrightarrow{p} P_0 \longrightarrow M \longrightarrow 0$ gives an exact sequence of left A-modules

$$0 \longrightarrow M^t \longrightarrow P_0^t \xrightarrow{p^t} P_1^t \longrightarrow \mathrm{Tr}M \longrightarrow 0.$$

Applying the standard duality $\mathrm{D} = \mathrm{Hom}_{\mathbf{k}}(-, \mathbf{k})$, we get an exact sequence of right A-modules.

$$0 \longrightarrow \mathrm{D\,Tr}\,M \longrightarrow \mathrm{D}P_1^t \xrightarrow{\mathrm{D}p^t} \mathrm{D}P_0^t \longrightarrow \mathrm{D}M^t \longrightarrow 0.$$

Now, $D(-)^t = D\operatorname{Hom}_A(-, A) \cong \nu$. Thus, because of Lemma III.1.8, we have $D\operatorname{Tr} M = \tau M$. This shows that $D\operatorname{Tr}$ and τ coincide on objects. Similarly, they coincide on morphisms. That is, Tr and $D\tau$ coincide on objects and on morphisms. Consequently, Tr extends to the functor $D\tau\colon \underline{\operatorname{mod}} A \longrightarrow \overline{\operatorname{mod}} A \longrightarrow \underline{\operatorname{mod}} A^{op}$, which is a duality, because τ is an equivalence and D a duality. This proves the first part.

To prove the second part, we apply successively D and $(-)^t$ to a minimal injective copresentation $0 \longrightarrow M \longrightarrow I_0 \xrightarrow{j} I_1$, getting the exact sequence

$$0 \longrightarrow (DM)^t \longrightarrow (DI_0)^t \xrightarrow{(Dj)^t} (DI_1)^t \longrightarrow \operatorname{Tr} DM \longrightarrow 0$$

where we used the fact that D applied to a minimal injective copresentation yields a minimal projective presentation. Now we have functorial isomorphisms

$$(DX)^t = \operatorname{Hom}_{A^{op}}(DX, A) \cong \operatorname{Hom}_A(DA, X) \cong \nu^{-1}X$$

for every A-module X. Thus, the previous exact sequence is isomorphic to the following

$$0 \longrightarrow \nu^{-1}M \longrightarrow \nu^{-1}I_0 \xrightarrow{\nu^{-1}j} \nu^{-1}I_1 \longrightarrow \tau^{-1}M \longrightarrow 0.$$

Applying Lemma III.1.8 again yields $\operatorname{Tr} DM \cong \tau^{-1}M$. Thus, Tr D and τ^{-1} coincide on objects and similarly they coincide on morphisms. Therefore, Tr extends to a duality $\tau^{-1}D\colon \overline{\operatorname{mod}} A \longrightarrow \underline{\operatorname{mod}} A^{op}$. □

III.1.4 Properties of the Auslander–Reiten translations

The following proposition records the most immediate properties of the translations.

Proposition III.1.10. *Let M be an indecomposable A-module.*

(a) *If M is projective, then $\tau M = 0$. If M is not projective, then τM is indecomposable noninjective and $\tau^{-1}\tau M \cong M$.*
(b) *If M is injective, then $\tau^{-1}M = 0$. If M is not injective, then $\tau^{-1}M$ is indecomposable nonprojective and $\tau\tau^{-1}M \cong M$.*

Proof. We only prove (a), because the proof of (b) is dual.

It follows from the definition that, if M is projective, then $\operatorname{Tr} M = 0$ and so $\tau M = D\operatorname{Tr} M = 0$.

Assume, thus, that M is indecomposable nonprojective. A minimal projective presentation $P_1 \xrightarrow{p} P_0 \longrightarrow M \longrightarrow 0$ yields an exact sequence

$$0 \longrightarrow \tau M \longrightarrow \nu P_1 \xrightarrow{\nu p} \nu P_0,$$

which is actually a minimal injective copresentation. Note that τM is not injective, because otherwise $\nu p = 0$ and so $p = 0$, a contradiction to the hypothesis that M is nonprojective. Applying ν^{-1} yields a commutative diagram with exact rows

$$\begin{array}{ccccccccc} 0 \longrightarrow \nu^{-1}\tau M & \longrightarrow & \nu^{-1}\nu P_1 & \xrightarrow{\nu^{-1}\nu p} & \nu^{-1}\nu P_0 & \longrightarrow & \tau^{-1}\tau M & \longrightarrow & 0 \\ & & \downarrow \cong & & \downarrow \cong & & & & \\ & & P_1 & \xrightarrow{p} & P_0 & \longrightarrow & M & \longrightarrow & 0 \end{array}$$

from which we deduce $\tau^{-1}\tau M \cong M$. Finally, this relation together with the indecomposability of M yield the indecomposability of τM. □

As an unexpected, but useful, dividend, we get a characterisation of modules of projective or injective dimension at most one.

Proposition III.1.11. *Let M be an indecomposable A-module. Then,*

(a) pd $M \leq 1$ *if and only if* $\mathrm{Hom}_A(DA, \tau M) = 0$.
(b) id $M \leq 1$ *if and only if* $\mathrm{Hom}_A(\tau^{-1}M, A) = 0$.

Proof. We only prove (a), because the proof of (b) is dual.

Let $P_1 \xrightarrow{p} P_0 \longrightarrow M \longrightarrow 0$ be a minimal projective presentation. Then pd $M \leq 1$ if only if $\mathrm{Ker}\, p = 0$. Now, we have an exact sequence

$$0 \longrightarrow \tau M \longrightarrow \nu P_1 \xrightarrow{\nu p} \nu P_0.$$

Applying $\nu^{-1} = \mathrm{Hom}_A(DA, -)$ yields a commutative diagram with exact rows

$$\begin{array}{ccccccc} 0 & \longrightarrow & \mathrm{Hom}_A(DA, \tau M) & \longrightarrow & \nu^{-1}\nu P_1 & \xrightarrow{\nu^{-1}\nu p} & \nu^{-1}\nu P_0 \\ & & & & \downarrow \cong & & \downarrow \cong \\ 0 & \longrightarrow & \mathrm{Ker}\, p & \longrightarrow & P_1 & \longrightarrow & P_0 \end{array}$$

Thus, $\mathrm{Ker}\, p \cong \mathrm{Hom}_A(DA, \tau M)$. The statement follows at once. □

The previous result is sometimes stated in the following equivalent form: let M be an indecomposable A-module, then:

(a) pd $M \leq 1$ if and only if for every indecomposable injective A-module I, we have $\mathrm{Hom}_A(I, \tau M) = 0$.
(b) id $M \leq 1$ if and only if for every indecomposable projective A-module P, we have $\mathrm{Hom}_A(\tau^{-1}M, P) = 0$.

Example III.1.12. Let A be given by the quiver

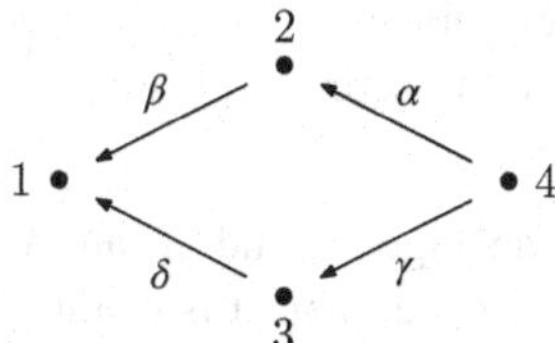

bound by $\alpha\beta = \gamma\delta$. We wish to compute the Auslander–Reiten translates of the simple nonprojective module S_2. Clearly, $S_2 \cong P_2/S_1$, so that we have a minimal projective presentation

$$0 \longrightarrow P_1 \xrightarrow{\ j\ } P_2 \xrightarrow{\ p\ } S_2 \longrightarrow 0$$

where j is the inclusion of $S_1 = P_1$ as the radical of P_2 and $p \colon P_2 \longrightarrow S_2$ is the projection. Applying ν yields an exact sequence

$$0 \longrightarrow \tau S_2 \longrightarrow I_1 = \nu P_1 \xrightarrow{\ \nu j\ } I_2 = \nu P_2.$$

Now, up to scalars, there exists a unique nonzero morphism from $I_1 = \begin{smallmatrix} 4 \\ 2\,3 \\ 1 \end{smallmatrix}$ to $I_2 = \begin{smallmatrix} 4 \\ 2 \end{smallmatrix}$, and its kernel is $\begin{smallmatrix} 3 \\ 1 \end{smallmatrix}$. Then, $\tau S_2 = \begin{smallmatrix} 3 \\ 1 \end{smallmatrix}$.

Similarly, we have a minimal injective copresentation

$$0 \longrightarrow S_2 \xrightarrow{\ i\ } I_2 \xrightarrow{\ q\ } I_4 \longrightarrow 0$$

where i is the inclusion and q the projection onto $I_4 \cong I_2/S_2$. Applying ν^{-1} yields an exact sequence

$$P_2 = \nu^{-1} I_2 \xrightarrow{\ \nu^{-1} q\ } P_4 = \nu^{-1} I_4 \longrightarrow \tau^{-1} S_2 \longrightarrow 0.$$

Again, there is a unique nonzero morphism from $\begin{smallmatrix} 2 \\ 1 \end{smallmatrix} = P_2$ to $P_4 = \begin{smallmatrix} 4 \\ 2\,3 \\ 1 \end{smallmatrix}$ (up to scalars) and we deduce that $\tau^{-1} S_2 \cong \begin{smallmatrix} 4 \\ 3 \end{smallmatrix}$.

Exercises for Section III.1

Exercise III.1.1. Let $\mathscr{C}$ be a $\mathbf{k}$-linear category and X an object in $\mathscr{C}$. Prove that the following conditions are equivalent.

(a) X is isomorphic to the zero object.
(b) The identity 1_X is equal to the zero morphism.
(c) $\mathrm{End}_{\mathscr{C}}\, X = 0$.

Exercise III.1.2. Let A be an algebra and X an A-module. Prove that, in the quotient category $\mathrm{mod}\, A/\langle \mathrm{add}\, X\rangle$, an object is isomorphic to the zero object if and only if it lies in $\mathrm{add}\, X$.

Exercise III.1.3. Let A be one of the following bound quiver Nakayama algebras and M an indecomposable A-module. Compute τM for every M nonprojective and $\tau^{-1}M$ for every M noninjective.

(a) $3 \xleftarrow{\ \beta\ } 2 \xleftarrow{\ \alpha\ } 1$ $\qquad \alpha\beta = 0$

(b) $4 \xleftarrow{\ \gamma\ } 3 \xleftarrow{\ \beta\ } 2 \xleftarrow{\ \alpha\ } 1$ $\qquad \alpha\beta\gamma = 0$

(c) $1 \underset{\beta}{\overset{\alpha}{\rightleftarrows}} 2$ $\qquad \alpha\beta = 0$

(d) $1 \underset{\beta}{\overset{\alpha}{\rightleftarrows}} 2$ $\qquad \alpha\beta\alpha = 0,\ \beta\alpha\beta = 0$

Exercise III.1.4. Let $f: M \longrightarrow N$ be a morphism of A-modules.

(a) Show that the following conditions are equivalent:

(i) For every epimorphism $h: L \longrightarrow N$, there exists $g: M \longrightarrow L$ such that $f = hg$.
(ii) For every epimorphism $p: P \longrightarrow N$ with P projective, there exists $g: M \longrightarrow P$ such that $f = pg$.
(iii) $f \in \mathscr{P}(M, N)$.

(b) Show that the following conditions are equivalent:

(i) For every monomorphism $h: M \longrightarrow L$, there exists $g: L \longrightarrow N$ such that $f = gh$.
(ii) For every monomorphism $j: M \longrightarrow I$ with I injective, there exists $g: I \longrightarrow M$ such that $f = gj$.
(iii) $f \in \mathscr{I}(M, N)$.

Exercise III.1.5. Let M, N be A-modules.

(a) If M, N have no projective direct summands, then $M \cong N$ in $\mathrm{mod}\, A$ if and only if $M \cong N$ in $\underline{\mathrm{mod}}\, A$.
(b) If M, N have no injective direct summands, then $M \cong N$ in $\mathrm{mod}\, A$ if and only if $M \cong N$ in $\overline{\mathrm{mod}}\, A$.

Exercise III.1.6. Prove that the functor $(-)^t$ induces a duality between $\mathrm{proj}\, A$ and $\mathrm{proj}\, A^{op}$.

Exercise III.1.7. Let M be an indecomposable nonprojective A-module, and $P_1 \xrightarrow{p} P_0 \longrightarrow M \longrightarrow 0$ a minimal projective presentation in $\operatorname{mod} A$. Prove that $P_0^t \xrightarrow{p^t} P_1^t \longrightarrow \operatorname{Tr} M \longrightarrow 0$ is a minimal projective presentation of $\operatorname{Tr} M$ in $\operatorname{mod} A^{op}$.

Exercise III.1.8. Let M, N be indecomposable nonprojective A-modules. Prove that

(a) $M \cong N$ if and only if $\operatorname{Tr} M \cong \operatorname{Tr} N$.
(b) $\operatorname{Tr}(M \oplus N) \cong \operatorname{Tr} M \oplus \operatorname{Tr} N$.
(c) $\operatorname{Tr}(\operatorname{Tr} M) \cong M$.

Exercise III.1.9. Let M, N be indecomposable A-modules. Prove that

(a) If M, N are nonprojective, then

  (i) $M \cong N$ if and only if $\tau M \cong \tau N$.
  (ii) $\tau(M \oplus N) \cong \tau M \oplus \tau N$.

(b) If M, N are noninjective, then

  (i) $M \cong N$ if and only if $\tau^{-1} M \cong \tau^{-1} N$.
  (ii) $\tau^{-1}(M \oplus N) \cong \tau^{-1} M \oplus \tau^{-1} N$.

Exercise III.1.10. Let M be an indecomposable A-module. Prove that

(a) If $P_1 \xrightarrow{p} P_0 \longrightarrow M \longrightarrow 0$ is a minimal projective presentation, then $\operatorname{soc} \tau M \cong P_1 / \operatorname{rad} P_1$.
(b) If $0 \longrightarrow M \longrightarrow I_0 \longrightarrow I_1$ is a minimal injective copresentation, then $\tau^{-1} M / \operatorname{rad} \tau^{-1} M \cong \operatorname{soc} I_1$.

III.2 The Auslander–Reiten formulae

III.2.1 Preparatory lemmata

Our motivation for defining the Auslander–Reiten translations was to express functorially the relation between the end terms of an almost split sequence. To do it, we first prove the Auslander–Reiten formulae, which express the extension groups between modules as stable homomorphism groups between one of these modules and the Auslander–Reiten translate of the other. As we see in Subsection III.2.3 below, they allow us to prove that, as expected, each of the end terms of an almost split sequence can be deduced from the other by an Auslander–Reiten translation. As a consequence, we obtain a second existence proof for almost split sequences. In this first subsection, we only prove lemmata, that are essentially used in the proof of the Auslander–Reiten formulae.

Lemma III.2.1. *Let M, N be A-modules. The functorial morphism $\varphi_{M,N}: N \otimes_A M^t \longrightarrow \operatorname{Hom}_A(M, N)$ defined by $y \otimes f \mapsto (x \mapsto yf(x))$ (with $x \in M, y \in N, f \in M^t$) satisfies the following properties:*

(a) *If M is projective, then $\varphi_{M,N}$ is an isomorphism.*
(b) $\operatorname{Coker} \varphi_{M,N} = \underline{\operatorname{Hom}}_A(M, N)$.

Proof.

(a) Setting $M = A_A$, we see that $\varphi_{A,N}$ equals the composition of the three well-known isomorphisms $N \otimes_A A^t \longrightarrow N \otimes A$ defined by $y \otimes f \mapsto y \otimes f(1)$, $N \otimes_A A \longrightarrow N$ defined by $y \otimes a \mapsto ya$ and $N \longrightarrow \operatorname{Hom}_A(A, N)$ defined by $y \mapsto (a \mapsto ya)$ (for $y \in N$, $f \in A^t$ and $a \in A$). The statement then follows from the fact that we are dealing with **k**-functors.
(b) It suffices to prove that $\operatorname{Im} \varphi_{M,N} = \mathscr{P}(M, N)$. Assume first that $f \in \operatorname{Im} \varphi_{M,N}$, then, there exist $y_1, \ldots, y_n \in N$ and $f_1, \ldots, f_n \in M^t$ such that

$$f = \varphi_{M,N}\left(\sum_{i=1}^{n} y_i \otimes f_i\right).$$

Thus, for $x \in M$, we have

$$f(x) = \sum_{i=1}^{n} y_i f_i(x) = (y_1, \ldots, y_n) \begin{pmatrix} f_1 \\ \vdots \\ f_n \end{pmatrix} (x).$$

Now, $\begin{pmatrix} f_1 \\ \vdots \\ f_n \end{pmatrix}$ is a morphism from M to A^n, whereas the morphism $(y_1, \ldots, y_n) : A^n \longrightarrow N$ is defined by left multiplication by the $y_i \in N$; thus, f factors through A^n, which is projective (even free). Therefore, $f \in \mathscr{P}(M, N)$.

Conversely, let $g \in \mathscr{P}(M, N)$. There exist a projective module P and morphisms $g_1: M \longrightarrow P$, $g_2: P \longrightarrow N$ such that $g = g_2 g_1$. Because N is finitely generated, there exist $m > 0$ and an epimorphism $p: A^m \longrightarrow N$. The projectivity of P yields $g_2': P \longrightarrow A^m$ such that $g_2 = p g_2'$ and hence $g = p g_2' g_1$. Let $\{e_1, \ldots, e_m\}$ denote the canonical basis of the free module A^m; thus, $p: A^m \longrightarrow N$ can be considered as a row matrix of elements of N of the form $p = (p(e_1), \ldots, p(e_m))$ where each $p(e_i)$ acts by left multiplication on an element of A. Also, $g_2' g_1: M \longrightarrow A^m$ can be expressed as a column matrix $\begin{pmatrix} f_1 \\ \vdots \\ f_n \end{pmatrix}$ where each f_i is the composition of $g_2' g_1$ with the ith projection from A^m to A. Therefore,

$$
\begin{aligned}
g &= pg_2'g_1 \\
&= (p(e_1), \dots, p(e_m)) \begin{pmatrix} f_1 \\ \vdots \\ f_m \end{pmatrix} \\
&= \varphi_{M,N}\left(\sum_{i=1}^{m} p(e_i) \otimes f_i\right) \in \operatorname{Im} \varphi_{M,N}.
\end{aligned}
$$

□

Corollary III.2.2. *There exists a functorial morphism* $\psi_{M,N} : \mathrm{D}\operatorname{Hom}_A(M, N) \longrightarrow \operatorname{Hom}_A(N, \nu M)$ *that is an isomorphism whenever* M *is projective.*

Proof. Indeed, the morphism $\varphi_{M,N} \colon N \otimes_A M^t \longrightarrow \operatorname{Hom}_A(M, N)$ of Lemma III.2.1 induces a morphism $\mathrm{D}\varphi_{M,N} \colon \mathrm{D}\operatorname{Hom}_A(M, N) \longrightarrow \mathrm{D}(N \otimes_A M^t)$, which can be composed with the adjunction isomorphism $\eta_{M,N}$:

$$
\begin{aligned}
\mathrm{D}(N \otimes_A M^t) &= \operatorname{Hom}_{\mathbf{k}}(N \otimes_A M^t, \mathbf{k}) \\
&\cong \operatorname{Hom}_A(N, \operatorname{Hom}_{\mathbf{k}}(M^t, \mathbf{k})) \\
&= \operatorname{Hom}_A(N, \mathrm{D}M^t) \\
&= \operatorname{Hom}_A(N, \nu M),
\end{aligned}
$$

thus yielding the required morphism

$$\psi_{M,N} = \eta_{M,N}\mathrm{D}\varphi_{M,N} \colon \mathrm{D}\operatorname{Hom}_A(M, N) \longrightarrow \operatorname{Hom}_A(N, \nu M).$$

The last statement follows from part (a) of Lemma III.2.1. □

We also need an easy diagram chasing lemma.

Lemma III.2.3. *Assume that we have a commutative diagram in* mod A

$$
\begin{array}{ccccc}
L & \xrightarrow{f} & M & \xrightarrow{g} & N \\
\downarrow{\scriptstyle u} & & \downarrow{\scriptstyle v} & & \downarrow{\scriptstyle w} \\
L' & \xrightarrow{f'} & M' & \xrightarrow{g'} & N'
\end{array}
$$

where v *is an isomorphism, the upper row is exact and the lower row satisfies* $g'f' = 0$. *Then the restriction of* gv^{-1} *to* $\operatorname{Ker} g'$ *defines a morphism* $\varphi \colon \operatorname{Ker} g' \longrightarrow \operatorname{Ker} w$ *such that*

(a) *If g is surjective, then so is φ.*
(b) *If u is surjective, then* $\operatorname{Ker}\varphi = \operatorname{Im} f'$.
(c) *If both g and u are surjective, then* $\operatorname{Ker} w \cong \frac{\operatorname{Ker} g'}{\operatorname{Im} f'}$.

Proof. Because v is an isomorphism, to each $x' \in \operatorname{Ker} g'$ corresponds a unique $x \in M$ such that $x' = v(x)$. Then, $\varphi(x') = g(x)$ and $\varphi(x') \in \operatorname{Ker} w$ because $w\varphi(x') = wg(x) = g'v(x) = g'(x') = 0$. This defines the required morphism φ.

(a) Let $y \in \operatorname{Ker} w$. Because g is surjective, there exists $x \in M$ such that $y = g(x)$. Then, $x' = v(x)$ belongs to $\operatorname{Ker} g'$ because $g'(x') = g'v(x) = wg(x) = w(y) = 0$, and we clearly have $\varphi(x') = g(x) = y$. Thus, φ is surjective.
(b) Assume $x' \in \operatorname{Im} f'$. There exists a unique $x \in M$ such that $x' = v(x)$. Because u is surjective, there exists $z \in L$ such that $x' = f'u(z) = vf(z)$. Then, $v(x) = vf(z)$ implies $x = f(z)$ and $\varphi(x') = g(x) = gf(z) = 0$. So $x' \in \operatorname{Ker}\varphi$ and $\operatorname{Im} f' \subseteq \operatorname{Ker}\varphi$.

    Conversely, let $x' \in \operatorname{Ker}\varphi$ and $x \in M$ be such that $x' = v(x)$. Then, $g(x) = \varphi(x') = 0$ and so $x \in \operatorname{Ker} g$. Because the upper row is exact, there exists $z \in L$ such that $x = f(z)$. Therefore, $x' = v(x) = vf(z) = f'u(z) \in \operatorname{Im} f'$. This completes the proof that $\operatorname{Ker}\varphi = \operatorname{Im} f'$.
(c) This follows easily from (a), (b) and the isomorphism theorem.

□

Observe that in (b), we did not need the surjectivity of u to prove that $\operatorname{Ker}\varphi \subseteq \operatorname{Im} f'$. Also, $g'f' = 0$ says that the lower row is a complex. We have computed in (c) its cohomology group at the middle term.

III.2.2 Proof of the formulae

We are now able to express the first extension space between two modules as (the dual of) a stable Hom-space. These are the Auslander–Reiten formulae.

Theorem III.2.4 (The Auslander–Reiten formulae). *Let M, N be A-modules. Then, there exist isomorphisms*

$$\operatorname{Ext}^1_A(M, N) \cong D\underline{\operatorname{Hom}}_A(\tau^{-1}N, M) \cong D\overline{\operatorname{Hom}}_A(N, \tau M)$$

that are functorial in both variables.

Proof. We prove only the first isomorphism, because the second is proved in the same way. As our functors are **k**-functors, it suffices to prove the statement assuming N indecomposable noninjective. Because of Proposition III.1.10, there exists an indecomposable nonprojective A-module L such that $N = \tau L$ and $L = \tau^{-1}N$. Let

$$P_1 \xrightarrow{p_1} P_0 \xrightarrow{p_0} L \longrightarrow 0$$

be a minimal projective presentation. Because of Lemma III.1.8, we have an exact sequence

$$0 \longrightarrow \tau L \longrightarrow \nu P_1 \xrightarrow{\nu p_1} \nu P_0 \xrightarrow{\nu p_0} \nu L \longrightarrow 0.$$

Applying the functor $\mathrm{Hom}_A(M, -)$ yields a complex

$$0 \longrightarrow \mathrm{Hom}_A(M, \tau L) \longrightarrow \mathrm{Hom}_A(M, \nu P_1) \xrightarrow{p_1^*} \mathrm{Hom}_A(M, \nu P_0) \xrightarrow{p_0^*} \mathrm{Hom}_A(M, \nu L)$$

where $p_i^* = \mathrm{Hom}_A(M, \nu p_i)$ for $i = 0, 1$.

We know that νP_1 and νP_0 are injective so that $\nu p_1 \colon \nu P_1 \longrightarrow \nu P_0$ is the beginning of an injective coresolution of τL. To compute $\mathrm{Ext}_A^1(M, \tau L) = \mathrm{Ext}_A^1(M, N)$, we need one more injective term. Let $j \colon \nu L \longrightarrow J$ be a monomorphism with J injective. We get an exact sequence

$$0 \longrightarrow \tau L \longrightarrow \nu P_1 \xrightarrow{\nu p_1} \nu P_0 \xrightarrow{j \nu p_0} J$$

and, by definition,

$$\mathrm{Ext}_A^1(M, \tau L) \cong \frac{\mathrm{Ker}\,\mathrm{Hom}_A(M, j\nu p_0)}{\mathrm{Im}\,\mathrm{Hom}_A(M, \nu p_1)}.$$

But now $\mathrm{Hom}_A(M, j\nu p_0) = \mathrm{Hom}_A(M, j)\,\mathrm{Hom}_A(M, \nu p_0)$. Because j is a monomorphism, so is $\mathrm{Hom}_A(M, j)$ and hence $\mathrm{Ker}\,\mathrm{Hom}_A(M, j\nu p_0) = \mathrm{Ker}\,\mathrm{Hom}_A(M, \nu p_0) = \mathrm{Ker}\, p_0^*$. We have thus proved that $\mathrm{Ext}_A^1(M, \tau L) \cong \mathrm{Ker}\, p_0^* / \mathrm{Im}\, p_1^*$.

On the other hand, applying the right exact functor $\mathrm{D}\,\mathrm{Hom}_A(-, M)$ to the given minimal projective presentation of L yields an exact sequence

$$\mathrm{D}\,\mathrm{Hom}_A(P_1, M) \xrightarrow{\widetilde{p_1}} \mathrm{D}\,\mathrm{Hom}_A(P_0, M) \xrightarrow{\widetilde{p_0}} \mathrm{D}\,\mathrm{Hom}_A(L, M) \longrightarrow 0$$

where $\widetilde{p_i} = \mathrm{D}\,\mathrm{Hom}_A(p_i, M)$ for $i = 0, 1$.

To relate the latter exact sequence with the complex found before, we use the functorial morphism ψ of Corollary III.2.2 and get a commutative diagram with the upper row exact and the lower row a complex

$$\begin{array}{ccccccc}
\mathrm{D}\,\mathrm{Hom}_A(P_1,M) & \xrightarrow{\widetilde{p_1}} & \mathrm{D}\,\mathrm{Hom}_A(P_0,M) & \xrightarrow{\widetilde{p_0}} & \mathrm{D}\,\mathrm{Hom}_A(L,M) & \longrightarrow & 0 \\
\downarrow {\scriptstyle \psi_{P_1,M}\ \cong} & & \downarrow {\scriptstyle \psi_{P_0,M}\ \cong} & & \downarrow {\scriptstyle \psi_{L,M}} & & \\
\mathrm{Hom}_A(M,\nu P_1) & \xrightarrow{p_1^*} & \mathrm{Hom}_A(M,\nu P_0) & \xrightarrow{p_0^*} & \mathrm{Hom}_A(M,\nu L) & &
\end{array}$$

Because of Corollary III.2.2, $\psi_{P_1,M}$ and $\psi_{P_0,M}$ are isomorphisms. Applying Lemma III.2.3, we get

$$\begin{aligned}
\mathrm{Ext}_A^1(M, \tau L) &\cong \frac{\mathrm{Ker}\, p_0^*}{\mathrm{Im}\, p_1^*} \\
&\cong \mathrm{Ker}\, \psi_{L,M} \\
&= \mathrm{Ker}(\eta_{L,M} \mathrm{D}\varphi_{L,M}) \\
&= \mathrm{Ker}(\mathrm{D}\varphi_{L,M}) \\
&\cong \mathrm{D}(\mathrm{Coker}\, \varphi_{L,M}).
\end{aligned}$$

Because of Lemma III.2.1(b) we have $\mathrm{Coker}\, \varphi_{L,M} \cong \underline{\mathrm{Hom}}_A(L, M)$; hence,

$$\begin{aligned}
\mathrm{Ext}_A^1(M, N) &= \mathrm{Ext}_A^1(M, \tau L) \\
&\cong \mathrm{D}(\mathrm{Coker}\, \varphi_{L,M}) \\
&\cong \mathrm{D}\underline{\mathrm{Hom}}_A(\tau^{-1} N, M).
\end{aligned}$$

□

III.2.3 Application to almost split sequences

We next show how the Auslander–Reiten formulae provide the relation between the end terms of an almost split sequence, and also a second existence proof for almost split sequences.

Lemma III.2.5. *Let M be an indecomposable A-module.*

(a) *If M is nonprojective, then the right and the left socles of the* $\mathrm{End}\, M - \mathrm{End}\, M$-*bimodule* $\mathrm{D}(\underline{\mathrm{End}}\ M)$ *are simple and coincide.*
(b) *If M is noninjective, then the right and the left socles of the* $\mathrm{End}\, M - \mathrm{End}\, M$-*bimodule* $\mathrm{D}(\overline{\mathrm{End}} M)$ *are simple and coincide.*

Proof. We only prove (a), because the proof of (b) is dual.

Clearly, $\underline{\mathrm{End}}\ M$ has a natural $\mathrm{End}\, M - \mathrm{End}\, M$-bimodule structure. We claim that its top as a left $\mathrm{End}\, M$-module and its top as a right $\mathrm{End}\, M$-module are simple.

Because M is indecomposable, $\mathrm{End}\, M$ is a local algebra. We claim that the nonprojectivity of M implies that $\mathscr{P}(M, M) \subseteq \mathrm{rad}\, \mathrm{End}\, M$: indeed, let $h \in \mathscr{P}(M, M)$, then there exist a projective module P and morphisms $f : M \longrightarrow P$ and $g : P \longrightarrow M$ such that $h = gf$. Now, if $h = gf \notin \mathrm{rad}\, \mathrm{End}\, M$, then h is invertible and g a retraction, so that M is projective, a contradiction. Therefore, $h \in \mathrm{rad}(\mathrm{End}\, M)$, as required.

This implies that

$$\mathrm{rad}(\underline{\mathrm{End}}\ M) = \mathrm{rad}\left(\frac{\mathrm{End}\, M}{\mathscr{P}(M, M)}\right) = \frac{\mathrm{rad}(\mathrm{End}\, M)}{\mathscr{P}(M, M)}$$

and so

$$\frac{\underline{\mathrm{End}}\, M}{\mathrm{rad}(\underline{\mathrm{End}}\, M)} \cong \frac{\mathrm{End}\, M/\mathscr{P}(M, M)}{\mathrm{rad}(\mathrm{End}\, M/\mathscr{P}(M, M))} \cong \frac{\mathrm{End}\, M}{\mathrm{rad}\,\mathrm{End}\, M}$$

is a skew field because End M is local. Now, a skew field is generated by each nonzero element; hence, $\underline{\mathrm{End}}\ M/\,\mathrm{rad}(\underline{\mathrm{End}}\ M)$ is simple as a left, or as a right, $\underline{\mathrm{End}}\ M$-module. Because the canonical projection End $M \longrightarrow \underline{\mathrm{End}}\ M$ is surjective, it is also simple as a left, or as a right, End M-module.

We have shown that $\mathrm{D}(\underline{\mathrm{End}}\ M)$ has a simple socle either as a left, or as a right, End M-module. These two socles coincide because they both correspond to the skew field $\underline{\mathrm{End}}\ M/\,\mathrm{rad}(\underline{\mathrm{End}}\ M)$. □

Lemma III.2.6. *Let M be an indecomposable A-module.*

(a) *If M is nonprojective and* $v\colon V \longrightarrow M$ *is a radical morphism, then, for every element* φ *of the socle of* $\mathrm{D}(\underline{\mathrm{End}}\ M)$ *we have* $\mathrm{D}\underline{\mathrm{Hom}}_A(M, v)(\varphi) = 0$.
(b) *If M is noninjective and* $u\colon M \longrightarrow U$ *is a radical morphism, then, for every element* φ *of the socle of* $\mathrm{D}(\overline{\mathrm{End}}M)$ *we have* $\mathrm{D}\overline{\mathrm{Hom}}_A(M, v)(\varphi) = 0$.

Proof. We only prove (a), because the proof of (b) is dual.

The morphism $v\colon V \longrightarrow M$ induces a morphism

$$\underline{\mathrm{Hom}}_A(M, v)\colon \underline{\mathrm{Hom}}_A(M, V) \longrightarrow \underline{\mathrm{Hom}}_A(M, M) = \underline{\mathrm{End}}\ M$$

and hence a morphism

$$\mathrm{D}\underline{\mathrm{Hom}}_A(M, v)\colon \mathrm{D}(\underline{\mathrm{End}}\ M) \longrightarrow \mathrm{D}\underline{\mathrm{Hom}}_A(M, V)$$

as follows: if $\varphi \in \mathrm{D}(\underline{\mathrm{End}}\ M)$ then

$$\mathrm{D}\underline{\mathrm{Hom}}_A(M, v)(\varphi) = \varphi\underline{\mathrm{Hom}}_A(M, v).$$

We need to prove that, if $\varphi \in \mathrm{soc}\,\mathrm{D}(\underline{\mathrm{End}}\ M)$, then $\varphi\underline{\mathrm{Hom}}_A(M, v) = 0$. Let $\underline{u} \in \underline{\mathrm{Hom}}_A(M, V)$. Then, $\underline{\mathrm{Hom}}_A(M, v)(\underline{u}) = \underline{vu} \in \mathrm{rad}(\underline{\mathrm{End}}\ M)$ because $v \in \mathrm{rad}_A(V, M)$. Now, φ belonging to the socle yields

$$\varphi\underline{vu} \in \mathrm{soc}\,\mathrm{D}(\underline{\mathrm{End}}\ M)\,\mathrm{rad}(\underline{\mathrm{End}}\ M) = \mathrm{rad}(\mathrm{soc}\,\mathrm{D}(\underline{\mathrm{End}}\ M)) = 0$$

because the socle of $\mathrm{D}(\underline{\mathrm{End}}\ M)$ is simple. Thus, $\varphi\underline{vu} = 0$. □

The reason for our interest in the End M – End M-bimodules $\mathrm{D}(\underline{\mathrm{End}}\ M)$ and $\mathrm{D}(\overline{\mathrm{End}}M)$ with M indecomposable comes from the Auslander–Reiten formulae, which imply that:

(a) If M is nonprojective, then $\mathrm{Ext}^1_A(M, \tau M) \cong \mathrm{D}(\underline{\mathrm{End}}\ M)$.
(b) If M is noninjective, then $\mathrm{Ext}^1_A(\tau^{-1}M, M) \cong \mathrm{D}(\overline{\mathrm{End}}M)$.

The nonzero elements of the socle of $\mathrm{D}(\underline{\mathrm{End}}\ M)$ and $\mathrm{D}(\overline{\mathrm{End}}M)$, such as, for instance, the (duals of the) identity morphism 1_M, correspond to nonsplit extensions. We now prove that these extensions are almost split sequences.

Corollary III.2.7. *Let M be an indecomposable A-module.*

(a) *If M is nonprojective and $\xi : 0 \longrightarrow \tau M \xrightarrow{f} E \xrightarrow{g} M \longrightarrow 0$ represents a nonzero element of the socle of* $\mathrm{Ext}^1_A(M, \tau M)$*, then the sequence ξ is almost split.*

(b) *If M is noninjective and $\xi : 0 \longrightarrow M \xrightarrow{f} E \xrightarrow{g} \tau^{-1}M \longrightarrow 0$ represents a nonzero element of the socle of* $\mathrm{Ext}^1_A(\tau^{-1}M, M)$*, then the sequence ξ is almost split.*

Proof. We only prove (a), because the proof of (b) is dual.

Because M is indecomposable nonprojective, the module τM is indecomposable owing to Proposition III.1.10. Applying Corollary II.3.13, it suffices to prove that the morphism g is right almost split. Because the given sequence is not split, g is not a retraction and therefore is a radical morphism. Let $v : V \longrightarrow M$ be radical. The functoriality in the Auslander–Reiten formulae yields a commutative square

$$
\begin{array}{ccc}
\mathrm{D}(\underline{\mathrm{End}}_A M) & \xrightarrow{\cong} & \mathrm{Ext}^1_A(M,\tau M) \\
{\scriptstyle \mathrm{D}\underline{\mathrm{Hom}}_A(M,v)}\downarrow & & \downarrow{\scriptstyle \mathrm{Ext}^1_A(v,\tau M)} \\
\mathrm{D}\underline{\mathrm{Hom}}_A(M,V) & \xrightarrow{\cong} & \mathrm{Ext}^1_A(V,\tau M)
\end{array}
$$

Because the given sequence ξ is a nonzero element of the socle, Lemma III.2.6 above gives $\mathrm{Ext}^1_A(v, \tau M)(\xi) = 0$. This means that, if we take the fibered product of the morphisms g and v, the upper sequence in the commutative diagram with exact rows

$$
\begin{array}{ccccccccc}
0 & \longrightarrow & \tau M & \xrightarrow{f'} & E' & \xrightarrow{g'} & V & \longrightarrow & 0 \\
 & & \| & & \downarrow{\scriptstyle u} & & \downarrow{\scriptstyle v} & & \\
0 & \longrightarrow & \tau M & \xrightarrow{f} & E & \xrightarrow{g} & M & \longrightarrow & 0
\end{array}
$$

is split. Let $g'' : V \longrightarrow E'$ be such that $g'g'' = 1_V$. Then, ug'' satisfies $g(ug'') = (gu)g'' = vg'g'' = v$. This proves that g is right almost split. □

This corollary is obviously a second existence proof for almost split sequences. But also, because of uniqueness, see Corollary II.2.32, it implies that, if $0 \longrightarrow L \longrightarrow M \longrightarrow N \longrightarrow 0$ is almost split, then $L \cong \tau N$ and $N \cong \tau^{-1}L$.

III.2.4 Starting to compute almost split sequences

This subsection is devoted to three easy lemmata that are useful in practical computations.

Lemma III.2.8. *Let M be an indecomposable module.*

(a) *There exists a right minimal almost split morphism $g\colon E \longrightarrow M$. Also, $E = 0$ if and only if M is simple projective.*
(b) *There exists a left minimal almost split morphism $f\colon M \longrightarrow E$. Also, $E = 0$ if and only if M is simple injective.*

Proof. We only prove (a), because the proof of (b) is dual.

The first statement is just Corollary II.3.13. In particular, if M is projective and $g\colon E \longrightarrow M$ is right minimal almost split, then it is isomorphic to the inclusion of $\operatorname{rad} M$ into M. Also, $\operatorname{rad} M = 0$ if and only if M is simple projective. On the other hand, if M is nonprojective and $g\colon E \longrightarrow M$ is right minimal almost split, then g is surjective, and so $E \neq 0$. □

We now prove that every irreducible morphism whose target (or source) is indecomposable nonprojective (or noninjective) corresponds to an irreducible morphism starting (or ending, respectively) at the translate of this module.

Lemma III.2.9.

(a) *Let N be indecomposable nonprojective. There exists an irreducible morphism $f\colon X \longrightarrow N$ if and only if there exists an irreducible morphism $f'\colon \tau N \longrightarrow X$.*
(b) *Let L be indecomposable noninjective. There exists an irreducible morphism $g\colon L \longrightarrow Y$ if and only if there exists an irreducible morphism $g'\colon Y \longrightarrow \tau^{-1}L$.*

Proof. We only prove (a) because the proof of (b) is dual.

Assume that there exists an irreducible morphism $f\colon X \longrightarrow N$. Because of Theorem II.2.24, there exists $h\colon Z \longrightarrow N$ such that $(f, h)\colon X \oplus Z \longrightarrow N$ is right minimal almost split. Because N is indecomposable nonprojective, there exists an almost split sequence

$$0 \longrightarrow \tau N \xrightarrow{\begin{pmatrix} f' \\ h' \end{pmatrix}} X \oplus Z \xrightarrow{(f\ h)} N \longrightarrow 0.$$

Therefore, $f'\colon \tau N \longrightarrow X$ is irreducible. The proof is similar, if one starts with f' instead of f. □

Finally, irreducible morphisms starting at a simple projective module have projective targets, and dually.

Lemma III.2.10.

(a) *Let S be simple projective and M indecomposable. If $f: S \longrightarrow M$ is irreducible, then M is projective.*
(b) *Let S be simple injective and M indecomposable. If $f: M \longrightarrow S$ is irreducible, then M is injective.*

Proof. We only prove (a), because the proof of (b) is dual.

Assume that M is nonprojective. Because of Lemma III.2.9, there exists an irreducible morphism $f': \tau M \longrightarrow S$. But S is simple; hence, f' is surjective. Now, S is projective; hence, f' is a retraction and this contradicts the fact that it is irreducible. □

We show in an example how to apply these lemmata.

Example III.2.11. Let A be given by the quiver

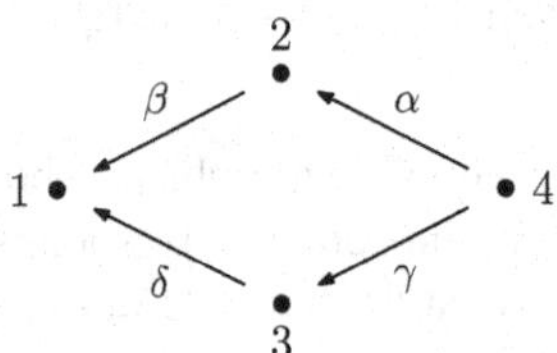

bound by $\alpha\beta = \gamma\delta$. The simple module S_1 is projective and noninjective. Because of Lemma III.2.10, every indecomposable target P of an irreducible morphism $S_1 \longrightarrow P$ is projective. Actually, the morphism $S_1 \longrightarrow P$ is a monomorphism whose image is a direct summand of rad P. So, to find P, we need to find those indecomposable projective A-modules that have S_1 as a summand of the radical. Now,

$$P_1 = S_1 \quad P_2 = {}^{2}_{1} \quad P_3 = {}^{3}_{1} \quad P_4 = \begin{smallmatrix} & 4 & \\ 2 & & 3 \\ & 1 & \end{smallmatrix}.$$

Thus, $S_1 = \text{rad}\, P_2 = \text{rad}\, P_3$. This shows that, up to scalars, there exist exactly two irreducible morphisms starting with S_1; namely, the inclusions $j_2: S_1 \longrightarrow P_2$ and $j_3: S_1 \longrightarrow P_3$. But then, because of Theorem II.2.24, the morphism $\begin{pmatrix} j_2 \\ j_3 \end{pmatrix}: S_1 \longrightarrow P_2 \oplus P_3$ is left minimal almost split. Therefore, there exists an almost split sequence

$$0 \longrightarrow S_1 \xrightarrow{\begin{pmatrix} j_2 \\ j_3 \end{pmatrix}} P_2 \oplus P_3 \xrightarrow{(f_2\ f_3)} \tau^{-1} S_1 \longrightarrow 0.$$

This allows to compute $\tau^{-1}S_1$. Indeed

$$\tau^{-1} S_1 = \text{Coker} \begin{pmatrix} j_2 \\ j_3 \end{pmatrix} \cong \frac{P_2 \oplus P_3}{S_1} \cong \begin{smallmatrix} 2 & & 3 \\ & 1 & \end{smallmatrix}.$$

Next, we wish to find the almost split sequence starting with P_2. Assume that X is indecomposable and we have an irreducible morphism $P_2 \longrightarrow X$. Either X is projective and then P_2 is isomorphic to a direct summand of its radical, or else there exists an irreducible morphism $\tau X \longrightarrow P_2$. The first case is easily discarded: there is no indecomposable projective A-module X that has P_2 isomorphic to a direct summand of its radical. Therefore, X is nonprojective. Because $\operatorname{rad} P_2 = S_1$, the only irreducible morphism of target P_2 is (up to scalars) the inclusion $S_1 \longrightarrow P_2$. Then, $\tau X \cong S_1$ and $X \cong \tau^{-1} S_1$. Because of Theorem II.2.24, this proves that the morphism $f_2 \colon P_2 \longrightarrow \tau^{-1} S_1$ is left minimal almost split. Thus, we get an almost split sequence

$$0 \longrightarrow P_2 \xrightarrow{f_2} \tau^{-1} S_1 \xrightarrow{g_2} \frac{\tau^{-1} S_1}{P_2} \cong S_3 \longrightarrow 0.$$

Similarly, we have an almost split sequence

$$0 \longrightarrow P_3 \xrightarrow{f_3} \tau^{-1} S_1 \xrightarrow{g_3} \frac{\tau^{-1} S_1}{P_3} \cong S_2 \longrightarrow 0.$$

We finally compute the almost split sequence starting with $\tau^{-1} S_1 = \begin{smallmatrix} 2 & & 3 \\ & 1 & \end{smallmatrix}$. This module $\tau^{-1} S_1$ is the (indecomposable) radical of P_4 so we have an irreducible morphism $j \colon \tau^{-1} S_1 \longrightarrow P_4$. This is easily seen to be the only irreducible morphism, up to scalars, from $\tau^{-1} S_1$ to an indecomposable projective module. Otherwise, if X is indecomposable nonprojective and there exists an irreducible morphism $\tau^{-1} S_1 \longrightarrow X$, then there exists an irreducible morphism $\tau X \longrightarrow \tau^{-1} S_1$. But, because of the (already known) almost split sequence ending with $\tau^{-1} S_1$, such an irreducible morphism can only be a scalar multiple of f_2 or f_3. Then, $\tau X \cong P_2$ or $\tau X \cong P_3$ so that $X \cong \tau^{-1} P_2 \cong S_3$ or $X \cong \tau^{-1} P_3 \cong S_2$ respectively. Hence, the morphism $\begin{pmatrix} j \\ g_2 \\ g_3 \end{pmatrix} \colon \tau^{-1} S_1 \longrightarrow P_4 \oplus S_3 \oplus S_2$ is left minimal almost split and we have an almost split sequence

$$0 \longrightarrow \tau^{-1} S_1 \xrightarrow{\begin{pmatrix} j \\ g_2 \\ g_3 \end{pmatrix}} P_4 \oplus S_3 \oplus S_2 \longrightarrow \begin{smallmatrix} & 4 & \\ 2 & & 3 \end{smallmatrix} \longrightarrow 0.$$

The alert reader will detect in this example an inductive method for computing almost split sequences. We shall return to it later.

Exercises for Section III.2

Exercise III.2.1. Prove that the morphism $\varphi_{M,N} \colon N \otimes_A M^t \longrightarrow \operatorname{Hom}_A(M, N)$ defined in Lemma III.2.1 is an isomorphism whenever N is projective.

Exercise III.2.2.

(a) Let M, N be indecomposable nonprojective. Prove that

$$\underline{\mathrm{Hom}}_A(M, N) \cong \overline{\mathrm{Hom}}_A(\tau M, \tau N).$$

(b) Let M, N be indecomposable noninjective. Prove that

$$\overline{\mathrm{Hom}}_A(M, N) \cong \underline{\mathrm{Hom}}_A(\tau^{-1} M, \tau^{-1} N).$$

Exercise III.2.3. Let M, N be indecomposable modules. Prove that

(a) If $\mathrm{pd}\, M \leq 1$, then $\mathrm{Ext}^1_A(M, N) \cong \mathrm{D}\,\mathrm{Hom}_A(N, \tau M)$.
(b) If $\mathrm{id}\, N \leq 1$, then $\mathrm{Ext}^1_A(M, N) \cong \mathrm{D}\,\mathrm{Hom}_A(\tau^{-1} N, M)$.
(c) If $\mathrm{pd}\, M \leq 1$ and $\mathrm{id}\, N \leq 1$, M is nonprojective and N is noninjective, then

$$\mathrm{Hom}_A(N, \tau M) \cong \mathrm{Hom}_A(\tau^{-1} N, M).$$

(d) If $\mathrm{pd}\, M \leq 1$, $\mathrm{id}\, \tau N \leq 1$ and N is nonprojective, then

$$\mathrm{Hom}_A(\tau N, \tau M) \cong \mathrm{Hom}_A(N, M).$$

(e) If $\mathrm{pd}\, \tau^{-1} M \leq 1$, $\mathrm{id}\, N \leq 1$ and M is noninjective, then

$$\mathrm{Hom}_A(\tau^{-1} N, \tau^{-1} M) \cong \mathrm{Hom}_A(N, M).$$

Exercise III.2.4. Let A be a hereditary algebra. Prove the following.

(a) We have an isomorphism of functors $\tau \cong \mathrm{Ext}^1_A(-, A)$.
(b) $\underline{\mathrm{mod}}\, A$ is equivalent to the full subcategory of $\mathrm{mod}\, A$ consisting of all modules that have no projective direct summand.
(c) $\overline{\mathrm{mod}}\, A$ is equivalent to the full subcategory of $\mathrm{mod}\, A$ consisting of all modules that have no injective direct summand.

Exercise III.2.5. Let M be an indecomposable nonprojective A-module. Prove that the functors $\underline{\mathrm{Hom}}_A(M, -)$ and $\mathrm{Tor}^A_1(-, \mathrm{Tr}\, M)$ from $\mathrm{mod}\, A$ to $\mathrm{mod}\, \mathbf{k}$ are isomorphic.

Exercise III.2.6. Let M be an indecomposable module. Prove that

(a) If M is nonprojective, then $\underline{\mathrm{End}} M$ is a skew field if and only if $\overline{\mathrm{End}} \tau M$ is a skew field and, in this case, any nonsplit short exact sequence $0 \longrightarrow \tau M \longrightarrow E \longrightarrow M \longrightarrow 0$ is almost split.
(b) If M is noninjective, then $\overline{\mathrm{End}} M$ is a skew field if and only if $\underline{\mathrm{End}} \tau^{-1} M$ is a skew field and, in this case, any nonsplit short exact sequence $0 \longrightarrow M \longrightarrow F \longrightarrow \tau^{-1} M \longrightarrow 0$ is almost split.

Exercise III.2.7. Let $0 \longrightarrow L \longrightarrow M \longrightarrow N \longrightarrow 0$ be an almost split sequence and P a nonzero projective module. Prove that the following assertions are equivalent:

(a) P is isomorphic to a direct summand of M.
(b) There exists an irreducible morphism $P \longrightarrow N$.
(c) There exists an irreducible morphism $L \longrightarrow P$.
(d) L is isomorphic to a direct summand of rad P.
(e) There is an indecomposable direct summand R of rad P such that $N \cong \tau^{-1}R$.
(f) If $v\colon V \longrightarrow N$ is a radical epimorphism, then P is isomorphic to a direct summand of V.

Exercise III.2.8. Let $0 \longrightarrow L \xrightarrow{f} M \xrightarrow{g} N \longrightarrow 0$ be a short exact sequence. Prove that the following are equivalent:

(a) It is almost split.
(b) $L \cong \tau N$ and g is right almost split.
(c) $N \cong \tau^{-1}L$ and f is left almost split.

Exercise III.2.9. Let $0 \longrightarrow L \xrightarrow{f} M \xrightarrow{g} N \longrightarrow 0$ be an almost split sequence. Prove that:

(a) M is projective if and only if g is a projective cover.
(b) M is injective if and only if f is an injective envelope.

Exercise III.2.10. Let S be a simple module. Prove that:

(a) If S is projective noninjective, then $\operatorname{pd}(\tau^{-1}S) = 1$ and $\operatorname{End}(\tau^{-1}S)$ is a skew field.
(b) If S is injective nonprojective, then $\operatorname{id}(\tau S) = 1$ and $\operatorname{End}(\tau S)$ is a skew field.

Exercise III.2.11.

(a) Let M be indecomposable nonprojective and X arbitrary. Prove that $f\colon \tau M \longrightarrow X$ is a section provided that the induced morphism $\operatorname{Ext}^1_A(M, f)\colon \operatorname{Ext}^1_A(M, \tau M) \longrightarrow \operatorname{Ext}^1_A(M, X)$ is a monomorphism.
(b) Let M be indecomposable noninjective and Y arbitrary. Prove that $g\colon Y \longrightarrow \tau^{-1}M$ is a retraction provided that the induced morphism $\operatorname{Ext}^1_A(g, M)\colon \operatorname{Ext}^1_A(\tau^{-1}M, M) \longrightarrow \operatorname{Ext}^1_A(Y, M)$ is a monomorphism.

Exercise III.2.12. Let I be an indecomposable injective, and P an indecomposable projective A-module. Prove that there exists no irreducible morphism $I \longrightarrow P$.

Exercise III.2.13. For each bound quiver below, consider the corresponding algebra and

(a) Compute the almost split sequences starting at P_i and at $\tau^{-1}P_i$ for $i = 1, 2, 3, 4$.
(b) Compute the almost split sequences ending at I_i and at τI_i for $i = 1, 2, 3, 4$.

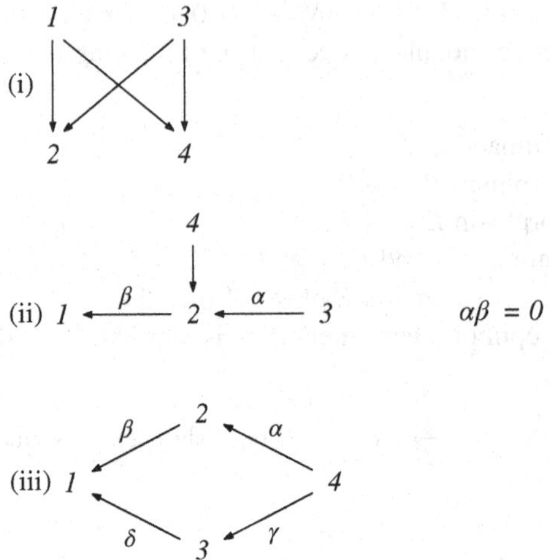

III.3 Examples of constructions of almost split sequences

III.3.1 The general case

In general, constructing an almost split sequence is a difficult exercise. There is, however, a technique that is implicit in the proofs of Theorem II.3.10 and Proposition II.3.11. Namely, assume that we want to construct an almost split sequence ending with a given indecomposable nonprojective A-module N. We start by constructing a minimal projective presentation

$$P_1 \xrightarrow{p_1} P_0 \xrightarrow{p_0} N \longrightarrow 0$$

Let $\xi_N : \operatorname{Hom}_A(-, N) \longrightarrow \mathrm{D}\operatorname{Hom}_A(N, -)$ be the unique functorial morphism that has the simple functor S_N as its image. We denote by

$$\alpha_{P_0} : \mathrm{D}\operatorname{Hom}_A(P_0, -) \longrightarrow \operatorname{Hom}_A(-, \nu P_0)$$

the functorial isomorphism induced by the Nakayama functor, see Lemma I.1.19. Let $u : N \longrightarrow \nu P_0$ be a morphism making the following diagram commutative

$$\begin{array}{ccccccc}
 & & & & \mathrm{Hom}_A(-,N) & & \\
 & & & \overset{\mathrm{Hom}_A(-,u)}{\swarrow} & \downarrow{\scriptstyle \xi_N} & & \\
\mathrm{Hom}_A(-,\nu P_0) & \xrightarrow[\alpha_{P_0}^{-1}]{\cong} & \mathrm{D}\,\mathrm{Hom}_A(P_0,-) & \xrightarrow[\mathrm{Hom}_A(p_0,-)]{} & \mathrm{D}\,\mathrm{Hom}_A(N,-) & \longrightarrow & 0.
\end{array}$$

The existence of such a morphism u follows from the projectivity of $\mathrm{Hom}_A(-,N)$ in the category Fun A, and from Yoneda's lemma II.3.1. We then construct the fibered product M of the morphisms $u\colon N \longrightarrow \nu P_0$ and $\nu p_1\colon \nu P_1 \longrightarrow \nu P_0$.

The upper row in the commutative diagram with exact rows

$$\begin{array}{ccccccccc}
0 & \longrightarrow & L & \xrightarrow{f} & M & \xrightarrow{g} & N & \longrightarrow & 0 \\
 & & \| & & \downarrow{\scriptstyle g'} & & \downarrow{\scriptstyle u} & & \\
0 & \longrightarrow & L & \longrightarrow & \nu P_1 & \xrightarrow{\nu p_1} & \nu P_0 & &
\end{array}$$

is an almost split sequence ending with N, as we now state.

Lemma III.3.1. *If the morphism u is chosen as shown above, then the short exact sequence*

$$0 \longrightarrow L \xrightarrow{f} M \xrightarrow{g} N \longrightarrow 0$$

obtained by taking M as the fibered product of the morphisms u and νp_1 is almost split.

Proof. This follows directly from the proofs of Theorem II.3.10 and Proposition II.3.11. □

The fibered product M and the morphism g can be computed as the kernel term in the exact sequence

$$0 \longrightarrow M \xrightarrow{\binom{g}{g'}} N \oplus \nu P_1 \xrightarrow{(u\ \nu p_1)} \nu P_0.$$

However, once one computes the module M and the morphism g, then it may become necessary to decompose M into its indecomposable summands (and hence g into its restriction to the summands). Carrying out this decomposition is either very difficult or tedious. No general technique seems to be known, and it can only be done efficiently in "small enough" examples. Our objective in this section is to present some instances where the computation of almost split sequences is reasonably easy.

III.3.2 Projective–injective middle term

We look here at the situation where an almost split sequence has a projective–injective module as summand of its middle term. Our main result in this subsection is the following.

Proposition III.3.2. *Let* $0 \longrightarrow L \xrightarrow{f} M \xrightarrow{g} N \longrightarrow 0$ *be an almost split sequence, and write* $M = \oplus_{i=1}^{t} M_i$ *with the* M_i *indecomposable. Then,*

(a) *If* M_i *is projective and* M_j *is injective, then* $i = j$ *and* M_i *is projective–injective.*
(b) *At most one of the indecomposable summands of* M *is projective–injective.*
(c) *If* M *admits an indecomposable projective–injective module* P *as a direct summand, then this sequence is isomorphic to*

$$0 \longrightarrow \operatorname{rad} P \xrightarrow{\binom{u}{p}} P \oplus \frac{\operatorname{rad} P}{\operatorname{soc} P} \xrightarrow{(q\ v)} \frac{P}{\operatorname{soc} P} \longrightarrow 0$$

where u, v *are the inclusions and* p, q *the projections. In addition, the summands of* M *other than* P *are neither projective nor injective.*

Proof.

(a) Assume that $i \neq j$. Comparing lengths, we have $l(L) < l(M_i)$ and $l(N) < l(M_j)$. Therefore,

$$l(L) + l(N) < l(M_i) + l(M_j) \leq \sum_{k=1}^{t} l(M_k) = l(L) + l(N),$$

a contradiction. Hence, $i = j$ and $M_i = M_j$ is projective–injective.
(b) Suppose that M_i, M_j are projective–injective. Because M_i is projective and M_j is injective, the reasoning made in (a) yields $M_i = M_j$.
(c) Write the sequence as

$$0 \longrightarrow L \xrightarrow{\binom{f_1}{f_2}} P \oplus M' \xrightarrow{(g_1\ g_2)} N \longrightarrow 0$$

with M' in general decomposable. The projectivity of P implies that L is a direct summand of its radical. However, because P is also indecomposable injective, it has a simple socle, and therefore its radical $\operatorname{rad} P$ also has a simple socle. This implies that $\operatorname{rad} P$ is indecomposable and thus $L \cong \operatorname{rad} P$, and the morphism $f_1 \colon L \longrightarrow P$ is isomorphic to the inclusion $u;\ \operatorname{rad} P \longrightarrow P$. Dually, $N \cong P/\operatorname{soc} P$ has simple top and the morphism $g_1 \colon P \longrightarrow N$ is isomorphic to the projection $q \colon P \longrightarrow P/\operatorname{soc} P$. Now we have

$$\begin{aligned} l(M') &= l(L) + l(N) - l(P) \\ &= l(\operatorname{rad} P) + l(P/\operatorname{soc} P) - l(P) \\ &= l(P) - 2, \end{aligned}$$

because $l(\operatorname{rad} P) = l(P/\operatorname{soc} P) = l(P) - 1$. In particular, $f_2 \colon L \longrightarrow M'$ is surjective. But now, because $\operatorname{rad} P$ has a simple socle, then it has, up to isomorphism, exactly one quotient of length $l(P) - 2 = l(\operatorname{rad} P) - 1$, namely its quotient $\operatorname{rad} P/\operatorname{soc} P$ by its simple socle. Therefore, $M' \cong \operatorname{rad} P/\operatorname{soc} P$ and f_2 is isomorphic to the projection $p \colon \operatorname{rad} P \longrightarrow \operatorname{rad} P/\operatorname{soc} P$. Similarly, $g_2 \colon M' \longrightarrow N$ is isomorphic to the inclusion v of $\operatorname{rad} P/\operatorname{soc} P$ as the unique maximal submodule of $P/\operatorname{soc} P$.

The last statement follows from the fact that, because of (a), $M' = \operatorname{rad} P/\operatorname{soc} P$ has neither projective nor injective summands. □

We point out that although $\operatorname{rad} P$ and $P/\operatorname{soc} P$ are indecomposable, the term $\operatorname{rad} P/\operatorname{soc} P$ is, in general, decomposable and may in fact have an arbitrary number of direct summands, see Exercise VI.4.2 below.

Example III.3.3. Let A be given by the quiver

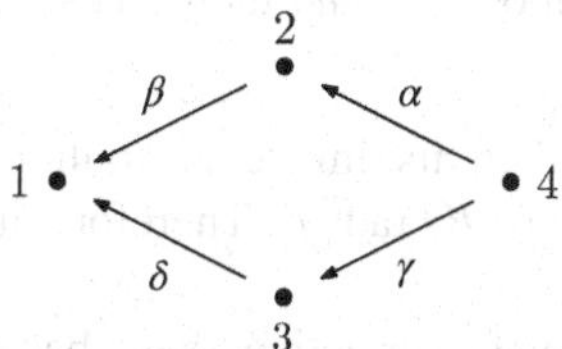

bound by $\alpha\beta = \gamma\delta$. Here,

$$P_4 = I_1 = \begin{smallmatrix} & 4 & \\ 2 & & 3 \\ & 1 & \end{smallmatrix}$$

is indecomposable projective–injective. The radical of P_4 is $\operatorname{rad} P_4 = \begin{smallmatrix} 2 & & 3 \\ & 1 & \end{smallmatrix}$ whereas its quotient by the socle is $P_4/\operatorname{soc} P_4 = \begin{smallmatrix} & 4 & \\ 2 & & 3 \end{smallmatrix}$. Finally, $\operatorname{rad} P_4/\operatorname{soc} P_4 = S_2 \oplus S_3$, so we get an almost split sequence

$$0 \longrightarrow \begin{smallmatrix} 2 & & 3 \\ & 1 & \end{smallmatrix} \longrightarrow \begin{smallmatrix} & 4 & \\ 2 & & 3 \\ & 1 & \end{smallmatrix} \oplus 2 \oplus 3 \longrightarrow \begin{smallmatrix} & 4 & \\ 2 & & 3 \end{smallmatrix} \longrightarrow 0$$

where the morphisms are either inclusions or projections. This is the sequence we obtained at the end of Example III.2.11.

III.3.3 Almost split sequences for Nakayama algebras

We recall from Subsection I.2.4 that an algebra A is a Nakayama algebra if all its indecomposable projective and injective modules are uniserial. We have also seen that, if N is an indecomposable A-module, then there exist an indecomposable projective A-module P and an integer $t \geq 0$ such that $N \cong P/\operatorname{rad}^t P$. We now describe the almost split sequence ending with N.

Proposition III.3.4. *Let A be a Nakayama algebra and $N = P/\operatorname{rad}^t P$ an indecomposable nonprojective A-module. Then the almost split sequence ending with N is isomorphic to the sequence*

$$0 \longrightarrow \frac{\operatorname{rad} P}{\operatorname{rad}^{t+1} P} \xrightarrow{\binom{i}{p}} \frac{P}{\operatorname{rad}^{t+1} P} \oplus \frac{\operatorname{rad} P}{\operatorname{rad}^t P} \xrightarrow{(q\ j)} \frac{P}{\operatorname{rad}^t P} \longrightarrow 0$$

where i, j are the inclusions and p, q the projections.

Proof. Clearly, the sequence in the statement is exact and nonsplit. In addition, both end terms are uniserial and hence indecomposable. Because of Corollary II.3.13, it suffices to prove that the morphism $(q\ j)$ is right almost split.

Let $v\colon V \longrightarrow P/\operatorname{rad}^t P$ be a radical morphism. We may assume without loss of generality that V is indecomposable (and then v is simply a nonisomorphism). We consider two cases.

(a) If v is not surjective, then its image is contained in the unique maximal submodule $\operatorname{rad} P/\operatorname{rad}^t P$ of $P/\operatorname{rad}^t P$. Therefore, in this case, v factors through the middle term.
(b) If, on the other hand, v is surjective then, because V is also uniserial, it must have the same top as $P/\operatorname{rad}^t P$. It follows from the description of the indecomposable A-modules, see Theorem I.2.25, that $V \cong P/\operatorname{rad}^s P$ with $s > t$. But then, v clearly factors through $P/\operatorname{rad}^{t+1} P$ and we are done.

□

There is a similarity between almost split sequences over Nakayama algebras and almost split sequences with projective–injective middle terms. Indeed, letting $U = P/\operatorname{rad}^{t+1} P$, the almost split sequence in the proposition can be written as

$$0 \longrightarrow \operatorname{rad} U \longrightarrow U \oplus \frac{\operatorname{rad} U}{\operatorname{soc} U} \longrightarrow \frac{U}{\operatorname{soc} U} \longrightarrow 0,$$

see Exercise III.3.4 below.

Corollary III.3.5. *Let A be a Nakayama algebra and N an indecomposable nonprojective A-module. Then, $l(\tau N) = l(N)$.*

Proof. If $N = P/\operatorname{rad}^t P$ for some projective A-module P and some $t > 0$, then $\tau N \cong \operatorname{rad} P/\operatorname{rad}^{t+1} P$ so that $l(\tau N) = t = l(N)$. □

In particular, if S is simple, then so are τS, $\tau^2 S$, etc.

Example III.3.6. Let A be given by the quiver

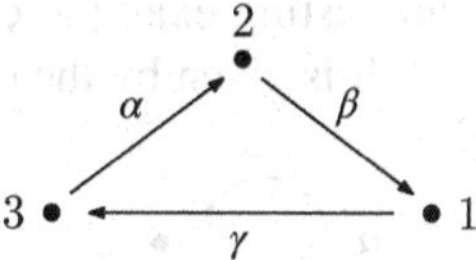

bound by $\alpha\beta\gamma = 0$. The indecomposable projective modules are

$$P_1 = \begin{smallmatrix}1\\2\\3\end{smallmatrix} \qquad P_2 = \begin{smallmatrix}2\\3\\1\\2\\3\end{smallmatrix} \qquad P_3 = \begin{smallmatrix}3\\1\\2\\3\end{smallmatrix}$$

We deduce a complete list of the isoclasses of indecomposable nonprojective A-modules

$$\frac{P_1}{\operatorname{rad} P_1} = 1 \qquad \frac{P_2}{\operatorname{rad} P_2} = 2 \qquad \frac{P_3}{\operatorname{rad} P_3} = 3$$

$$\frac{P_1}{\operatorname{rad}^2 P_1} = \begin{smallmatrix}1\\2\end{smallmatrix} \qquad \frac{P_2}{\operatorname{rad}^2 P_2} = \begin{smallmatrix}2\\3\end{smallmatrix} \qquad \frac{P_3}{\operatorname{rad}^2 P_3} = \begin{smallmatrix}3\\1\end{smallmatrix}$$

$$\frac{P_2}{\operatorname{rad}^3 P_2} = \begin{smallmatrix}2\\3\\1\end{smallmatrix} \qquad \frac{P_3}{\operatorname{rad}^3 P_3} = \begin{smallmatrix}3\\1\\2\end{smallmatrix} \qquad \frac{P_2}{\operatorname{rad}^4 P_2} = \begin{smallmatrix}2\\3\\1\\2\end{smallmatrix}$$

Proposition III.3.4 above gives all almost split sequences in mod A

$$0 \longrightarrow 2 \longrightarrow \begin{smallmatrix}1\\2\end{smallmatrix} \longrightarrow 1 \longrightarrow 0 \qquad 0 \longrightarrow 3 \longrightarrow \begin{smallmatrix}2\\3\end{smallmatrix} \longrightarrow 2 \longrightarrow 0$$

$$0 \longrightarrow 1 \longrightarrow \begin{smallmatrix}3\\1\end{smallmatrix} \longrightarrow 3 \longrightarrow 0 \qquad 0 \longrightarrow \begin{smallmatrix}2\\3\end{smallmatrix} \longrightarrow \begin{smallmatrix}1\\2\\3\end{smallmatrix} \oplus 2 \longrightarrow \begin{smallmatrix}1\\2\end{smallmatrix} \longrightarrow 0$$

$$0 \longrightarrow \begin{smallmatrix}3\\1\end{smallmatrix} \longrightarrow \begin{smallmatrix}2\\3\\1\end{smallmatrix} \oplus 3 \longrightarrow \begin{smallmatrix}2\\3\end{smallmatrix} \longrightarrow 0 \qquad 0 \longrightarrow \begin{smallmatrix}1\\2\end{smallmatrix} \longrightarrow \begin{smallmatrix}3\\1\\2\end{smallmatrix} \oplus 1 \longrightarrow \begin{smallmatrix}3\\1\end{smallmatrix} \longrightarrow 0$$

$$0 \longrightarrow \begin{smallmatrix}3\\1\\2\end{smallmatrix} \longrightarrow \begin{smallmatrix}2\\3\\1\\2\end{smallmatrix} \oplus \begin{smallmatrix}3\\1\end{smallmatrix} \longrightarrow \begin{smallmatrix}2\\3\\1\end{smallmatrix} \longrightarrow 0 \qquad 0 \longrightarrow \begin{smallmatrix}1\\2\\3\end{smallmatrix} \longrightarrow \begin{smallmatrix}3\\1\\2\\3\end{smallmatrix} \oplus \begin{smallmatrix}1\\2\end{smallmatrix} \longrightarrow \begin{smallmatrix}3\\1\\2\end{smallmatrix} \longrightarrow 0$$

$$0 \longrightarrow \begin{smallmatrix}3\\1\\2\\3\end{smallmatrix} \longrightarrow \begin{smallmatrix}2\\3\\1\\2\\3\end{smallmatrix} \oplus \begin{smallmatrix}3\\1\\2\end{smallmatrix} \longrightarrow \begin{smallmatrix}2\\3\\1\\2\end{smallmatrix} \longrightarrow 0$$

where the morphisms are either inclusions or projections. The reader will notice that $P_2 = I_3$ is projective–injective; thus, the last sequence provides an example of an almost split sequence with a projective–injective middle term.

Example III.3.7. A particularly interesting example of a Nakayama algebra is the algebra $A = \mathbf{k}[t]/\langle t^n \rangle$, with $n \geq 2$. It is given by the quiver

bound by $\alpha^n = 0$. In this case, a complete list of the isoclasses of indecomposable A-modules is obtained as follows. Let $I = \langle t \rangle / \langle t^n \rangle$ be the unique maximal ideal of A and thus its radical. Then, because A is indecomposable projective, the list is $A, A/I, A/I^2, \ldots, A/I^{n-1}$. The almost split sequence ending with A/I^t, where t is such that $1 \leq t \leq n-1$, is of the form

$$0 \longrightarrow \frac{A}{I^t} \longrightarrow \frac{A}{I^{t+1}} \oplus \frac{A}{I^{t-1}} \longrightarrow \frac{A}{I^t} \longrightarrow 0.$$

If $t = n-1$, then the middle term is indecomposable and equal to A/I^{n-2}, whereas, if $t = 1$, then the middle term has the projective–injective module A_A as a direct summand. In particular, for every indecomposable nonprojective A-module M, we have $\tau M \cong M$.

III.3.4 Examples of almost split sequences over bound quiver algebras

Example III.3.8. Let A be given by the quiver

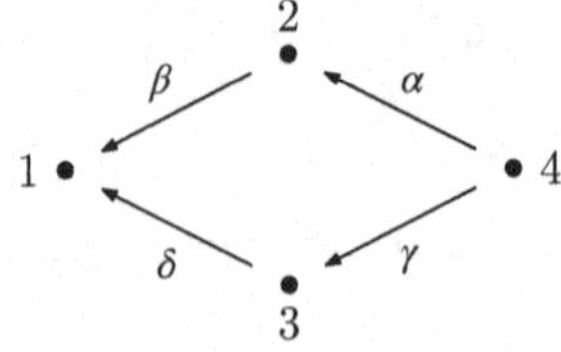

bound by $\alpha\beta = \gamma\delta$. We wish to compute the almost split sequences ending and starting in the simple module S_3, which is neither projective nor injective.

We first compute τS_3. The projective cover morphism $q\colon P_3 \longrightarrow S_3$ has as a kernel $S_1 = P_1$, so that a minimal projective presentation of S_3 is

$$0 \longrightarrow P_1 \xrightarrow{p} P_3 \xrightarrow{q} S_3 \longrightarrow 0$$

where p is the inclusion morphism. In particular, $\operatorname{pd} S_3 \leq 1$. Applying the Nakayama functor yields an exact sequence

$$0 \longrightarrow \tau S_3 \longrightarrow I_1 \xrightarrow{\nu p} I_3.$$

Now

$$I_1 = \begin{smallmatrix} & 4 & \\ 2 & & 3 \\ & 1 & \end{smallmatrix} \quad \text{and} \quad I_3 = \begin{smallmatrix} 4 \\ 3 \end{smallmatrix}.$$

Up to scalars, there is exactly one nonzero morphism $I_1 \longrightarrow I_3$: it is surjective with kernel $P_2 = \begin{smallmatrix} 2 \\ 1 \end{smallmatrix}$. Thus, $\tau S_3 \cong P_2$. The middle term M of the almost split sequence ending with S_3 is the fibered product of the morphisms

$$\begin{array}{ccc} & & S_3 \\ & & \downarrow u \\ I_1 & \xrightarrow{\nu p} & I_3 \end{array}$$

In our case, there is exactly one nonzero morphism $S_3 \longrightarrow I_3$ up to scalars, namely the inclusion of S_3 as the socle of I_3. Then, M is the kernel term in the short exact sequence

$$0 \longrightarrow M \longrightarrow I_1 \oplus S_3 \xrightarrow{(\nu p, u)} I_3 \longrightarrow 0.$$

Indeed, $(\nu p, u)$ is surjective because so is νp. We claim that $M = \begin{smallmatrix} 2 & & 3 \\ & 1 & \end{smallmatrix}$. Clearly, the composition factors of M are S_1, S_2, S_3. Moreover, $P_2 = \begin{smallmatrix} 2 \\ 1 \end{smallmatrix}$ lies in the kernel of νp and is a proper submodule of M. On the other hand, $(\nu p, u)$ maps diagonally the direct sum of the two composition factors of $I_1 \oplus S_3$ isomorphic to S_3 onto the socle of I_3. Therefore, the kernel must contain a copy of S_3 located above S_1 in its composition series. This may be viewed in the following sequence

$$0 \longrightarrow \begin{smallmatrix} 2 & & 3 \\ & 1 & \end{smallmatrix} \longrightarrow \begin{smallmatrix} & 4 & \\ 2 & & 3 \\ & 1 & \end{smallmatrix} \oplus 3 \longrightarrow \begin{smallmatrix} 4 \\ 3 \end{smallmatrix} \longrightarrow 0.$$

Thus, the required almost split sequence is:

$$0 \longrightarrow \begin{smallmatrix} 2 \\ 1 \end{smallmatrix} \longrightarrow \begin{smallmatrix} 2 & & 3 \\ & 1 & \end{smallmatrix} \longrightarrow 3 \longrightarrow 0.$$

Dually, to compute $\tau^{-1} S_3$, we start with a minimal injective copresentation

$$0 \longrightarrow S_3 \xrightarrow{i} I_3 \xrightarrow{j} I_4 \longrightarrow 0$$

where i is the inclusion and j is the projection. Applying ν^{-1} yields an exact sequence

$$P_3 \xrightarrow{\nu^{-1}j} P_4 \longrightarrow \tau^{-1}S_3 \longrightarrow 0.$$

Now

$$P_3 = \begin{smallmatrix} 3 \\ 1 \end{smallmatrix}, \; P_4 = \begin{smallmatrix} & 4 & \\ 2 & & 3 \\ & 1 & \end{smallmatrix} \text{ and } P_3 = \begin{smallmatrix} 3 \\ 1 \end{smallmatrix}.$$

Up to scalars, there is exactly one nonzero morphism $P_3 \longrightarrow P_4$: it is injective with cokernel $\begin{smallmatrix} 4 \\ 2 \end{smallmatrix} = I_2$. Thus, $\tau^{-1}S_3 \cong I_2$. The middle term N of the almost split sequence starting with S_3 is the amalgamated sum of the morphisms

$$\begin{array}{ccc} P_3 & \xrightarrow{\nu^{-1}j} & P_4 \\ \big\downarrow{\scriptstyle \nu} & & \\ S_3 & & \end{array}$$

In our case, there is, up to scalars, exactly one nonzero morphism $P_3 \longrightarrow S_3$. This is the projection of P_3 onto its top. So N is the cokernel term in the short exact sequence

$$0 \longrightarrow P_3 \xrightarrow{\binom{\nu}{\nu^{-1}j}} S_3 \oplus P_4 \longrightarrow N \longrightarrow 0$$

where we have used the injectivity of $\nu^{-1}j$. Reasoning as above, we get

$$0 \longrightarrow \begin{smallmatrix} 3 \\ 1 \end{smallmatrix} \longrightarrow 3 \oplus \begin{smallmatrix} & 4 & \\ 2 & & 3 \\ & 1 & \end{smallmatrix} \longrightarrow \begin{smallmatrix} & 4 & \\ 2 & & 3 \end{smallmatrix} \longrightarrow 0$$

So $N = \begin{smallmatrix} & 4 & \\ 2 & & 3 \end{smallmatrix}$ and the required almost split sequence is

$$0 \longrightarrow 3 \longrightarrow \begin{smallmatrix} & 4 & \\ 2 & & 3 \end{smallmatrix} \longrightarrow \begin{smallmatrix} 4 \\ 2 \end{smallmatrix} \longrightarrow 0.$$

In the previous example, we used essentially the fact that, between two of the modules under consideration, the morphism space is at most one dimensional. Obviously, this is not the case in general. To perform the calculation, we need, given an indecomposable nonprojective module M, and a minimal projective presentation

$$P_1 \xrightarrow{p} P_0 \longrightarrow M \longrightarrow 0,$$

to compute the image of the morphism $p\colon P_1 \longrightarrow P_0$ under the action of the Nakayama functor ν, because $\tau M = \mathrm{Ker}(\nu p)$. Dually, if M is indecomposable noninjective and has a minimal injective copresentation

$$0 \longrightarrow M \longrightarrow I_1 \xrightarrow{j} I_1,$$

we need to compute $\nu^{-1} j$, because $\tau^{-1} M = \mathrm{Coker}(\nu^{-1} j)$.

We show how to solve the first problem, leaving the dual case to the reader, see Exercise III.3.5. Given projective modules $P_1 = e_1 A$, $P_2 = e_2 A$ (with e_1, e_2 idempotents), we have

$$\mathrm{Hom}_A(P_2, P_1) \cong \mathrm{Hom}_A(e_2 A, e_1 A) \cong e_1 A e_2$$

and the isomorphism is given by left multiplication by an element $w \in e_1 A e_2$. That is, every morphism $e_2 A \longrightarrow e_1 A$ is of the form $e_2 a \mapsto wa \in e_1 A$, for some $w \in e_1 A e_2$. In particular, the image of e_2 is exactly w.

We also remind the reader that, given an idempotent $e \in A$, we have canonical isomorphisms $\mu : eA \otimes_A \mathrm{D}A \longrightarrow e\mathrm{D}A$ given by $e \otimes f \longmapsto ef$ and $\varphi : e\mathrm{D}A \longrightarrow \mathrm{D}(Ae)$ given by $ef \longmapsto f(\cdot e)$ where the latter is the linear form $ae \longmapsto f(ae)$.

Lemma III.3.9. *Let $p\colon e_2 A \longrightarrow e_1 A$ be given by left multiplication by $w \in e_1 A e_2$. Then, $\nu p\colon \mathrm{D}(Ae_2) \longrightarrow \mathrm{D}(Ae_1)$ is the morphism $f \mapsto (ae_1 \mapsto f(aw))$.*

Proof. Let $\eta\colon \mathrm{D}(Ae_2) \longrightarrow \mathrm{D}(Ae_1)$ be defined by $f \mapsto (ae_1 \mapsto f(aw))$. Because $\nu = - \otimes_A \mathrm{D}A$, it suffices to prove that the following square commutes.

$$\begin{array}{ccc}
e_2 A \otimes_A \mathrm{D}A & \xrightarrow{p \otimes \mathrm{D}A} & e_1 A \otimes_A \mathrm{D}A \\
\mu_2 \downarrow \cong & & \cong \downarrow \mu_1 \\
e_2 \mathrm{D}A & & e_1 \mathrm{D}A \\
\varphi_2 \downarrow \cong & & \cong \downarrow \varphi_1 \\
\mathrm{D}(Ae_2) & \xrightarrow{\eta} & \mathrm{D}(Ae_1)
\end{array}$$

where the μ_i, φ_i are the canonical isomorphisms. For this, let $e_2 \otimes f \in e_2 A \otimes \mathrm{D}A$, then

$$\eta\varphi_2\mu_2(e_2 \otimes f) = \eta\varphi_2(e_2 f) = \eta f(\cdot e_2)$$

is the linear form mapping $ae_1 \in Ae_1$ to $f(awe_2) = f(aw)$. On the other hand,

$$\varphi_1\mu_1(p \otimes \mathrm{D}A)(e_2 \otimes f) = \varphi_1\mu_1(p(e_2) \otimes f) = \varphi_1\mu_1(w \otimes f) = \varphi_1(wf) = (wf)(\cdot e_1)$$

is the linear form that maps $ae_1 \in Ae_1$ to $(wf)(ae_1) = f(aw)$. The proof is complete. $\square$

Example III.3.10. Let A be given by the quiver

$$1 \bullet \underset{\beta}{\overset{\alpha}{\rightleftarrows}} \bullet\, 2$$

bound by $\beta\alpha\beta\alpha = 0$. We want to compute τS_1. For this purpose, we first need a minimal projective presentation. The projective cover of S_1 is clearly

$$P_1 = \begin{smallmatrix} 1 \\ 2 \\ 1 \\ 2 \\ 1 \end{smallmatrix}$$

with kernel equal to the indecomposable projective P_2. Thus, the required minimal projective presentation is

$$0 \longrightarrow P_2 \xrightarrow{\ p\ } P_1 \longrightarrow S_1 \longrightarrow 0,$$

where p is the inclusion. Identifying paths to their classes modulo the binding ideal, the basis of A as a $\mathbf{k}$-vector space is $\{e_1, e_2, \alpha, \beta, \alpha\beta, \beta\alpha, \alpha\beta\alpha, \beta\alpha\beta, \alpha\beta\alpha\beta\}$, that of P_1 is $\{e_1, \alpha, \alpha\beta, \alpha\beta\alpha, \alpha\beta\alpha\beta\}$ and that of P_2 is $\{e_2, \beta, \beta\alpha, \beta\alpha\beta\}$. We see easily that the morphism p is given by left multiplication by α. Applying the Nakayama functor ν yields an exact sequence

$$0 \longrightarrow \tau S_1 \longrightarrow I_2 \xrightarrow{\ \nu p\ } I_1.$$

Denoting by $\{e_1^\vee, e_2^\vee, \alpha^\vee, \beta^\vee, (\alpha\beta)^\vee, (\beta\alpha)^\vee, (\alpha\beta\alpha)^\vee, (\beta\alpha\beta)^\vee, (\alpha\beta\alpha\beta)^\vee\}$ the dual basis to the above basis of A, we see that the basis of I_2 is $\{e_2^\vee, \alpha^\vee, (\beta\alpha)^\vee, (\alpha\beta\alpha)^\vee\}$, see Lemma I.2.17. Because of the previous lemma, νp maps $f \in \mathrm{D}(Ae_2)$ to the linear form $ae_1 \mapsto f(a\alpha)$. Thus,

$$\begin{aligned}
(\nu p)(e_2^\vee)(ae_1) &= e_2^\vee(a\alpha) = 0 && \text{for every } a \in A \\
(\nu p)(\alpha^\vee)(ae_1) &= \alpha^\vee(a\alpha) \neq 0 && \text{because } \alpha^\vee(\alpha) = 1 \\
(\nu p)(\beta\alpha)^\vee)(ae_1) &= (\beta\alpha)^\vee(a\alpha) \neq 0 && \text{because } (\beta\alpha)^\vee(\beta\alpha) = 1
\end{aligned}$$

and similarly $(\nu p)(\alpha\beta\alpha)^\vee(ae_1) \neq 0$. This proves that $\tau S_1 = \mathrm{Ker}(\nu p)$ is the one-dimensional vector space spanned by $e_2^\vee$; thus, $\tau S_1 \cong S_2$.

Example III.3.11. Let A be the Kronecker algebra with quiver

$$1 \bullet \underset{\beta}{\overset{\alpha}{\leftleftarrows}} \bullet\, 2$$

and consider the indecomposable nonprojective module

$$M = \begin{smallmatrix} 2 \\ \downarrow \alpha \\ 1 \end{smallmatrix}$$

We wish to construct the almost split sequence ending with M.

A minimal projective presentation of M is

$$0 \longrightarrow P_1 \xrightarrow{p} P_2 \longrightarrow M \longrightarrow 0$$

where the morphism p is given by left multiplication by β. Applying the Nakayama functor ν yields an exact sequence

$$0 \longrightarrow \tau S_1 \longrightarrow I_1 \xrightarrow{\nu p} I_2.$$

Denote by $\{e_1^\vee, e_2^\vee, \alpha^\vee, \beta^\vee\}$ the dual basis to the $\mathbf{k}$-basis of A given by $\{e_1, e_2, \alpha, \beta\}$, we have $I_2 = \mathrm{D}(Ae_2) = S_2$, whereas $I_1 = \mathrm{D}(Ae_1)$ has a basis $\{e_1^\vee, \alpha^\vee, \beta^\vee\}$ and is in fact the module

$$\begin{smallmatrix} 2 & & 2 \\ & \alpha^\vee \searrow \ 1 \ \swarrow \beta^\vee & \end{smallmatrix}$$

The morphism $\nu p \colon I_1 \longrightarrow I_2$ maps $f \in \mathrm{D}(Ae_2)$ to the linear form $ae_1 \mapsto f(a\beta)$. Thus, we see that

$$\begin{aligned} (\nu p)(e_2^\vee)(ae_1) &= e_2^\vee(a\beta) = 0 && \text{for every } a \in A \\ (\nu p)(\alpha^\vee)(ae_1) &= \alpha^\vee(a\beta) = 0 && \text{for every } a \in A \\ \text{whereas } (\nu p)(\beta^\vee)(ae_1) &= \beta^\vee(a\beta) \neq 0 && \text{because } \beta^\vee(\beta) = 1. \end{aligned}$$

A basis of τM is $\{e_2^\vee, \alpha^\vee\}$ so that

$$\tau M = \begin{smallmatrix} 2 \\ \downarrow \alpha \\ 1 \end{smallmatrix} .$$

In particular, $\tau M \cong M$.

Next, the middle term H of the almost split sequence

$$0 \longrightarrow \tau M \longrightarrow H \longrightarrow M \longrightarrow 0$$

is the fibered product of the morphisms

$$\begin{array}{ccc} & & M \\ & & \downarrow{\scriptstyle u} \\ I_1 & \xrightarrow{\mathrm{D}p_1'} & I_2 \end{array}$$

where u is as in Subsection III.3.1. Again, up to scalars, there is a unique morphism $u: M \longrightarrow I_2 = S_2$: it is the surjection with kernel S_1. Hence, H is the kernel term in the short exact sequence

$$0 \longrightarrow H \longrightarrow I_1 \oplus M \xrightarrow{(\mathrm{D}p_1', u)} S_2 \longrightarrow 0.$$

We get that H is the module

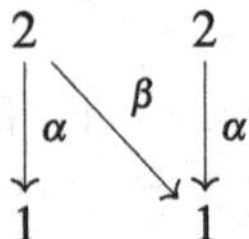

and, in particular, is indecomposable. This module H will appear again below in Subsection IV.4.1.

Exercises for Section III.3

Exercise III.3.1. State and prove the dual of Lemma III.3.9.

Exercise III.3.2. Let A be given by the quiver

$$1 \underset{\beta}{\overset{\alpha}{\rightleftarrows}} 2$$

bound by $\beta\alpha\beta = 0$.

(a) Applying the techniques of this subsection, for each indecomposable nonprojective A-module M, compute τM by exhibiting in each case a $\mathbf{k}$-basis of this module.

(b) Same question for $\tau^{-1}M$, where M is indecomposable noninjective.

Exercise III.3.3. Same exercise if the quiver is bound by $\beta\alpha\beta\alpha\beta = 0$.

Exercise III.3.4. Let A be a Nakayama algebra and U an indecomposable A-module.

(a) Prove that the **support** of U, namely the full subquiver

$$\operatorname{supp} U = \{x \in (Q_A)_0 \mid Ue_x \neq 0\}$$

is a nonzero path w in Q_A.

(b) Let α, β be the unique arrows (if they exist) such that $t(\alpha) = s(w)$ and $s(\beta) = t(w)$. Prove that the quotient of A by the ideal I generated by the classes of the paths αw and $w\beta$ is a Nakayama algebra.

(c) Prove that U is a projective–injective A/I-module.

Exercise III.3.5. Let A be an algebra, and e_1, e_2 idempotents of A. Given a morphism $j : \mathrm{D}(e_1A) \longrightarrow \mathrm{D}(e_2A)$, compute the morphism $\nu^{-1}j : e_1A \longrightarrow e_2A$.

Exercise III.3.6. For each of the following Nakayama algebras given by their bound quivers, give a complete list of all the almost split sequences.

(a) $1 \xleftarrow{\varepsilon} 2 \xleftarrow{\delta} 3 \xleftarrow{\gamma} 4 \xleftarrow{\beta} 5 \xleftarrow{\alpha} 6 \qquad \alpha\beta = 0, \beta\gamma\delta\epsilon = 0$

(b) $$\begin{array}{ccc} 1 & \xrightarrow{\gamma} & 3 \\ & {\scriptstyle\beta}\nwarrow \quad \swarrow{\scriptstyle\alpha} & \\ & 3 & \end{array} \qquad \alpha\beta = 0$$

(c) $1 \underset{\beta}{\overset{\alpha}{\rightleftarrows}} 2 \qquad \alpha\beta\alpha\beta = 0$

(d) $$\begin{array}{ccc} 1 & \xrightarrow{\alpha} & 2 \\ {\scriptstyle\delta}\uparrow & & \downarrow{\scriptstyle\beta} \\ 4 & \xleftarrow[\gamma]{} & 3 \end{array} \qquad \alpha\beta = 0, \beta\gamma = 0$$

Exercise III.3.7. An algebra A is called selfinjective if the module A_A is injective. Prove that each of the following bound quiver algebras A is selfinjective and compute the almost split sequences with a projective–injective middle term.

(a) $$\begin{array}{ccc} & 2 & \\ {\scriptstyle\beta}\swarrow & & \nwarrow{\scriptstyle\alpha} \\ 1 & \xrightarrow{\varepsilon} & 4 \\ {\scriptstyle\delta}\nwarrow & & \swarrow{\scriptstyle\gamma} \\ & 3 & \end{array} \qquad \alpha\beta = \gamma\delta, \delta\varepsilon\alpha = 0, \beta\varepsilon\gamma = 0$$

(b) $1 \overset{\varepsilon}{\curvearrowright} 2$, $\;1 \underset{\gamma}{\overset{\beta}{\leftleftarrows}} 2 \xleftarrow{\alpha} 3$, $\;1 \underset{\delta}{\curvearrowright} 3$

$$\alpha\beta = 0, \varepsilon\gamma = 0, \varepsilon\beta = \delta\alpha\gamma, \beta\delta = 0,$$
$$\gamma\varepsilon = 0, \beta\varepsilon = \gamma\delta\alpha, \alpha\gamma\,\varepsilon = 0, \varepsilon\gamma\delta = 0$$

(c) $1 \underset{\beta}{\overset{\alpha}{\leftrightarrows}} 2 \underset{\delta}{\overset{\gamma}{\rightleftarrows}} 3 \qquad \alpha\beta = \gamma\delta, \beta\gamma = 0, \delta\alpha = 0$

(d) 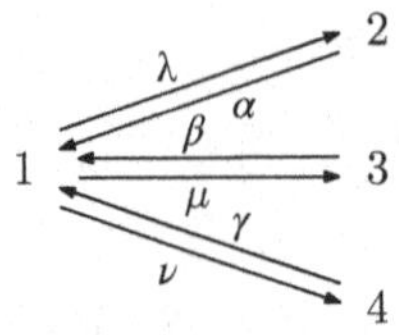 $\lambda\alpha = \mu\beta = \nu\gamma, \alpha\mu = 0, \alpha\nu = 0, \beta\lambda = 0,$
$\beta\nu = 0, \gamma\lambda = 0, \gamma\mu = 0$

III.4 Almost split sequences over quotient algebras

III.4.1 The change of rings functors

Let A be a finite dimensional algebra, E a two-sided ideal in A and $B = A/E$. Our aim in this section is to derive relations between the Auslander–Reiten translations τ_A in $\operatorname{mod} A$ and τ_B in $\operatorname{mod} B$. In this situation, we have the four classical change of rings functors, namely the functors $- \otimes_A B_B$ and $\operatorname{Hom}_A({}_BB_A, -)$ from $\operatorname{mod} A$ to $\operatorname{mod} B$, and the corresponding functors $- \otimes_B A_A$ and $\operatorname{Hom}_B({}_AA_B, -)$ from $\operatorname{mod} B$ to $\operatorname{mod} A$. There exist several adjunction relations between these functors, but moreover we have the following lemma.

Lemma III.4.1. *There exist isomorphisms of functors*

(a) $- \otimes_B A_A \otimes_A B_B \cong 1_{\operatorname{mod} B}$.
(b) $\operatorname{Hom}_A({}_BB_A, \operatorname{Hom}_B({}_AA_B, -)) \cong 1_{\operatorname{mod} B}$.

Proof. The proof of (a) is clear.

For (b), we observe that, for every B-module M, we have functorial isomorphisms

$$\operatorname{Hom}_A({}_BB_A, \operatorname{Hom}_B({}_AA_B, M)) \cong \operatorname{Hom}_B({}_BB_A \otimes_A A_B, M) \cong \operatorname{Hom}_B({}_BB_B, M) \cong M_B.$$

□

It is important to observe that the reverse compositions of these functors are not isomorphic to the identity in $\operatorname{mod} A$. Another important observation is that, because $B = A/E$, there exist a surjective morphism of algebras $\varphi : A \longrightarrow B$, so that every B-module M can be viewed as an A-module when one defines multiplication as follows

$$xa = x\varphi(a)$$

for $x \in M$ and $a \in A$. Then, $\operatorname{mod} B$ is fully embedded in $\operatorname{mod} A$ and can actually be identified with the full subcategory of $\operatorname{mod} A$ consisting of all the A-modules M that are annihilated by E, that is, such that $ME = 0$, see Exercise III.4.2.

Corresponding to an arbitrary A-module M, there are two modules that are clearly annihilated by E, namely, the quotient module $M^* = M/ME$ and the submodule $M_* = \{x \in M : xE = 0\}$. Thus, M^* and M_* are B-modules. We prove that these constructions are other versions of using the change of rings functors.

Lemma III.4.2. *Let M be an A-module. We have functorial isomorphisms:*

(a) $M \otimes_A B \cong M^*$.
(b) $\mathrm{Hom}_A(B, M) \cong M_*$.

Proof.

(a) Applying $M \otimes_A -$ to the short exact sequence of left A-modules $0 \longrightarrow E \xrightarrow{i} A \xrightarrow{p} B \longrightarrow 0$ yields a commutative diagram with exact rows

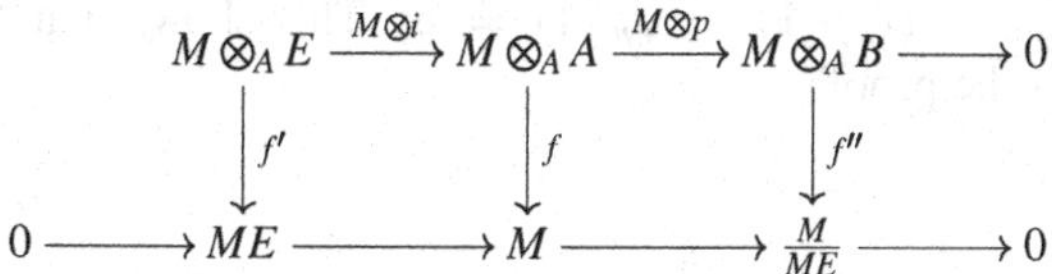

$$\begin{array}{ccccccccc} & & M\otimes_A E & \xrightarrow{M\otimes i} & M\otimes_A A & \xrightarrow{M\otimes p} & M\otimes_A B & \longrightarrow & 0 \\ & & \downarrow f' & & \downarrow f & & \downarrow f'' & & \\ 0 & \longrightarrow & ME & \longrightarrow & M & \longrightarrow & \frac{M}{ME} & \longrightarrow & 0 \end{array}$$

where i and p are respectively the inclusion and the projection, $f : M \otimes_A A \longrightarrow M$ is the well-known functorial isomorphism given by the multiplication map $m \otimes a \mapsto ma$ (for $m \in M, a \in A$), $f' : M \otimes_A E \longrightarrow ME$ is the induced morphism defined also by $m \otimes a \mapsto ma$ (for $m \in M, a \in E$) and f'' is obtained by passing to cokernels. In particular, the surjectivity of p and the fact that f is an isomorphism imply that f'' is surjective. On the other hand, ME is, by definition, generated by all products ma with $m \in M$ and $a \in E$. Therefore, f' is also surjective. But then the snake lemma gives that f'' is injective and is thus an isomorphism.

(b) Let $p : A_A \longrightarrow B_A$ denote, as in (a), the canonical projection. There is a well-known functorial isomorphism $f : \mathrm{Hom}_A(A, M) \longrightarrow M$ given by $f : u \longmapsto u(1)$ for $u \in \mathrm{Hom}_A(A, M)$. We claim that the image of the composition of f with $\mathrm{Hom}_A(p, M) : \mathrm{Hom}_A(B, M) \longrightarrow \mathrm{Hom}_A(A, M)$ lies in M_*. Indeed, if $\varphi \in \mathrm{Hom}_A(B, M)$, then

$$f\,\mathrm{Hom}_A(p, M)(\varphi) = f(\varphi p) = \varphi p(1).$$

For every $a \in E$ we have

$$\varphi p(1)a = \varphi p(a) = 0$$

because $E = \mathrm{Ker}\, p$. This shows that $\varphi p(1) \in M_*$ as required, implying the existence of a morphism $f' : \mathrm{Hom}_A(B, M) \longrightarrow M_*$ making the following square commutative

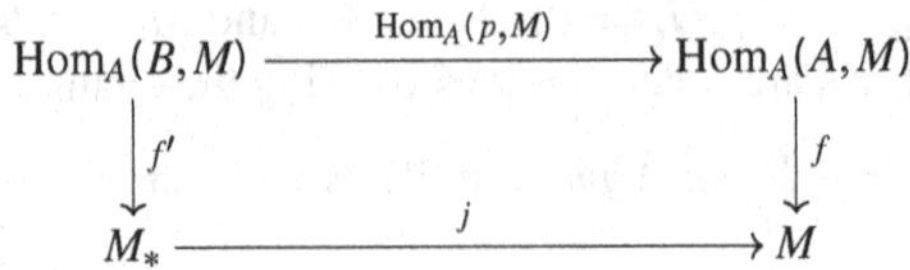

where j is the inclusion. In particular, the injectivity of f and that of $\mathrm{Hom}_A(p, M)$ imply that of f'. There remains to show that f' is also surjective. Let $m \in M_*$, then $m \in M$ and $mE = 0$. Because $m \in M$, there exists $\varphi_m \in \mathrm{Hom}_A(A, M)$ such that $\varphi_m(1) = m$. But then $\varphi_m(E) = \varphi_m(1)E = mE = 0$. Because $E = \mathrm{Ker}\, p$, there exists $\varphi'_m : B \longrightarrow M$ such that $\varphi'_m p = \varphi_m$. Hence, $f'(\varphi'_m) = \varphi'_m p(1) = \varphi_m(1) = m$. This shows that f' is surjective and completes the proof.

□

III.4.2 The embedding of mod B inside mod A

Using the assumptions and notation of the previous subsection, we prove that every B-module, which, when considered as an A-module annihilated by E, is A-projective, must also be B-projective (and dually).

Lemma III.4.3. *Let M be a B-module.*

(a) *If M is projective in* mod A*, then it is also projective in* mod B.
(b) *If M is injective in* mod A*, then it is also injective in* mod B.

Proof. We only prove (a), because the proof of (b) is dual.

If $f : L \longrightarrow L'$ is an epimorphism in mod B, then it is also an epimorphism in mod A. Because M is projective, $\mathrm{Hom}_A(M, f) : \mathrm{Hom}_A(M, L) \longrightarrow \mathrm{Hom}_A(M, L')$ is surjective. Because mod B is a full subcategory of mod A, we have $\mathrm{Hom}_A(M, N) = \mathrm{Hom}_B(M, N)$ for every B-module N. Hence, $\mathrm{Hom}_B(M, f) = \mathrm{Hom}_A(M, f)$ is surjective. □

We now see how, starting from a right minimal almost split morphism in mod A ending in a B-module, we can construct the corresponding right minimal almost split morphism in mod B ending in the same module.

Lemma III.4.4.

(a) *Let N be an indecomposable B-module and $g : M \longrightarrow N$ right minimal almost split in* mod A*. Then its composition g_* with the inclusion $M_* \longrightarrow M$ is right almost split in* mod B*. In addition, if N is nonprojective, then g_* is surjective.*
(b) *Let L be an indecomposable B-module and $f : L \longrightarrow M$ left minimal almost split in* mod A*. Then, its composition f^* with the projection $M \longrightarrow M^*$ is left almost split in* mod B*. In addition, if L is noninjective, then f^* is injective.*

Proof. We only prove (a), because the proof of (b) is dual.

We have a commutative diagram in mod A

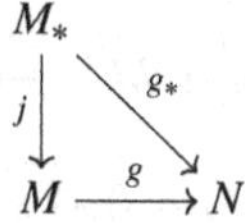

in which M_* and N are B-modules.

First, we prove that g_* is a radical morphism in mod B. If it is not, then it is a retraction and there exists $g'\colon N \longrightarrow M_*$ such that $g_*g' = 1_N$. But then $gjg' = 1_N$; thus, g is itself a retraction, a contradiction.

Let $v\colon V \longrightarrow N$ be a radical morphism in mod B. Then, v is also radical in mod A (for, otherwise, v would be a retraction in mod A and hence in mod B) and there exists $v'\colon V \longrightarrow M$ such that $v = gv'$. But V is a B-module; hence, for every $x \in V$ we have $v'(xE) = 0$. Therefore, the image of v' lies in M_*, providing a morphism $v''\colon V \longrightarrow M_*$ such that $v = gjv'' = g_*v''$. This completes the proof that g_* is right almost split in mod B.

Because of Corollary II.2.22, the morphism g_* is isomorphic to a morphism of the form $(g_0, 0)\colon M_0 \oplus L \longrightarrow N$ with $g_0\colon M_0 \longrightarrow N$ right minimal almost split in mod B. Clearly, if N is nonprojective in mod B, then g_0 is surjective. Hence, so is g_*. □

In the situation of (a) above, if N is projective, then g_0 is the (proper) inclusion of a summand of the radical of N into N. In this case g_* is not surjective.

Let N be an indecomposable nonprojective B-module. Because of Lemma III.4.3 above, it is also indecomposable nonprojective in mod A. Therefore, there exist two almost split sequences ending with N, one in mod A and the other in mod B. We explain the relation between them.

Proposition III.4.5. *Let* $0 \longrightarrow L \xrightarrow{f} M \xrightarrow{g} N \longrightarrow 0$ *be an almost split sequence in* mod A.

(a) *Assume that* N *is an indecomposable nonprojective* B*-module, then there exists a short exact sequence in* mod B

$$0 \longrightarrow L_* \xrightarrow{f_*} M_* \xrightarrow{g_*} N \longrightarrow 0$$

that is isomorphic to the direct sum of an almost split sequence

$$0 \longrightarrow L_0 \xrightarrow{f_0} M_0 \xrightarrow{g_0} N \longrightarrow 0$$

in mod B *with a sequence of the form* $0 \longrightarrow X \longrightarrow X \longrightarrow 0 \longrightarrow 0$.

(b) *Assume that* L *is an indecomposable noninjective* B*-module. Then, there exists a short exact sequence in* mod B

$$0 \longrightarrow L \xrightarrow{f^*} M^* \xrightarrow{g^*} N^* \longrightarrow 0$$

that is isomorphic to the direct sum of an almost split sequence

$$0 \longrightarrow L \xrightarrow{f^0} M^0 \xrightarrow{g^0} N^0 \longrightarrow 0$$

in mod B *with a sequence of the form* $0 \longrightarrow 0 \longrightarrow Y \longrightarrow Y \longrightarrow 0$.

Proof. We only prove (a), because the proof of (b) is dual.

First, $N_* = N$, because N itself is a B-module. Applying the functor $\mathrm{Hom}_A(B, -)$ to the given almost split sequence in mod A yields a left exact sequence

$$0 \longrightarrow L_* \xrightarrow{f_*} M_* \xrightarrow{g_*} N_* = N,$$

where $f_* = \mathrm{Hom}_A(B, f)$ and $g_* = \mathrm{Hom}_A(B, g)$. Because N is nonprojective in mod B, Lemma III.4.4 yields that g_* is surjective and right almost split. In particular, it is not a retraction; thus, we have a nonsplit short exact sequence

$$0 \longrightarrow L_* \xrightarrow{f_*} M_* \xrightarrow{g_*} N \longrightarrow 0.$$

In addition, as seen in the proof of Lemma III.4.4, g_* is isomorphic to a morphism of the form $(g_0, 0): M_0 \oplus X \longrightarrow N$ with g_0 right minimal almost split in mod B. This shows that the latter short exact sequence is actually isomorphic to a sequence of the form

$$0 \longrightarrow L_0 \oplus X \xrightarrow{\begin{pmatrix} f_0 & 0 \\ 0 & 1_X \end{pmatrix}} M_0 \oplus X \xrightarrow{(g_0,0)} N \longrightarrow 0.$$

In particular, it is isomorphic to the direct sum of an exact sequence of the form $0 \longrightarrow X \longrightarrow X \longrightarrow 0 \longrightarrow 0$ and another of the form

$$0 \longrightarrow L_0 \xrightarrow{f_0} M_0 \xrightarrow{g_0} N \longrightarrow 0$$

with g_0 right minimal almost split in mod B. Then, because of Theorem II.2.31, the latter sequence is almost split in mod B. □

It is possible to prove that the module X in (a) above is actually a projective B-module, whereas the module Y in (b) is an injective B-module. This is Exercise III.4.1 below. Recall that we denote by τ_A and τ_B the Auslander–Reiten translations in mod A and mod B respectively.

Corollary III.4.6.

(a) *If N is an indecomposable nonprojective B-module, then $\tau_B N$ is isomorphic to a submodule of $\tau_A N$.*

(b) *If* L *is an indecomposable noninjective* A*-module, then* $\tau_B^{-1}L$ *is isomorphic to a quotient of* $\tau_A^{-1}L$.

Proof. We only prove (a), because the proof of (b) is dual.

Using the notations in the proof of Proposition III.4.5, the statement follows from the inclusion morphisms

$$\tau_B N \cong L_0 \hookrightarrow L_0 \oplus X = L_* \hookrightarrow L = \tau_A N.$$

□

Corollary III.4.7. *Let* M, N *be indecomposable* B*-modules. If there exists an irreducible morphism* $f: M \longrightarrow N$ *in* mod A, *then this morphism remains irreducible in* mod B.

Proof. Because of Theorem II.2.24, there exists a right minimal almost split morphism in mod A of the form $(f, g): M \oplus X \longrightarrow N$. Now we have $(M \oplus X)_* \cong \mathrm{Hom}_A(B, M \oplus X) \cong \mathrm{Hom}_A(B, M) \oplus \mathrm{Hom}_A(B, X) \cong M_* \oplus X_* \cong M \oplus X_*$ because M is a B-module. For the same reason, $f_* = f: M \longrightarrow N$ and it is a nonzero morphism. Therefore, there exists a direct sum decomposition $X_* \cong X_0 \oplus Y$ such that $(f, g)_*: M \oplus X_* \longrightarrow N$ is of the form $(f, g_0, 0): M \oplus X_0 \oplus Y \longrightarrow N$ with (f, g_0) right minimal almost split in mod B. This implies the statement. □

Example III.4.8. Let A be given by the quiver

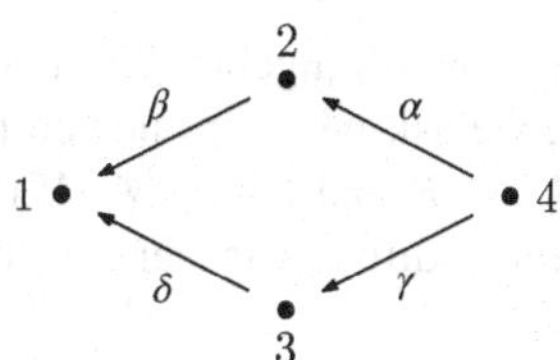

bound by $\alpha\beta = \gamma\delta$. Let $E = Ae_3A$ be the two-sided ideal of A generated by the idempotent e_3. As a vector space, E has the basis $\{e_3, \gamma, \delta, \gamma\delta\}$ whereas A has the basis $\{e_1, e_2, e_3, e_4, \alpha, \beta, \gamma, \delta, \alpha\beta = \gamma\delta\}$. Therefore, $B = A/E$ is given by the quiver

$$\overset{1}{\bullet} \longleftarrow \overset{2}{\bullet} \longleftarrow \overset{4}{\bullet}$$

bound by $\alpha\beta = 0$. Indeed, this relation comes from the fact that in A we have $\alpha\beta = \gamma\delta$ and $\gamma\delta \in E$.

We have previously computed in Example III.2.11 the almost split sequence ending in S_2 in mod A

$$0 \longrightarrow \begin{smallmatrix}3\\1\end{smallmatrix} \longrightarrow \begin{smallmatrix}2\,3\\1\end{smallmatrix} \longrightarrow 2 \longrightarrow 0.$$

Because S_2 is also a simple B-module, the almost split sequence in mod B ending with S_2 is a direct summand of the short exact sequence

$$0 \longrightarrow \left(\begin{smallmatrix}3\\1\end{smallmatrix}\right)_* \longrightarrow \left(\begin{smallmatrix}2\,3\\1\end{smallmatrix}\right)_* \longrightarrow 2 \longrightarrow 0.$$

Because the functor $(-)_*$ is actually $\mathrm{Hom}_A(B, -)$, a direct calculation gives

$$\left(\begin{smallmatrix}3\\1\end{smallmatrix}\right)_* = 1 \quad \text{and} \quad \left(\begin{smallmatrix}2\,3\\1\end{smallmatrix}\right)_* = \begin{smallmatrix}2\\1\end{smallmatrix}.$$

Thus, the corresponding almost split sequence in mod B is

$$0 \longrightarrow 1 \longrightarrow \begin{smallmatrix}2\\1\end{smallmatrix} \longrightarrow 2 \longrightarrow 0.$$

In this example, the module X of Proposition III.4.5(a) is equal to zero.

III.4.3 Split-by-nilpotent extensions

Again, let A be a finite dimensional **k**-algebra, E a two-sided ideal of A and $B = A/E$. Given a B-module M, we ask whether one can relate the Auslander–Reiten translates of the modules $M \otimes_A B$ and $\mathrm{Hom}_B(A, M)$ in mod A with those of M itself in mod B. This problem is difficult in general, but there is one case where computation is actually possible.

Let B be a finite dimensional algebra and E a B–B-bimodule, finite dimensional over **k**. We wish to consider the case where elements of E may be multiplied together. We say that E is equipped with an **associative product** if there exists a morphism of B–B-bimodules $E \otimes_B E \longrightarrow E$, denoted as $x \otimes y \longmapsto xy$, for $x, y \in E$, such that $x(yz) = (xy)z$ for all $x, y, z \in E$.

Definition III.4.9. Let B be an algebra and E a B–B-bimodule equipped with an associative product. The **k**-vector space

$$A = B \oplus E = \{(b, x) \mid b \in B, x \in E\}$$

together with the multiplication defined by

$$(b, x)(b', x') = (bb', bx' + xb' + xx')$$

for $(b, x), (b', x') \in A$ is an algebra called a **split extension** of B by E. In addition, if E is nilpotent as an ideal in A, then A is called a **split-by-nilpotent extension**.

In this case, there is a short exact sequence of **k**-vector spaces

$$0 \longrightarrow E \xrightarrow{i} A \underset{q}{\overset{p}{\rightleftarrows}} B \longrightarrow 0$$

where the projection $p : (b, x) \longmapsto b$ is an algebra morphism having as a section the inclusion $q : b \longmapsto (b, 0)$, which is also an algebra morphism. Thus, the above sequence is split as a short exact sequence of B–B-bimodules, but not as a short exact sequence of (left or right) A-modules.

The assumption that E is nilpotent amounts to saying that $E \subseteq \operatorname{rad} A$. As an easy consequence, $\operatorname{rad} B = (\operatorname{rad} A)/E$: indeed, $(\operatorname{rad} A)/E$ is nilpotent as an ideal in $B = A/E$; in addition,

$$\frac{A/E}{(\operatorname{rad} A)/E} \cong \frac{A}{\operatorname{rad} A}$$

is semisimple. This establishes our claim. Incidentally, the isomorphism $A/\operatorname{rad} A \cong B/\operatorname{rad} B$ implies that the projection $p : A \longrightarrow B$ induces a bijection between the idempotents of A and those of B. In the sequel, we always assume that E is nilpotent.

Example III.4.10. Let A be an elementary algebra. As seen in Subsection I.2.2, $A = A/\operatorname{rad} A \oplus \operatorname{rad} A$. Therefore, A is a split extension of the semisimple algebra $A/\operatorname{rad} A$ by the nilpotent bimodule $\operatorname{rad} A$.

Example III.4.11. Assume $B = \mathbf{k}$ and $E = \mathbf{k}^2$ equipped with the (obviously associative) product

$$(b, c)(b', c') = (0, bb')$$

for $b, c, b', c' \in \mathbf{k}$. It is easy to prove that the split extension of B by E is isomorphic to the truncated polynomial algebra $A = \mathbf{k}[t]/\langle t^3 \rangle$, that is, the algebra given by the quiver

bound by $\alpha^3 = 0$.

Example III.4.12. Let A be given by the quiver

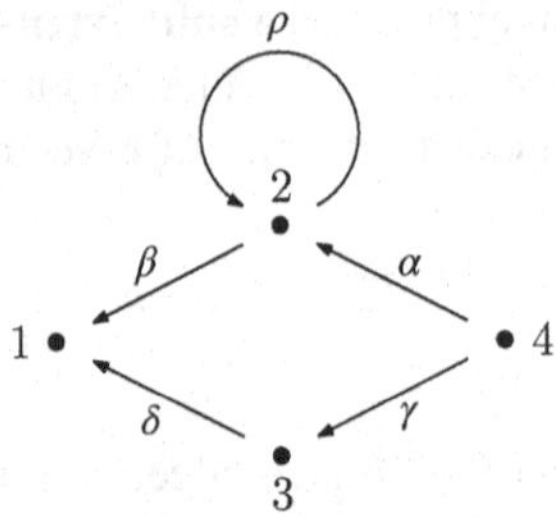

bound by $\alpha\beta = 0,\ \delta\gamma = 0,\ \rho^3 = 0$. Let $E = \langle \alpha, \rho \rangle$. Then, A is a split extension of the algebra B given by the quiver

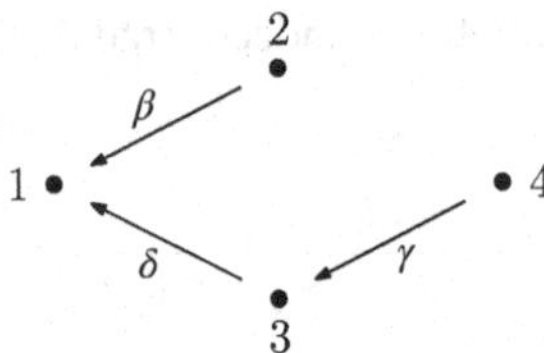

bound by $\delta\gamma = 0$.

As the previous examples show, if A is a split-by-nilpotent extension of B, passing from A to B may be thought of as "dropping arrows" (according to certain rules, which are not within the scope of these notes). On the other hand, the points of the quiver (that is, the idempotents) remain the same, as mentioned above.

Let A be a split extension of B by the nilpotent bimodule E and M an A-module. As seen in Subsection III.4.1, there exist a canonical epimorphism $p_M : M \longrightarrow M^*$ and a canonical monomorphism $j_M : M_* \longrightarrow M$. We also recall that $M^* \cong M \otimes_A B$ whereas $M_* \cong \mathrm{Hom}_A(B, M)$. We prove that p_M and j_M are respectively superfluous and essential, in the sense of Example II.2.17.

Lemma III.4.13. *Let M be an A-module. Then:*

(a) *The canonical epimorphism $p_M : M \longrightarrow M \otimes_A B$ is superfluous.*
(b) *The canonical monomorphism $j_M : \mathrm{Hom}_A(B, M) \longrightarrow M$ is essential.*

Proof.

(a) It follows from Nakayama's lemma that the canonical epimorphism $f : M \longrightarrow M/M \operatorname{rad} A$ is superfluous. Because $E \subseteq \operatorname{rad} A$, there exists a canonical epimorphism $g : M/ME \longrightarrow M/M \operatorname{rad} A$ such that $f = gp_M$ (here, we use that, because of Lemma III.4.1, we have $M/ME \cong M \otimes_A B$). Let $h : L \longrightarrow M$ be such that $p_M h$ is surjective, then so is $gp_M h = fh$. Now, f is superfluous. Hence, h is surjective.
(b) To show that j_M is essential, it suffices to prove that $M_* = \operatorname{Im} j_M$ intersects every nonzero submodule of M, see Exercise II.2.3. Let $L \subseteq M$ be a nonzero submodule. Because E is nilpotent, there exists $s \geq 1$ such that $LE^{s-1} \neq 0$ but

$LE^s = 0$. Let $l \in LE^{s-1}$ be a nonzero element. Then, $lE = 0$ yields $l \in M_*$. This shows that $M_* \cap L \neq 0$.

□

Corollary III.4.14. *Let M be an indecomposable B-module. Then,*

(a) *$M \otimes_B A$ is indecomposable in* mod A.
(b) $\mathrm{Hom}_B(A, M)$ *is indecomposable in* mod A.

Proof. We only prove (a), because the proof of (b) is dual.

Assume $M \otimes_B A \cong X_1 \oplus X_2$ in mod A. Then,

$$M \cong M \otimes_B A \otimes_A B \cong (X_1 \otimes_A B) \oplus (X_2 \otimes_A B).$$

Because M_B is indecomposable, we have either $X_1 \otimes_A B = 0$ or $X_2 \otimes_A B = 0$, say the former. Because $p_{X_1} : X_1 \longrightarrow X_1 \otimes_A B$ is superfluous, then $X_1 \otimes_A B = 0$ implies $X_1 = 0$. Therefore, $M \otimes_B A$ is indecomposable. □

Corollary III.4.15. *Let M be a B-module. Then, there exist bijections between the isoclasses of indecomposable summands of M in* mod B *and*

(a) *Those of $M \otimes_B A$ in* mod A, *given by $N \mapsto N \otimes_B A$.*
(b) *Those of* $\mathrm{Hom}_B(A, M)$ *in* mod A, *given by $N \mapsto$* $\mathrm{Hom}_B(A, N)$.

Proof. We only prove (a), because the proof of (b) is dual.

In view of Corollary III.4.14, it suffices to prove that $N_1 \cong N_2$ if and only if $N_1 \otimes_B A \cong N_2 \otimes_B A$. Indeed, if $N_1 \cong N_2$, then clearly $N_1 \otimes_B A \cong N_2 \otimes_B A$. Conversely, if $N_1 \otimes_B A \cong N_2 \otimes_B A$, then $N_1 \otimes_B A \otimes_A B \cong N_2 \otimes_B A \otimes_A B$ and the result follows from Lemma III.4.1. □

Applying Corollary III.4.15 to B_B and $B \otimes_B A \cong A_A$, we get a bijection between the isoclasses of indecomposable projective B- and A-modules given by $P_B \mapsto P \otimes_B A$. Dually, there is a bijection between the isoclasses of indecomposable injective B- and A-modules, given by $I_B \mapsto \mathrm{Hom}_B(A, I)$.

We now look at the correspondence between projective covers (injective envelopes) in mod B and mod A.

Lemma III.4.16. *Let M be a B-module.*

(a) *If $f : P \longrightarrow M$ is a projective cover in* mod B, *then*

$$f \otimes_B A : P \otimes_B A \longrightarrow M \otimes_B A$$

is a projective cover in mod A.

(b) *If $g : M \longrightarrow I$ is an injective envelope in* mod B, *then*

$$\mathrm{Hom}_B(A, g) : \mathrm{Hom}_B(A, M) \longrightarrow \mathrm{Hom}_B(A, I)$$

is an injective envelope in mod A.

Proof.

(a) Viewing P and M as A-modules, we have a commutative diagram of A-modules and epimorphisms

$$\begin{array}{ccc} P\otimes_B A & \xrightarrow{f\otimes_B A} & M\otimes_B A \\ {\scriptstyle p_{P\otimes_B A}}\downarrow & & \downarrow{\scriptstyle p_{M\otimes_B A}} \\ P & \xrightarrow{f} & M \end{array}$$

where the vertical morphisms are those of Lemma III.4.13(a). Because $P\otimes_B A$ is projective and $f\otimes_B A$ is an epimorphism, it suffices to prove that $f\otimes_B A$ is superfluous. Let $h\colon X\longrightarrow P\otimes_B A$ be such that $(f\otimes_B A)h$ is surjective, then so is $fp_{P\otimes_B A}h = p_{M\otimes_B A}(f\otimes_B A)h$. Because both f and $p_{P\otimes_B A}$ are superfluous, h is surjective.

(b) The morphism $\mathrm{D}g\colon \mathrm{D}I\longrightarrow \mathrm{D}M$ is a projective cover in mod B^{op}. Applying (a) yields that $A\otimes_B \mathrm{D}g\colon A\otimes_B \mathrm{D}I\longrightarrow A\otimes_B \mathrm{D}M$ is a projective cover in mod A^{op}. The result then follows from the commutative diagram

$$\begin{array}{ccc} A\otimes_B \mathrm{D}I & \xrightarrow{A\otimes_B \mathrm{D}g} & A\otimes_B \mathrm{D}M \\ \cong\downarrow & & \downarrow\cong \\ \mathrm{D}\,\mathrm{Hom}_B(A,I) & \xrightarrow{\mathrm{D}\,\mathrm{Hom}_B(A,g)} & \mathrm{D}\,\mathrm{Hom}_B(A,M) \end{array}$$

where the vertical maps are functorial isomorphisms.

□

Corollary III.4.17. *Let M be a B-module.*

(a) *If $P_1 \xrightarrow{p_1} P_0 \xrightarrow{p_0} M \longrightarrow 0$ is a projective presentation in* mod B, *then so is $P_1\otimes_B A \xrightarrow{p_1\otimes_B A} P_0\otimes_B A \xrightarrow{p_0\otimes_B A} M\otimes_B A \longrightarrow 0$ in* mod A. *In addition, if the first presentation is minimal, then so is the second.*

(b) *If $0\longrightarrow M \xrightarrow{j_0} I_0 \xrightarrow{j_1} I_1$ is an injective copresentation in* mod B, *then so is $0\longrightarrow \mathrm{Hom}_B(A,M) \xrightarrow{\mathrm{Hom}_B(A,j_0)} \mathrm{Hom}_B(A,I_0) \xrightarrow{\mathrm{Hom}_B(A,j_1)} \mathrm{Hom}_B(A,I_1)$ in* mod A. *In addition, if the first copresentation is minimal, then so is the second.*

Proof. We only prove (a), because the proof of (b) is dual.

The first statement is obvious. Assume that the first presentation is minimal. Because of Lemma III.4.16, $p_0\otimes_B A\colon P_0\otimes_B A\longrightarrow M\otimes_B A$ is a projective cover in mod A. Also, because $p_1\colon P_1\longrightarrow p_1(P_1)$ is a projective cover in mod B, then $p_1\otimes_B A\colon P_1\otimes_B A\longrightarrow (p_1\otimes_B A)(P_1\otimes_B A)\cong \mathrm{Ker}(p_0\otimes_B A)$ is a projective cover in mod A. □

We are now able to state and prove the result announced at the beginning of this subsection.

Proposition III.4.18. *Let M be a B-module. Then,*

(a) $\tau_A(M \otimes_B A) \cong \mathrm{Hom}_B(A, \tau_B M)$.
(b) $\tau_A^{-1} \mathrm{Hom}_B(A, M) \cong (\tau_B^{-1} M) \otimes_B A$.

Proof. We only prove (a), because the proof of (b) is dual.

If $e \in B$ is a primitive idempotent and $P = eB$, then we have functorial isomorphisms

$$\begin{aligned}\mathrm{Hom}_A(P \otimes_B A, A) &= \mathrm{Hom}_A(eB \otimes_B A, A) \cong \mathrm{Hom}_A(eA, A) \cong Ae \cong A \otimes_B Be \\ &\cong A \otimes_B \mathrm{Hom}_B(eB, B) = A \otimes_B \mathrm{Hom}_B(P, B).\end{aligned}$$

Therefore, for every projective B-module P, we have a functorial isomorphism

$$\varphi\colon \mathrm{Hom}_A(P \otimes_B A, A) \cong A \otimes_B \mathrm{Hom}_B(P, B).$$

Let $P_1 \xrightarrow{p_1} P_0 \xrightarrow{p_0} M \longrightarrow 0$ be a minimal projective presentation of M in $\mathrm{mod}\, B$. Because of Corollary III.4.17,

$$P_1 \otimes_B A \xrightarrow{p_1 \otimes_B A} P_0 \otimes_B A \xrightarrow{p_0 \otimes_B A} M \otimes_B A \longrightarrow 0$$

is a minimal projective presentation of $M \otimes_B A$ in $\mathrm{mod}\, A$. Applying $\mathrm{Hom}_A(-, A)$ yields the upper row in the commutative diagram with exact rows

$$\begin{array}{ccccccc}\mathrm{Hom}_A(P_0 \otimes_B A, A) & \longrightarrow & \mathrm{Hom}_A(P_1 \otimes_B A, A) & \longrightarrow & \mathrm{Tr}(M \otimes_B A) & \longrightarrow & 0 \\ \varphi_{P_0} \downarrow \cong & & \varphi_{P_1} \downarrow \cong & & & & \\ A \otimes \mathrm{Hom}_B(P_0, A) & \longrightarrow & A \otimes \mathrm{Hom}_B(P_0, A) & \longrightarrow & A \otimes_B \mathrm{Tr} M & \longrightarrow & 0.\end{array}$$

We deduce that $\mathrm{Tr}(M \otimes_B A) \cong A \otimes_B \mathrm{Tr}\, M$ in $\mathrm{mod}\, A^{op}$ and so we get, in $\mathrm{mod}\, A$,

$$\tau_A(M \otimes_B A) = \mathrm{D}\,\mathrm{Tr}(M \otimes_B A) \cong \mathrm{D}(A \otimes_B \mathrm{Tr}\, M) \cong \mathrm{Hom}_B(A, \mathrm{D}\,\mathrm{Tr}\, M) = \mathrm{Hom}_B(A, \tau_B M).$$

□

Example III.4.19. Let B be given by the quiver

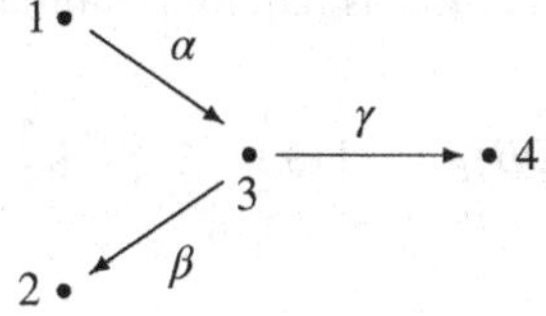

bound by $\alpha\gamma = 0$, and A by the quiver

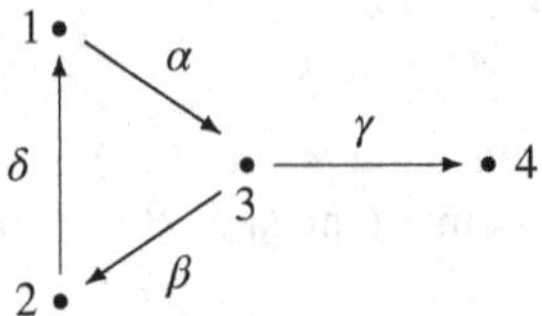

bound by $\alpha\gamma = 0$ and $\delta\alpha\beta\delta\alpha = 0$. It is easily seen that A is a split extension of the algebra B by the B–B-bimodule E generated by the arrow δ. In addition, one has

$$B_B = \begin{smallmatrix}1\\3\\2\end{smallmatrix} \oplus 2 \oplus \begin{smallmatrix}&3&\\2&&4\end{smallmatrix} \oplus 4$$

whereas

$$A_A = \begin{smallmatrix}1\\3\\2\\1\\3\\2\\1\end{smallmatrix} \oplus \begin{smallmatrix}2\\1\\3\\2\\1\end{smallmatrix} \oplus \begin{smallmatrix}&3&\\2&&4\\1&&\\3&&\\2&&\\1&&\end{smallmatrix} \oplus 4$$

from which we deduce

$$E_B = \left(\begin{smallmatrix}1\\3\\2\end{smallmatrix}\right)^3 \oplus (1)^3.$$

Indeed, writing A as a B-module amounts to deleting the arrow δ in the indecomposable projective A-modules, thus getting the direct sum decomposition

$$A_B = \left(\begin{smallmatrix}1\\3\\2\end{smallmatrix} \oplus \begin{smallmatrix}1\\3\\2\end{smallmatrix} \oplus 1\right) \oplus \left(2 \oplus \begin{smallmatrix}1\\3\\2\end{smallmatrix} \oplus 1\right) \oplus \left(\begin{smallmatrix}&3&\\2&&4\end{smallmatrix} \oplus \begin{smallmatrix}1\\3\\2\end{smallmatrix} \oplus 1\right) \oplus 4.$$

Writing $A_B = B_B \oplus E_B$, we get E_B as required. Similarly, one has

$$(DB)_B = 1 \oplus \begin{smallmatrix}1\\3\\2\end{smallmatrix} \oplus \begin{smallmatrix}1\\3\end{smallmatrix} \oplus \begin{smallmatrix}3\\4\end{smallmatrix}$$

whereas

$$(\mathrm{D}A)_A = \begin{smallmatrix}1\\3\\2\\1\\3\\2\\1\end{smallmatrix} \oplus \begin{smallmatrix}1\\3\\2\\1\\3\\2\end{smallmatrix} \oplus \begin{smallmatrix}1\\3\\2\\1\\3\end{smallmatrix} \oplus \begin{smallmatrix}3\\4\end{smallmatrix}$$

so that a calculation similar to the one above yields

$$(\mathrm{D}E)_B = \mathrm{D}({}_BE) = \left(\begin{smallmatrix}1\\3\\2\end{smallmatrix}\right)^4.$$

Now, in mod B, the simple projective $S_4 = 4$ is a radical summand of $P_3 = \begin{smallmatrix}&3&\\2&&4\end{smallmatrix}$ and of no other indecomposable projective module. Therefore, we have an almost split sequence in mod B.

$$0 \longrightarrow 4 \longrightarrow \begin{smallmatrix}&3&\\2&&4\end{smallmatrix} \longrightarrow \begin{smallmatrix}3\\2\end{smallmatrix} \longrightarrow 0$$

and, in particular, $\tau_B\left(\begin{smallmatrix}3\\2\end{smallmatrix}\right) = 4$. We wish to compute $\begin{smallmatrix}3\\2\end{smallmatrix} \otimes_B A$ and $\tau_A\left(\begin{smallmatrix}3\\2\end{smallmatrix} \otimes_B A\right)$. For this purpose, we consider the almost split sequence above as a minimal projective presentation of $M = \begin{smallmatrix}3\\2\end{smallmatrix}$ in mod B

$$P_4 \longrightarrow P_3 \longrightarrow \begin{smallmatrix}3\\2\end{smallmatrix} \longrightarrow 0.$$

We apply the right exact functor $- \otimes_B A$, obtaining an exact sequence

$$P_4 \otimes_B A \longrightarrow P_3 \otimes_B A \longrightarrow \begin{smallmatrix}3\\2\end{smallmatrix} \otimes_B A \longrightarrow 0.$$

Now, $P_4 \otimes_B A$ and $P_3 \otimes_B A$ are the indecomposable projective A-modules corresponding to the points 4 and 3 respectively. We thus get

$$4 \longrightarrow \begin{smallmatrix}&3&\\2&&4\\1&&\\3&&\\2&&\\1&&\end{smallmatrix} \longrightarrow \begin{smallmatrix}3\\2\end{smallmatrix} \otimes_B A \longrightarrow 0$$

and therefore $\begin{smallmatrix}3\\2\end{smallmatrix} \otimes_B A \cong \begin{smallmatrix}3\\2\\1\\3\\2\\1\end{smallmatrix}$.

To compute the Auslander–Reiten translate in mod A of the module ${\substack{3\\2}} \otimes_B A$, we recall that $\tau_B({\substack{3\\2}}) = 4$. Applying Proposition III.4.18, we need to find $\mathrm{Hom}_B(A,4)$. For this purpose we consider the minimal injective copresentation of 4

$$0 \longrightarrow 4 \longrightarrow {\substack{3\\4}} \longrightarrow {\substack{1\\3}}$$

in mod B, to which we apply the left exact functor $\mathrm{Hom}_B(A,-)$ obtaining an exact sequence

$$0 \longrightarrow \mathrm{Hom}_B(A,4) \longrightarrow \mathrm{Hom}_B(A,{\substack{3\\4}}) \longrightarrow \mathrm{Hom}_B(A,{\substack{1\\3}}).$$

Now

$$\mathrm{Hom}_B(A,{\substack{3\\4}}) = {\substack{3\\4}} \quad \text{and} \quad \mathrm{Hom}_B(A,{\substack{1\\3}}) = {\substack{1\\3\\2\\1\\3}}$$

from where we get

$$\tau_A({\substack{3\\2}} \otimes_B A) \cong \mathrm{Hom}_B(A,\tau_B{\substack{3\\2}}) \cong \mathrm{Hom}_B(A,4) \cong 4.$$

Exercises for Section III.4

Exercise III.4.1. Assume $B = A/E$, and $0 \longrightarrow L \longrightarrow M \longrightarrow N \longrightarrow 0$ is an almost split sequence in mod A. Prove that:

(a) If N is an indecomposable nonprojective B-module, then the short exact sequence $0 \longrightarrow L_* \longrightarrow M_* \longrightarrow N \longrightarrow 0$ is the direct sum of an almost split sequence $0 \longrightarrow L_0 \longrightarrow M_0 \longrightarrow N \longrightarrow 0$ and a sequence of the form $0 \longrightarrow P \longrightarrow P \longrightarrow 0 \longrightarrow 0$, where P is a projective B-module.

(b) If L is an indecomposable noninjective B-module, then the short exact sequence $0 \longrightarrow L \longrightarrow M^* \longrightarrow N^* \longrightarrow 0$ is the direct sum of an almost split sequence $0 \longrightarrow L \longrightarrow M^0 \longrightarrow N^0 \longrightarrow 0$ and a sequence of the form $0 \longrightarrow 0 \longrightarrow I \longrightarrow I \longrightarrow 0$, where I is an injective B-module.

Exercise III.4.2. Let $B = A/E$. Prove that there exist an equivalence between mod B and the full subcategory of mod A consisting of the modules M such that $ME = 0$.

Exercise III.4.3. Let $B = A/E$. Prove the following facts.

(a) A module X is projective in mod A if and only if:

(i) $X \otimes_A B$ is projective in mod B, and
(ii) $X \otimes_A B \otimes_B A \cong X$ in mod A.

In addition, X is indecomposable if and only if $X \otimes_A B$ is indecomposable too.

(b) A module Y is injective in mod A if and only if:

(i) $\mathrm{Hom}_A(B_A, Y)$ is injective in mod B, and
(ii) $\mathrm{Hom}_B(A_B, \mathrm{Hom}_A(B_A, Y)) \cong Y$ in mod A.

In addition, Y is indecomposable if and only if $\mathrm{Hom}_A(B, Y)$ is indecomposable too.

Exercise III.4.4. Let $B = A/E$. Prove that, for each A-module X, we have short exact sequences

(a) $0 \longrightarrow \mathrm{Tor}_1^A(X, B) \longrightarrow X \otimes_A E \longrightarrow XE \longrightarrow 0$,
(b) $0 \longrightarrow X/X_* \longrightarrow \mathrm{Hom}_A(E, X) \longrightarrow \mathrm{Ext}_A^1(B, X) \longrightarrow 0$.

Exercise III.4.5. Let A be a split extension of an algebra B by the nilpotent bimodule E. Prove that, for a B-module M

(a) $\mathrm{pd}(M \otimes_B A) \leq 1$ if and only if $\mathrm{pd}\, M_B \leq 1$ and $\mathrm{Hom}_B(\mathrm{D}E, \tau_B M) = 0$.
(b) $\mathrm{id}\,\mathrm{Hom}_B(A, M) \leq 1$ if and only if $\mathrm{id}\, M_B \leq 1$ and $\mathrm{Hom}_B(\tau_B^{-1} M, E) = 0$.

Exercise III.4.6. Let A be a split extension of an algebra B by the nilpotent bimodule E. Prove that, for a B-module M

(a) $\mathrm{pd}(M \otimes_B A) \leq 1$ and $\mathrm{Ext}_A^1(M \otimes_B A, M \otimes_B A) = 0$ if and only if $\mathrm{pd}\, M_B \leq 1$, $\mathrm{Ext}_B^1(M, M) = 0$, $\mathrm{Hom}_B(\mathrm{D}E, \tau_B M) = 0$ and $\mathrm{Hom}_B(M \otimes_B E, \tau_A M) = 0$.
(b) $\mathrm{id}\,\mathrm{Hom}_B(A, M) \leq 1$ and $\mathrm{Ext}_A^1(\mathrm{Hom}_B(A, M), \mathrm{Hom}_B(A, M)) = 0$ if and only if $\mathrm{id}\, M_B \leq 1$, $\mathrm{Ext}_B^1(M, M) = 0$, $\mathrm{Hom}_B(\tau_B^{-1} M, \mathrm{Hom}_B(E, M))) = 0$ and $\mathrm{Hom}_B(\tau_B^{-1} M, E) = 0$.

Exercise III.4.7. Let B be given by the quiver

$$\begin{array}{ccccc} & & 1 & & \\ & & \uparrow{\scriptstyle\beta} & & \\ 2 & \xleftarrow{\gamma} & 3 & \xleftarrow{\alpha} & 4 \end{array}$$

and A by the quiver

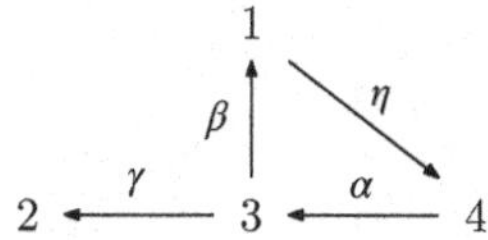

bound by $\beta\eta = 0$, $\eta\alpha\beta = 0$, $\eta\alpha\gamma = 0$. Prove that A is a split extension of B by a nilpotent bimodule E such that $E^2 = 0$. Compute the right and the left B-module structures of E. Compute $M \otimes_B A$, where

$$M_B = 1 \oplus \begin{smallmatrix} & 4 & \\ & 3 & \\ 2 & & 1 \end{smallmatrix} \oplus \begin{smallmatrix} 4 \\ 3 \\ 1 \end{smallmatrix} \oplus 4.$$

Exercise III.4.8. Let A be a split extension of an algebra B by a bimodule E. Prove that:

(a) The quiver Q_A of A has the same points as the quiver Q_B of B.
(b) The set of arrows from x to y in Q_A equals the set of arrows from x to y in Q_B plus

$$\dim_{\mathbf{k}} e_x \left(\frac{E}{E.\operatorname{rad} B + \operatorname{rad} B.E + E^2} \right) e_y$$

additional arrows.

Exercise III.4.9. A split extension by a bimodule E such that $E^2 = 0$ is called a **trivial extension**. Let A be a trivial extension of B by E, and $\mathscr{C}$ the category consisting of the pairs (M, φ_M) where M is a B–B-bimodule and $\varphi_M : M \otimes_B E \longrightarrow M$ is a morphism in $\operatorname{mod} B$ such that $\varphi_M(\varphi_M \otimes E) = 0$; a morphism $f : (M, \varphi_M) \longrightarrow (N, \varphi_N)$ is a morphism f in $\operatorname{mod} B$ from M to N such that $f\varphi_M = \varphi_N(f \otimes E)$. Prove that $\operatorname{mod} A$ is equivalent to $\mathscr{C}$.

Exercise III.4.10. Let $A = \begin{pmatrix} C & 0 \\ M & \mathbf{k} \end{pmatrix}$ be the one-point extension of an algebra C by a C-module M, see Exercise I.2.21. Prove that A is the split extension of $B = C \times \mathbf{k}$ by some bimodule E. Compute the right and the left B-module structures of E.

Exercise III.4.11. In each of the following examples, prove that A is the split extension of B by some bimodule E. Compute the right and the left B-module structures of E.

(a) B $\quad 1 \xleftarrow{\beta} 2 \xleftarrow{\alpha} 3$

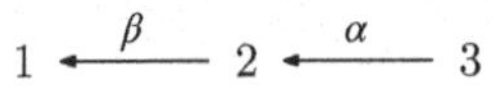

A $\quad 1 \xleftarrow{\beta} 2 \xleftarrow{\alpha} 3$, with an arrow γ from 1 to 3, $\quad \beta\gamma\alpha\beta\gamma = 0.$

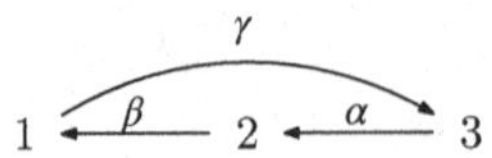

(b) B

$$\begin{array}{ccccc} 1 & \xleftarrow{\beta} & 3 & \xleftarrow{\alpha} & 4 \\ & & \downarrow{\scriptstyle\gamma} & & \\ & & 2 & & \end{array} \qquad \alpha\gamma = 0,$$

A

$$\begin{array}{ccccc} & & \overset{\delta}{\frown} & & \\ 1 & \xleftarrow{\beta} & 3 & \xleftarrow{\alpha} & 4 \\ & & \downarrow{\scriptstyle\gamma} & & \\ & & 2 & & \end{array} \qquad \alpha\gamma = 0, \delta\alpha\beta\,\delta\alpha = 0.$$

(where δ is an arrow $1 \to 4$)

(c) B

$$1 \xleftarrow{\beta} 2 \xleftarrow{\alpha} 3 \qquad \alpha\beta = 0,$$

A

$$1 \xleftarrow{\beta} 2 \underset{\alpha}{\overset{\gamma}{\rightleftarrows}} 3 \qquad \alpha\beta = 0, \gamma\alpha\gamma = 0.$$

(d) B

$$1 \xleftarrow{\gamma} 2 \xleftarrow{\beta} 3 \xleftarrow{\alpha} 4 \qquad \alpha\beta\gamma = 0,$$

A

$$1 \xleftarrow{\gamma} 2 \xleftarrow{\beta} 3 \xleftarrow{\alpha} 4, \quad 1 \xrightarrow{\delta} 4 \qquad \begin{array}{l} \alpha\beta\gamma = 0, \beta\gamma\delta = 0, \\ \gamma\delta\alpha = 0, \delta\alpha\beta = 0. \end{array}$$

Exercise III.4.12. Let A be as in Exercise III.4.11(c). For each indecomposable B-module M, compute $M \otimes_B A$ and $\tau_A(M \otimes_B A)$.

Exercise III.4.13. Let A be as in Exercise III.4.11(d). Let $M = \begin{smallmatrix} 3 \\ 2 \end{smallmatrix}$ and $N = \begin{smallmatrix} 3 \\ 1 \end{smallmatrix}$.

(a) Prove that the almost split sequence in mod B ending with M remains almost split in mod A.
(b) Prove that the almost split sequence in mod B ending with N does not remain almost split in mod A. In this case, compare the two sequences.

Chapter IV
The Auslander–Reiten quiver of an algebra

Let A be a finite dimensional $\mathbf{k}$-algebra. As seen in Corollary II.3.13, every indecomposable A-module is the source, and the target, of an almost split morphism, and thus fits into an almost split sequence. The knowledge of all almost split sequences implies the knowledge of all indecomposable A-modules, up to isomorphism, and all irreducible morphisms. We have seen several examples and some suggest the possibility of constructing all almost split sequences over an algebra, using a recursive procedure, see, for instance, Example III.2.11 or Example III.3.8. To carry out this recursion efficiently, it is practical to arrange the results in the form of a quiver. This is the Auslander–Reiten quiver of the algebra, which we define in the present chapter. We give a construction procedure for the simplest Auslander–Reiten quivers, and study the shape of some of their connected components. In the third section, we show how the Auslander–Reiten quiver can be used for computing radical morphisms, and in the fourth, we compute the Auslander–Reiten quiver of the Kronecker algebra.

As usual, we assume throughout that A is a finite dimensional $\mathbf{k}$-algebra, but here in this chapter as well as in the rest of the book, we assume that the base field $\mathbf{k}$ is algebraically closed. The reason is that, as we shall see, thanks to the assumption, several statements take on a simpler form and in particular, almost split sequences and irreducible morphisms are easier to visualise directly in the Auslander–Reiten quiver.

IV.1 The Auslander–Reiten quiver

IV.1.1 The space of irreducible morphisms

We know that the morphisms occurring in almost split sequences are just the irreducible morphisms between indecomposable modules. Now, if M, N are

I. Assem, F. U. Coelho, *Basic Representation Theory of Algebras*, Graduate Texts in Mathematics 283, https://doi.org/10.1007/978-3-030-35118-2_4

indecomposable A-modules, a morphism $f: M \longrightarrow N$ is irreducible if and only if it belongs to $\operatorname{rad}_A(M, N) \setminus \operatorname{rad}_A^2(M, N)$, see Lemma II.2.2. Therefore, the quotient vector space $\operatorname{rad}_A(M, N)/\operatorname{rad}_A^2(M, N)$ can be considered as a measure for the set of irreducible morphisms from M to N. This leads to the following definition.

Definition IV.1.1. Let M, N be indecomposable A-modules. The **space of irreducible morphisms** is the **k**-vector space

$$\operatorname{Irr}_A(M, N) = \frac{\operatorname{rad}_A(M, N)}{\operatorname{rad}_A^2(M, N)}.$$

Our first objective is to describe a basis of this space. Because each irreducible morphism can be completed to a left, and also to a right, minimal almost split morphism, it is reasonable to use the latter to construct the required basis.

We need some notation. Let $M = \oplus_{i=1}^t M_i^{m_i}$ be an A-module, with the M_i indecomposable and pairwise nonisomorphic (that is, $M_i \ncong M_j$ for $i \neq j$). For each i, with $1 \leq i \leq t$, we denote by $M_{i1}, \dots, M_{im_i}$ the different copies of M_i occurring in the above decomposition of M. In this notation, we have $M = \oplus_{i=1}^t (\oplus_{j=1}^{m_i} M_{ij})$, and $M_{ij} \ncong M_{kl}$ if and only if $i \neq k$. Then, a morphism $f: L \longrightarrow M$ induces, for each pair (i, j), a morphism $f_{ij}: L \longrightarrow M_{ij}$ obtained by composing f with the projection $M \longrightarrow M_{ij}$. Also, a morphism $g: M \longrightarrow N$ induces morphisms $g_{ij}: M_{ij} \longrightarrow N$ by composing the inclusion morphisms $M_{ij} \longrightarrow M$ with g.

Proposition IV.1.2. *Let L, N be indecomposable and M as above.*

(a) *A morphism $f: L \longrightarrow M$ is left minimal almost split if and only if*

   (i) *for each i, the set of residual classes*

$$\{f_{i1} + \operatorname{rad}_A^2(L, M_i), \dots, f_{im_i} + \operatorname{rad}_A^2(L, M_i)\}$$

   is a basis of $\operatorname{Irr}_A(L, M_i)$, *and*

   (ii) *if* $\operatorname{Irr}(L, M') \neq 0$ *with M' indecomposable, then there exists i such that $M' \cong M_i$.*

(b) *A morphism $g: M \longrightarrow N$ is right minimal almost split if and only if*

   (i) *for each i, the set of residual classes*

$$\{g_{i1} + \operatorname{rad}_A^2(M_i, N), \dots, g_{im_i} + \operatorname{rad}_A^2(M_i, N)\}$$

   is a **k***-basis of* $\operatorname{Irr}_A(M_i, N)$, *and*

   (ii) *if* $\operatorname{Irr}(M', N) \neq 0$ *with M' indecomposable, then there exists i such that $M' \cong M_i$.*

Proof. We only prove (a) because the proof of (b) is dual.

Necessity. Because of Corollary II.2.25, each f_{ij} is irreducible, so that $f_{ij} \in \mathrm{rad}_A(L, M_i) \setminus \mathrm{rad}^2_A(L, M_i)$. Because (ii) follows directly from Theorem II.2.24, it remains to prove that the classes $\overline{f_{ij}} = f_{ij} + \mathrm{rad}^2_A(L, M_i)$ constitute a basis of $\mathrm{Irr}_A(L, M_i)$.

We first show their linear independence. Suppose $\sum_{j=1}^{m_i} \lambda_{ij}\overline{f_{ij}} = 0$ in $\mathrm{Irr}_A(L, M_i)$, where the λ_{ij} are scalars. Assume that there exists j such that $\lambda_{ij} \neq 0$. Then, the morphism $\boldsymbol{\lambda} = (\lambda_{i1}, \ldots, \lambda_{im_i})\colon M_i^{m_i} \longrightarrow M_i$ is a retraction. Indeed, an associated section is $(0, \ldots, 0, \lambda_{ij}^{-1}, 0, \ldots, 0)^t\colon M_i \longrightarrow M_i^{m_i}$, where λ_{ij}^{-1} occurs in the coordinate j. Let $f_i = (f_{i1}, \ldots, f_{im_i})^t\colon L \longrightarrow M_i^{m_i}$, then $\boldsymbol{\lambda} f_i = \sum_{i=1}^{m_i} \lambda_{ij} f_{ij}$ is irreducible, because of Corollary II.2.25. But this contradicts the hypothesis that it belongs to $\mathrm{rad}^2_A(L, M_i)$. Therefore, $\lambda_{ij} = 0$ for all j, and the $\overline{f_{ij}}$ are linearly independent.

We next prove that the $\overline{f_{ij}}$ generate the $\mathbf{k}$-vector space $\mathrm{Irr}_A(L, M_i)$. Let $h \in \mathrm{rad}_A(L, M_i)$. Because $f\colon L \longrightarrow M$ is left almost split, there exists $u\colon M \longrightarrow M_i$ such that $h = uf$. Decomposing M into its indecomposable summands, this equality becomes

$$h = \sum_{l=1}^{t}\sum_{j=1}^{m_l} u_{lj} f_{lj}$$

where $u_{lj}\colon M_{lj} \longrightarrow M_i$ is the composition of u with the inclusion $M_{lj} \longrightarrow M$. First, assume $l \neq i$. Then, $M_{lj} \neq M_i$ and so $u_{lj} \in \mathrm{rad}_A(M_{lj}, M_i)$. Because $f_{lj} \in \mathrm{rad}_A(L, M_{lj})$, we have $u_{lj} f_{lj} \in \mathrm{rad}^2_A(L, M_i)$. On the other hand, if $l = i$, then u_{lj} is an endomorphism of M_i. Because $\mathrm{End}\, M_i$ is local, and the field $\mathbf{k}$ is algebraically closed, we have $\mathrm{End}\, M_i / \mathrm{rad}(\mathrm{End}\, M_i) \cong \mathbf{k}$. Therefore, there exist $\alpha_{ij} \in \mathbf{k}$ and $u'_{ij} \in \mathrm{rad}(\mathrm{End}\, M_i)$ such that, for every j with $1 \leq j \leq m_i$,

$$u_{ij} = \alpha_{ij} 1_{M_i} + u'_{ij}.$$

Because $f_{ij} \in \mathrm{rad}_A(L, M_i)$, we have $u'_{ij} f_{ij} \in \mathrm{rad}^2_A(L, M_i)$. Passing to residual classes, we have

$$\overline{h} = \sum_{l=1}^{t}\sum_{j=1}^{m_l} \overline{u_{lj}}\,\overline{f_{lj}} = \sum_{j=1}^{m_i} \alpha_{ij}\overline{f_{lj}}.$$

This completes the proof that the $\overline{f_{ij}}$ form a basis of $\mathrm{Irr}_A(L, M_i)$ and thus the proof of the necessity part.

Sufficiency. Let $f'\colon L \longrightarrow M'$ be left minimal almost split and $M' = \oplus_{j=1}^{s} M_j'^{\,m'_j}$ where the M'_j are indecomposable and pairwise nonisomorphic. Because of (ii), for each j, we have $M'_j \cong M_i$ for some i. Because $m'_j = \dim_{\mathbf{k}} \mathrm{Irr}_A(L, M'_j) = \dim_{\mathbf{k}} \mathrm{Irr}_A(L, M_i) = m_i$, we get that $M' \cong M$. On the other hand, $\overline{f_{ij}} \neq 0$ implies $f_{ij} \in \mathrm{rad}_A(L, M_i) \setminus \mathrm{rad}^2_A(L, M_i)$, that is, it is irreducible.

Therefore, f itself is irreducible and so is radical (because L is indecomposable, see Lemma II.2.2). Hence, there exists $h\colon M' \longrightarrow M$ such that $f = hf'$. Because f' is not a section, h must be a retraction and in particular, an epimorphism. But $M' \cong M$; hence, h must be an isomorphism. □

The hypothesis that $\mathbf{k}$ is algebraically closed was used essentially in the proof of the necessity part to show that the induced morphisms generate the space of irreducible morphisms. An immediate but useful consequence of this result is the following corollary.

Corollary IV.1.3. *Let* $0 \longrightarrow L \longrightarrow \oplus_{i=1}^{t} M_i^{m_i} \longrightarrow N \longrightarrow 0$ *be an almost split sequence with the* M_i *indecomposable and pairwise nonisomorphic. Then, for each* i*, we have*

$$\dim_{\mathbf{k}} \operatorname{Irr}_A(L, M_i) = m_i = \dim_{\mathbf{k}} \operatorname{Irr}_A(M_i, N).$$

Proof. This follows from parts (i) of both (a) and (b) of Proposition IV.1.2. □

We warn the reader that this corollary does not hold true if the base field $\mathbf{k}$ is not algebraically closed.

Example IV.1.4. Let A be the Kronecker algebra with quiver

$$1 \bullet \overset{\alpha}{\underset{\beta}{\leftleftarrows}} \bullet\, 2$$

Every irreducible morphism starting from the simple projective module P_1 has as a target a projective module, hence must be a morphism from P_1 to P_2 and therefore the inclusion of P_1 as a direct summand of rad P_2. Because rad $P_2 = P_1^2$, there exist exactly two linearly independent irreducible morphisms from P_1 to P_2, which are two embeddings f_1, f_2 of P_1 into rad P_2. Therefore, a basis of $\operatorname{Irr}_A(P_1, P_2)$ is given by the residual classes $\overline{f_1}, \overline{f_2}$ of f_1, f_2 respectively, modulo $\operatorname{rad}_A^2(P_1, P_2)$. Now, $\operatorname{rad}_A^2(P_1, P_2) = 0$. For, assume that $f = \alpha_1 f_1 + \alpha_2 f_2 \in \operatorname{rad}_A^2(P_1, P_2)$, with $\alpha_1, \alpha_2 \in \mathbf{k}$. Then, there exist a module M and radical morphisms $h\colon P_1 \longrightarrow M$, $g\colon M \longrightarrow P_2$ such that $f = gh$. Because g is radical, it is not surjective; hence, it factors through rad $P_2 = P_1^2$, that is, there exists a morphism $g'\colon M \longrightarrow P_1^2$ such that f is the composition of $g'h$ with the inclusion $P_1^2 \longrightarrow P_2$. But P_1 is simple; hence, $g'h\colon P_1 \longrightarrow P_1^2$ is injective, and every monomorphism $P_1 \longrightarrow P_1^2$ is a section. Therefore, h itself is a section; thus, it is not a radical morphism, a contradiction. In this example, we have $\operatorname{Irr}_A(P_1, P_2) \cong \operatorname{rad}_A(P_1, P_2) = \operatorname{Hom}_A(P_1, P_2)$ spanned by $\{f_1, f_2\}$. The almost split sequence starting with P_1 is given by

$$0 \longrightarrow P_1 \xrightarrow{\binom{f_1}{f_2}} P_2^2 \longrightarrow N \longrightarrow 0.$$

An easy calculation yields that $N = P_2^2/P_1$ is the indecomposable module represented as follows

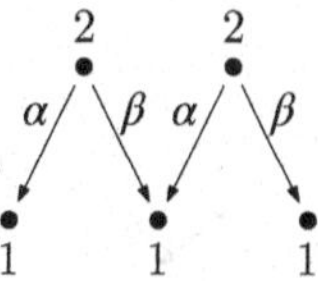

compare with Example II.4.3.

IV.1.2 Defining the Auslander–Reiten quiver

We define our main object of study in this chapter.

Definition IV.1.5. The **Auslander–Reiten quiver** $\Gamma(\operatorname{mod} A)$ of the algebra A is defined as follows

(a) The points of $\Gamma(\operatorname{mod} A)$ are the isoclasses of indecomposable A-modules. For an indecomposable A-module M, we denote simply its isoclass by M, thus identifying them.
(b) For points M, N, the arrows are in bijection with the vectors of a basis of the $\mathbf{k}$-vector space $\operatorname{Irr}_A(M, N)$. In particular, the number of these arrows equals $\dim_{\mathbf{k}} \operatorname{Irr}_A(M, N)$.

Remark IV.1.6.

(a) Let M, N be indecomposable A-modules. There exists an arrow from M to N in $\Gamma(\operatorname{mod} A)$ if and only if there exists an irreducible morphism from M to N, that is, if and only if $\operatorname{Irr}_A(M, N) \neq 0$.
(b) Because there are no irreducible morphisms from an indecomposable module to itself, the Auslander–Reiten quiver has no loops.
(c) Let N be an indecomposable nonprojective A-module (so that τN exists) and let M be indecomposable. It follows from Corollary IV.1.3 above that we have m arrows from M to N in $\Gamma(\operatorname{mod} A)$ if and only if we have m arrows from τN to M.
(d) Assume that we have an almost split sequence

$$0 \longrightarrow L \xrightarrow{f} \oplus_{i=1}^{t} M_i^{m_i} \xrightarrow{g} N \longrightarrow 0$$

with the M_i indecomposable and pairwise nonisomorphic. It follows from (c) that it induces in $\Gamma(\operatorname{mod} A)$ a so-called **mesh**

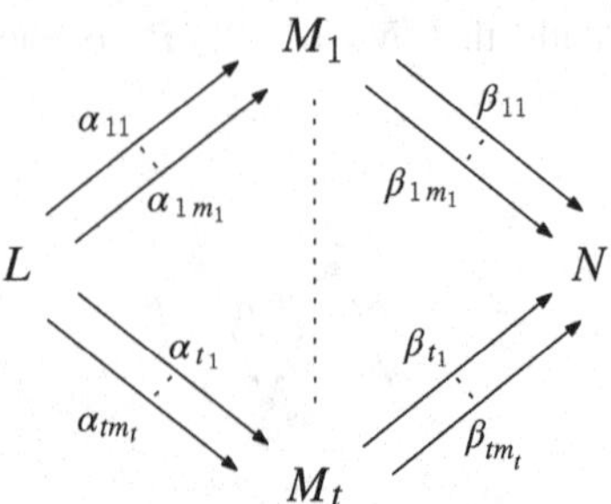

where the arrows $\alpha_{ij}: L \longrightarrow M_i$ (with $1 \leq j \leq m_i$) are in bijection with the morphisms induced from the compositions of f with the projections $M \longrightarrow M_i$, and the $\beta_{ij}: M_i \longrightarrow N$ are in bijection with the compositions of g with the inclusions $M_i \longrightarrow M$.

If N is indecomposable projective, then the right minimal almost split morphism ending at N is the inclusion $g\colon \operatorname{rad} N \longrightarrow N$. Setting $\operatorname{rad} N = \oplus_{i=1}^{t} M_i^{m_i}$ with the M_i indecomposable and pairwise nonisomorphic, we see that g induces a "half-mesh"

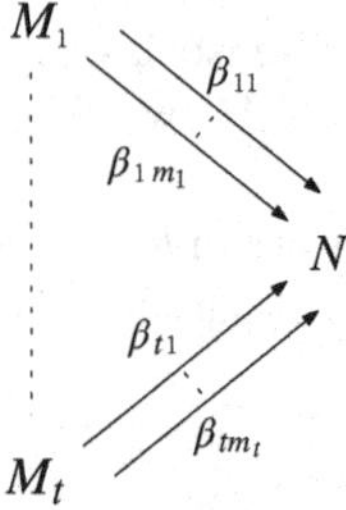

with the β_{ij} as before. Dually, if L is indecomposable injective and $L/\operatorname{soc} L = \oplus_{i=1}^{t} M_i^{m_i}$ with the M_i indecomposable and pairwise nonisomorphic, the left minimal almost split projection morphism $f: L \longrightarrow L/\operatorname{soc} L$ induces a "half-mesh"

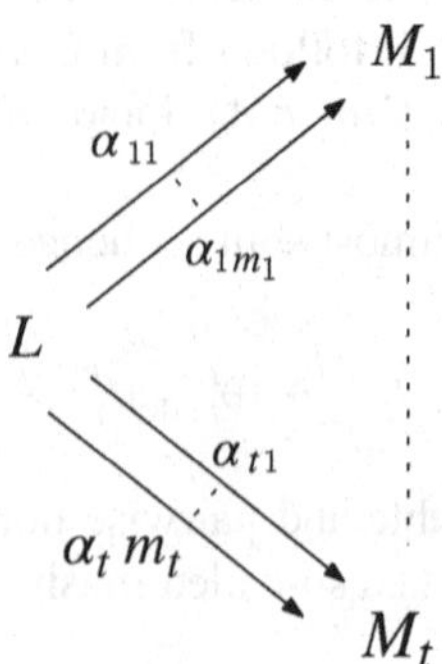

(e) The situation described in (d) above shows that the Auslander–Reiten quiver has a remarkable combinatorial structure. Every point, and every arrow, occurs in a mesh or in a half-mesh; therefore, $\Gamma(\text{mod}\, A)$ is the union of these meshes or half-meshes. In particular, every point of $\Gamma(\text{mod}\, A)$ is the source, and the target, of at most finitely many arrows, a situation expressed by saying that $\Gamma(\text{mod}\, A)$ is **locally finite**, see Definition IV.1.17 below. In general, an Auslander–Reiten quiver has infinitely many connected components. But the local finiteness of the quiver implies that each of these components has at most countably many points and countably many arrows. In particular, $\Gamma(\text{mod}\, A)$ is a finite quiver if and only if A is a representation-finite algebra. We prove later in Chapter VI that, if an Auslander–Reiten quiver admits one finite connected component, then this component is the totality of the quiver and therefore the algebra is representation-finite. This is a theorem due to Auslander, which we admit for the time being.

Before giving examples, we should mention that, in the important special case of a representation-finite algebra, the Auslander–Reiten quiver has no multiple arrows.

Proposition IV.1.7. *Let A be a representation-finite algebra and M, N indecomposable A-modules. Then*

$$\dim_{\mathbf{k}} \text{Irr}_A(M, N) \leq 1.$$

Proof. Assume that there exist indecomposable A-modules M, N such that $\dim_{\mathbf{k}} \text{Irr}_A(M, N) \geq 2$. Because every irreducible morphism is either a monomorphism or an epimorphism, we must have $\dim_{\mathbf{k}} M \neq \dim_{\mathbf{k}} N$. Without loss of generality, we assume that $\dim_{\mathbf{k}} M > \dim_{\mathbf{k}} N$. In particular, this implies that N is not projective. Therefore, there exists an almost split sequence:

$$0 \longrightarrow \tau N \longrightarrow M^2 \oplus X \longrightarrow N \longrightarrow 0.$$

But then,

$$\dim_{\mathbf{k}} \tau N = 2\dim_{\mathbf{k}} M + \dim_{\mathbf{k}} X - \dim_{\mathbf{k}} N \geq 2\dim_{\mathbf{k}} M - \dim_{\mathbf{k}} N > \dim_{\mathbf{k}} M$$

due to our hypothesis that $\dim_{\mathbf{k}} M > \dim_{\mathbf{k}} N$. In addition, Corollary IV.1.3 yields

$$\dim_{\mathbf{k}} \text{Irr}_A(\tau N, M) = \dim_{\mathbf{k}} \text{Irr}_A(M, N) \geq 2.$$

We may therefore repeat the procedure, replacing N by M and M by τN. Inductively, we get that none of the modules $\tau^i N$ or $\tau^i M$ is projective and we have

$$\dim_{\mathbf{k}} N < \dim_{\mathbf{k}} M < \dim_{\mathbf{k}} \tau N < \dim_{\mathbf{k}} \tau M < \ldots < \dim_{\mathbf{k}} \tau^i N < \dim_{\mathbf{k}} \tau^i M \ldots$$

In particular, all these indecomposable modules are nonisomorphic. This contradicts the assumption that A is representation-finite. □

Example IV.1.8. In the previous section, we proved that the Kronecker algebra A

$$1 \bullet \underset{\beta}{\overset{\alpha}{\leftleftarrows}} \bullet\, 2$$

has an almost split sequence of the form

$$0 \longrightarrow P_1 \longrightarrow P_2^2 \longrightarrow \tau^{-1}P_1 \longrightarrow 0$$

with $\tau^{-1}P_1 = \begin{smallmatrix} & 2 & & 2 & \\ 1 & & 1 & & 1 \end{smallmatrix}$. This yields a mesh of the form

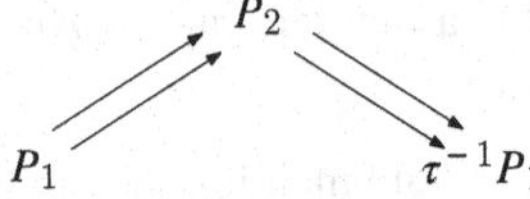

Because we have multiple arrows, the previous proposition says that A is representation-infinite, a fact already seen in Example II.4.3.

IV.1.3 Examples and construction procedures

In general, constructing an Auslander–Reiten quiver can be extremely difficult: indeed, its construction presupposes the knowledge of all (isoclasses of) indecomposable modules and all irreducible morphisms between them. We have this knowledge in the case of Nakayama algebras, which we illustrate in two examples.

Example IV.1.9. Let A be given by the quiver

$$\overset{1}{\bullet} \xleftarrow{\ \gamma\ } \overset{2}{\bullet} \xleftarrow{\ \beta\ } \overset{3}{\bullet} \xleftarrow{\ \alpha\ } \overset{4}{\bullet}$$

bound by $\alpha\beta\gamma = 0$. Applying Proposition III.3.4, we get all almost split sequences in mod A

$$0 \longrightarrow 3 \longrightarrow \begin{smallmatrix} 4 \\ 3 \end{smallmatrix} \longrightarrow 4 \longrightarrow 0 \qquad 0 \longrightarrow 2 \longrightarrow \begin{smallmatrix} 3 \\ 2 \end{smallmatrix} \longrightarrow 3 \longrightarrow 0$$

$$0 \longrightarrow 1 \longrightarrow \begin{smallmatrix} 2 \\ 1 \end{smallmatrix} \longrightarrow 2 \longrightarrow 0 \qquad 0 \longrightarrow \begin{smallmatrix} 2 \\ 1 \end{smallmatrix} \longrightarrow \begin{smallmatrix} 3 \\ 2 \\ 1 \end{smallmatrix} \oplus 2 \longrightarrow \begin{smallmatrix} 3 \\ 2 \end{smallmatrix} \longrightarrow 0$$

$$0 \longrightarrow \begin{smallmatrix} 3 \\ 2 \end{smallmatrix} \longrightarrow \begin{smallmatrix} 4 \\ 3 \\ 2 \end{smallmatrix} \oplus 3 \longrightarrow \begin{smallmatrix} 4 \\ 3 \end{smallmatrix} \longrightarrow 0.$$

This gives rise to five meshes, which we draw below

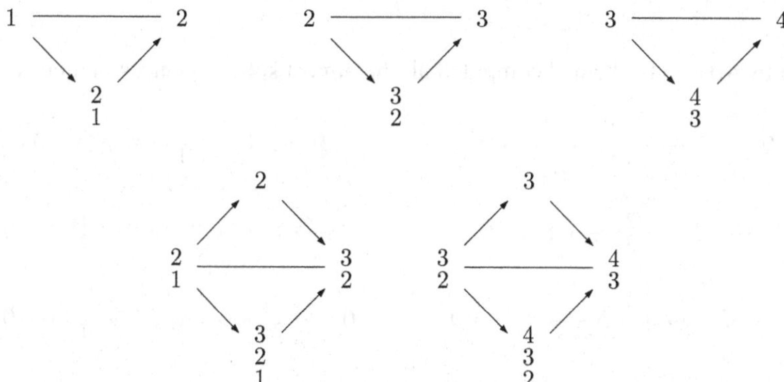

It is customary, when drawing (a mesh of) the Auslander–Reiten quiver, to put translates on the same horizontal line and to join them with a dashed line. Assembling the meshes above in the obvious way, we get the Auslander–Reiten quiver

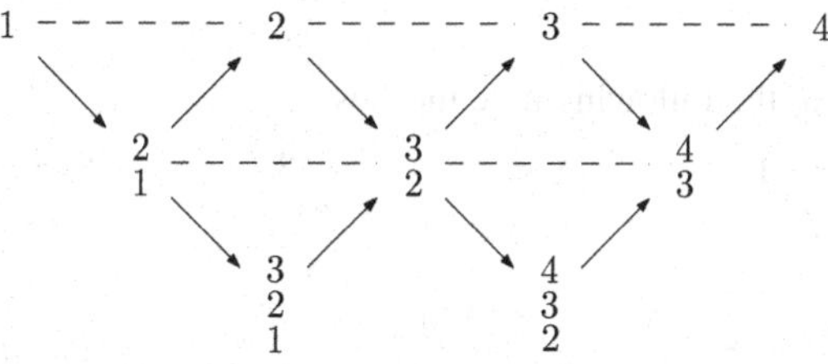

Observe that

$$\begin{matrix} 3 \\ 2 \\ 1 \end{matrix} \qquad \text{and} \qquad \begin{matrix} 4 \\ 3 \\ 2 \end{matrix},$$

though lying on the same horizontal line, are not joined by a dashed line: indeed, both are projective–injective and are not translates of each other.

Example IV.1.10. Let A be as in Example III.3.6, that is, A is given by the quiver

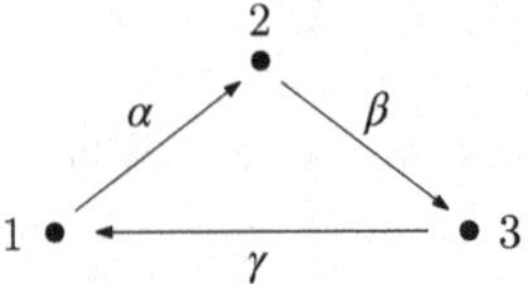

bound by $\alpha\beta\gamma = 0$. We had computed all the almost split sequences in mod A:

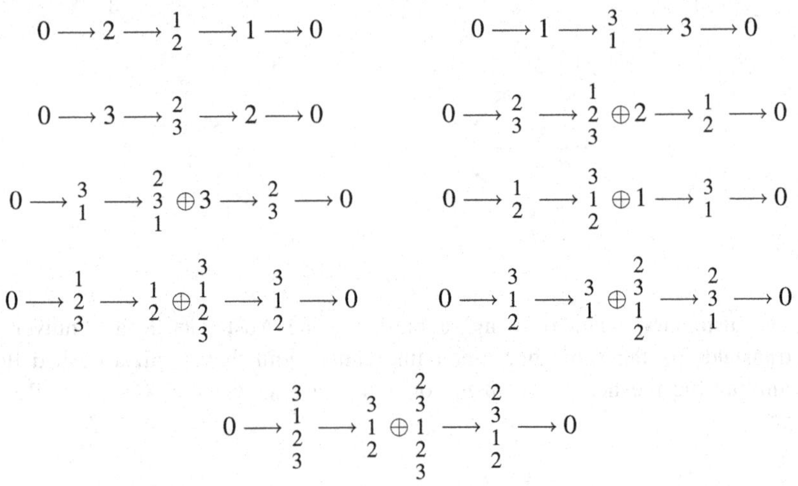

This gives rise to the following nine meshes

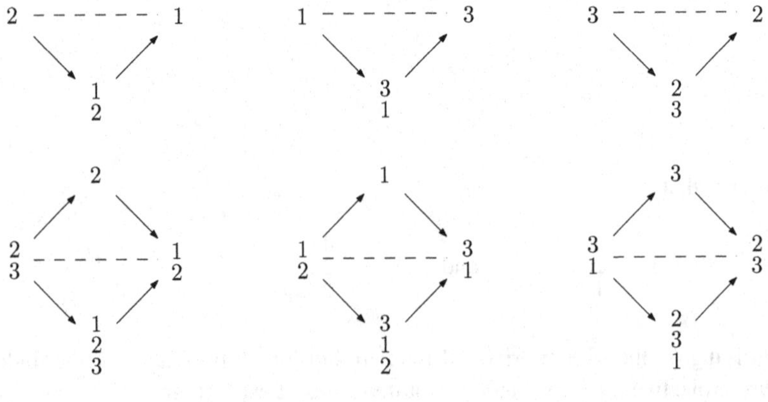

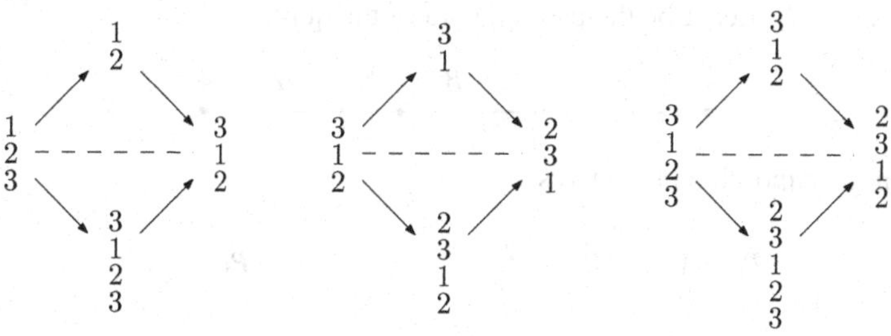

where again the dashed lines denote the Auslander–Reiten translations. Assembling these meshes, we get the Auslander–Reiten quiver

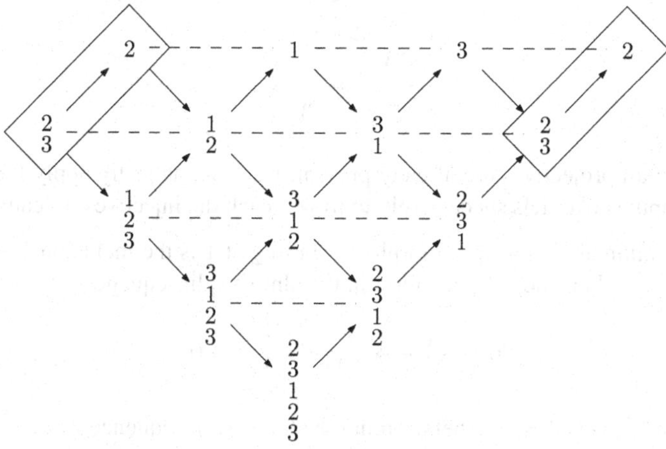

where one has to identify the two copies of the arrow $\begin{smallmatrix}2\\3\end{smallmatrix} \longrightarrow 2$ so that $\Gamma(\operatorname{mod} A)$ lies on a cylinder.

We show a simple technique, known as **knitting**, which consists of using in a systematic manner the following already proven facts and their duals.

1. The sources of $\Gamma(\operatorname{mod} A)$ are the simple projective modules, see Lemma III.2.8.
2. Every arrow in $\Gamma(\operatorname{mod} A)$ starting in a simple projective ends in a projective, see Lemma III.2.10.
3. Every arrow in $\Gamma(\operatorname{mod} A)$ ending in a projective starts at an indecomposable direct summand of its radical, see Example II.2.21 and Theorem II.2.24.
4. If an indecomposable module L is not injective and one knows the left minimal almost split morphism $f\colon L \longrightarrow M$, then $\tau^{-1}L \cong \operatorname{Coker} f$, and, for every indecomposable module X, we have n arrows $L \longrightarrow X$ if and only if there exist n arrows $X \longrightarrow \tau^{-1}L$, see Lemma III.2.9 and Corollary IV.1.3.

The knitting technique works perfectly well for all finite acyclic Auslander–Reiten quivers. The technique can be written as a formal algorithm, but we prefer to illustrate it using examples.

Example IV.1.11. Let A be the path algebra of the quiver

$$\overset{1}{\bullet} \xleftarrow{\ \gamma\ } \overset{2}{\bullet} \xrightarrow{\ \beta\ } \overset{3}{\bullet} \xleftarrow{\ \alpha\ } \overset{4}{\bullet}$$

The indecomposable projectives are

$$P_1 = 1 \qquad P_2 = \begin{smallmatrix}2\\1\,3\end{smallmatrix} \qquad P_3 = 3 \qquad P_4 = \begin{smallmatrix}4\\3\end{smallmatrix}.$$

One sees immediately that rad $P_2 = P_1 \oplus P_3$ whereas rad $P_4 = P_3$. Because of facts 1, 2 and 3 above, we have a full subquiver of $\Gamma(\text{mod}\,A)$ of the form

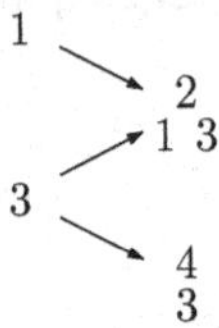

Because all projectives are already present, we systematically apply fact 4, that is, we compute cokernels successively until we reach the injectives. Because of fact 2, the left minimal almost split morphism starting at 1 is the inclusion $1 \longrightarrow \begin{smallmatrix}2\\1\,3\end{smallmatrix}$. Therefore, $\tau^{-1}(1)$ is the cokernel term in the almost split sequence

$$0 \longrightarrow 1 \longrightarrow \begin{smallmatrix}2\\1\,3\end{smallmatrix} \longrightarrow \begin{smallmatrix}2\\3\end{smallmatrix} \longrightarrow 0.$$

Similarly, $\tau^{-1}(3)$ is the cokernel term in the almost split sequence

$$0 \longrightarrow 3 \longrightarrow \begin{smallmatrix}2\\1\,3\end{smallmatrix} \oplus \begin{smallmatrix}4\\3\end{smallmatrix} \longrightarrow \begin{smallmatrix}2\,4\\1\,3\end{smallmatrix} \longrightarrow 0.$$

Thus, we get the meshes

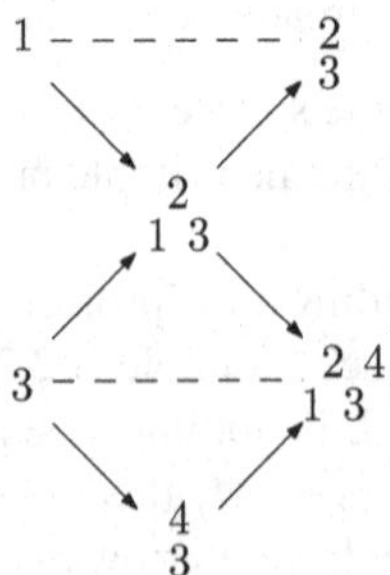

We claim that the morphism $\begin{smallmatrix}2\\1\,3\end{smallmatrix} \longrightarrow \begin{smallmatrix}2\\3\end{smallmatrix} \oplus \begin{smallmatrix}2\,4\\1\,3\end{smallmatrix}$ is left minimal almost split. Indeed, if $\begin{smallmatrix}2\\1\,3\end{smallmatrix} \longrightarrow M$ is irreducible with M indecomposable, then either M

is projective and $\begin{smallmatrix} 2 \\ 1\,3 \end{smallmatrix}$ is an indecomposable summand of its radical, or we have an irreducible morphism $\tau M \longrightarrow \begin{smallmatrix} 2 \\ 1\,3 \end{smallmatrix}$. Now, all the projectives have already appeared; thus, M is not projective and we are in the second case. However, the only irreducible morphisms of target $\begin{smallmatrix} 2 \\ 1\,3 \end{smallmatrix}$ and of indecomposable source are the inclusions $1 \longrightarrow \begin{smallmatrix} 2 \\ 1\,3 \end{smallmatrix}$ and $3 \longrightarrow \begin{smallmatrix} 2 \\ 1\,3 \end{smallmatrix}$. Therefore, either $M \cong \tau^{-1}(1) = \begin{smallmatrix} 2 \\ 3 \end{smallmatrix}$ or $M \cong \tau^{-1}(3) = \begin{smallmatrix} 2\,4 \\ 1\,3 \end{smallmatrix}$. This establishes our claim.

Consequently, $\tau^{-1}\left(\begin{smallmatrix} 2 \\ 1\,3 \end{smallmatrix}\right)$ is the cokernel term in the almost split sequence

$$0 \longrightarrow \begin{smallmatrix} 2 \\ 1\,3 \end{smallmatrix} \longrightarrow \begin{smallmatrix} 2 \\ 3 \end{smallmatrix} \oplus \begin{smallmatrix} 2\,4 \\ 1\,3 \end{smallmatrix} \longrightarrow \begin{smallmatrix} 2\;\;4 \\ 3 \end{smallmatrix} \longrightarrow 0.$$

Similarly, $\tau^{-1}\left(\begin{smallmatrix} 4 \\ 3 \end{smallmatrix}\right)$ is the cokernel term in the almost split sequence

$$0 \longrightarrow \begin{smallmatrix} 4 \\ 3 \end{smallmatrix} \longrightarrow \begin{smallmatrix} 2\,4 \\ 1\,3 \end{smallmatrix} \longrightarrow \begin{smallmatrix} 2 \\ 1 \end{smallmatrix} \longrightarrow 0.$$

Thus, we get two new meshes

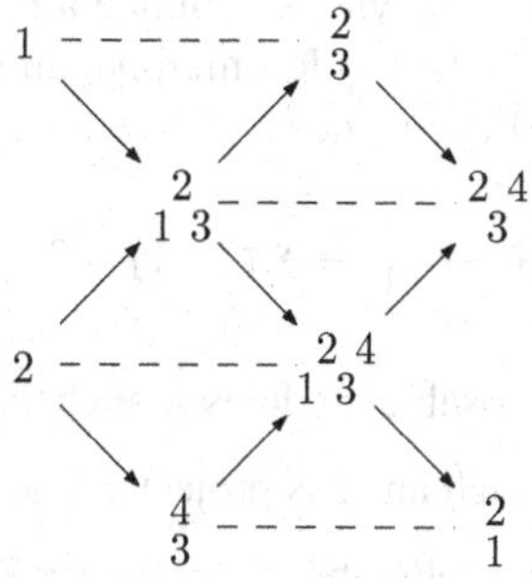

We repeat the procedure, finding two more meshes

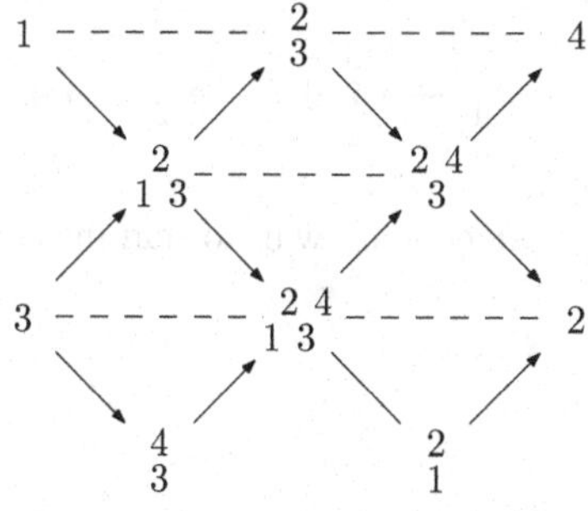

Now $I_1 = {\tiny\begin{matrix}2\\1\end{matrix}}$, $I_2 = 2$, $I_3 = {\tiny\begin{matrix}2\,4\\3\end{matrix}}$ and $I_4 = 4$. We have reached all the injectives and thus finished knitting a connected component of $\Gamma(\operatorname{mod} A)$.

Because of Auslander's theorem, see Remark IV.1.6(e) above, this component is the whole Auslander–Reiten quiver.

Example IV.1.12. Let A be given by the quiver

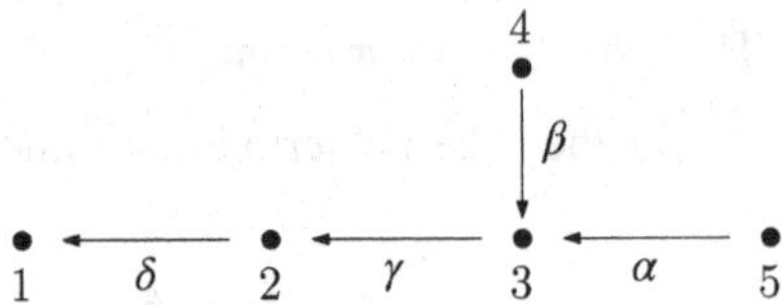

bound by $\alpha\gamma\delta = 0$, $\beta\gamma = 0$. The indecomposable projectives are

$$P_1 = 1 \qquad P_2 = {\tiny\begin{matrix}2\\1\end{matrix}} \qquad P_3 = \begin{matrix}3\\2\\1\end{matrix} \qquad P_4 = {\tiny\begin{matrix}4\\3\end{matrix}} \qquad P_5 = \begin{matrix}5\\3\\2\end{matrix}.$$

An acyclic quiver, such as the quiver of A, always has a sink; thus, A has at least one simple projective module. Here, $P_1 = 1$ is the only simple projective. Every arrow of source P_1 admits a projective as a target, and the target admits P_1 as a direct summand of its radical. This yields a unique arrow starting with P_1, namely the inclusion $P_1 \longrightarrow P_2$, which is thus left minimal almost split. Because P_1 is not injective, we get an almost split sequence

$$0 \longrightarrow 1 \longrightarrow {\tiny\begin{matrix}2\\1\end{matrix}} \longrightarrow \tau^{-1}(1) = 2 \longrightarrow 0.$$

Let us search for the indecomposable modules X such that there exists an irreducible morphism ${\tiny\begin{matrix}2\\1\end{matrix}} \longrightarrow X$. Either such an X is projective, and ${\tiny\begin{matrix}2\\1\end{matrix}}$ is a direct summand of its radical, or there exists an irreducible morphism $\tau X \longrightarrow {\tiny\begin{matrix}2\\1\end{matrix}}$. In the first case, $X \cong P_3 = \begin{matrix}3\\2\\1\end{matrix}$, and in the second $X \cong \tau^{-1}(1) = 2$; hence, the almost split sequence

$$0 \longrightarrow {\tiny\begin{matrix}2\\1\end{matrix}} \longrightarrow \begin{matrix}3\\2\\1\end{matrix} \oplus 2 \longrightarrow {\tiny\begin{matrix}3\\2\end{matrix}} \longrightarrow 0.$$

These two almost split sequences allow us to start the construction

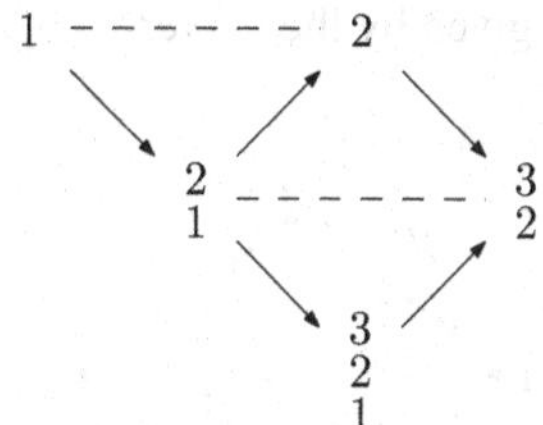

The module $P_3 = \begin{smallmatrix}3\\2\\1\end{smallmatrix} = I_1$ is projective–injective. On the other hand, the morphism $2 \longrightarrow \begin{smallmatrix}3\\2\end{smallmatrix}$ is left minimal almost split, so we have an almost split sequence

$$0 \longrightarrow 2 \longrightarrow \begin{smallmatrix}3\\2\end{smallmatrix} \longrightarrow 3 \longrightarrow 0.$$

On the other hand, rad $P_5 = \begin{smallmatrix}3\\2\end{smallmatrix}$ and rad $P_4 = 3$, from which we deduce the following full subquiver of $\Gamma(\operatorname{mod} A)$

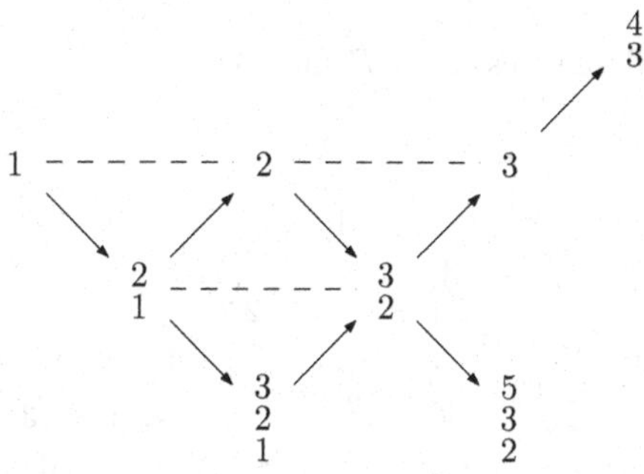

All projectives have now appeared. Therefore, the construction proceeds by constructing cokernels recursively until we reach the injectives.

This yields the whole Auslander–Reiten quiver

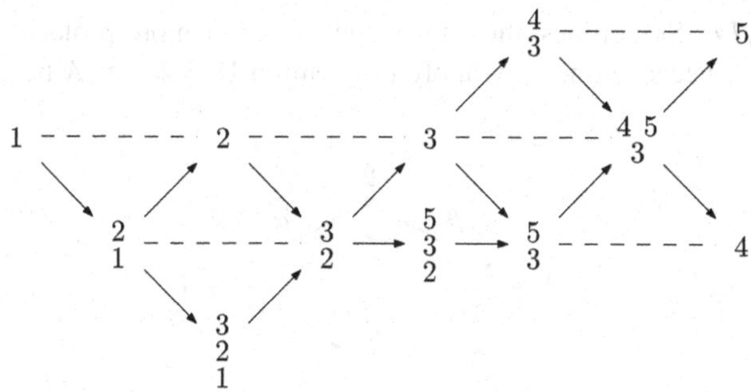

Example IV.1.13. Let A be given by the quiver

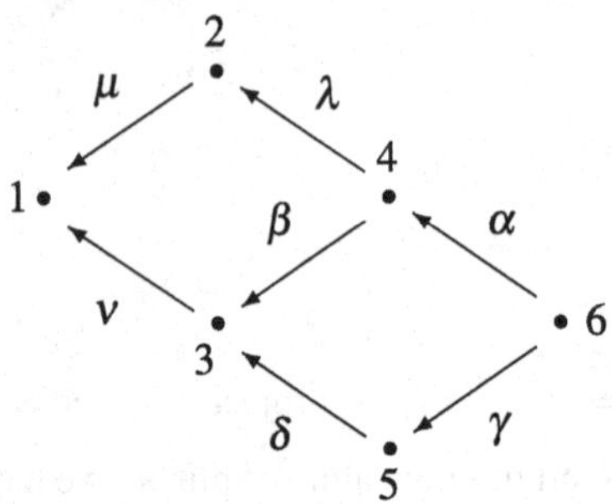

bound by $\alpha\beta = \gamma\delta$ and $\lambda\mu = \beta\nu$. Here,

$$P_1 = 1,\quad P_2 = \begin{smallmatrix}2\\1\end{smallmatrix},\quad P_3 = \begin{smallmatrix}3\\1\end{smallmatrix},\quad P_4 = \begin{smallmatrix}&4&\\2&&3\\&1&\end{smallmatrix},\quad P_5 = \begin{smallmatrix}5\\3\\1\end{smallmatrix}\quad \text{and}\quad P_6 = \begin{smallmatrix}&&6&\\&4&&5\\2&&3&\\&1&&\end{smallmatrix}.$$

The knitting procedure gives easily $\Gamma(\text{mod}\,A)$.

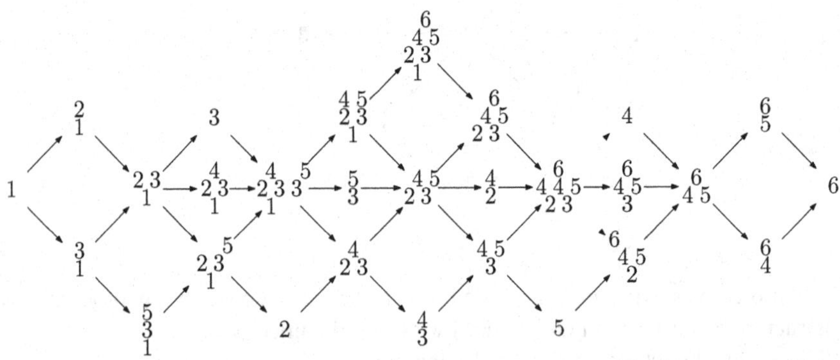

Example IV.1.14. Sometimes, the knowledge of one or more projective–injectives is very helpful, because one may apply Proposition III.3.2. Let A be given by the quiver

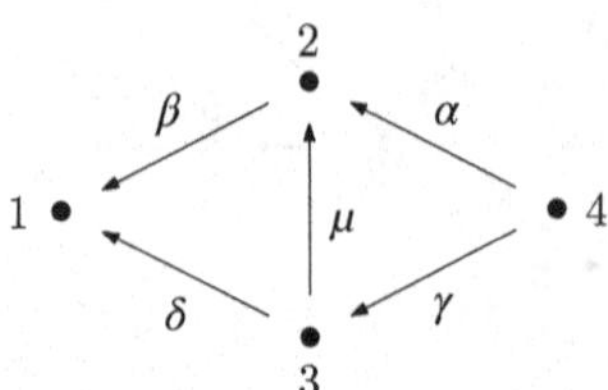

bound by $\alpha\beta = \gamma\delta$, $\gamma\mu = 0$ and $\mu\beta = 0$. Here we have

$$P_1 = 1, \quad P_2 = \begin{smallmatrix}2\\1\end{smallmatrix}, \quad P_3 = \begin{smallmatrix}&3&\\1&&2\end{smallmatrix} \quad \text{and} \quad P_4 = \begin{smallmatrix}&4&\\2&&3\\&1&\end{smallmatrix}$$

$$I_1 = \begin{smallmatrix}&4&\\2&&3\\&1&\end{smallmatrix} \quad I_2 = \begin{smallmatrix}3&&4\\&2&\end{smallmatrix}, \quad I_3 = \begin{smallmatrix}4\\3\end{smallmatrix} \quad \text{and} \quad I_4 = 4.$$

Because $P_4 = I_1$ is projective–injective, we have an almost split sequence

$$0 \longrightarrow \begin{smallmatrix}2&&3\\&1&\end{smallmatrix} \longrightarrow \begin{smallmatrix}&4&\\2&&3\\&1&\end{smallmatrix} \oplus 2 \oplus 3 \longrightarrow \begin{smallmatrix}&4&\\3&&2\end{smallmatrix} \longrightarrow 0$$

hence the mesh

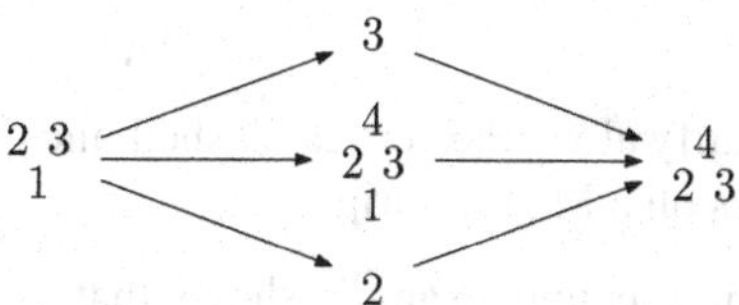

On the other hand, 2 is a direct summand of rad $P_3 = 2 \oplus 1$. The other summand 1 is also equal to rad P_2 so we have a full subquiver of $\Gamma(\text{mod}\, A)$

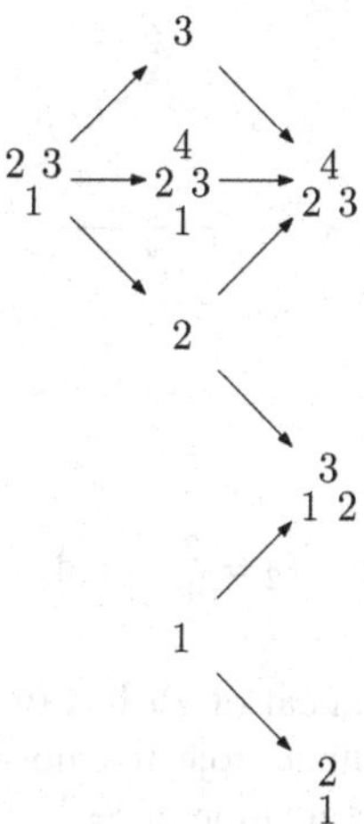

We now have all projectives. We may continue knitting and deduce $\Gamma(\text{mod}\, A)$

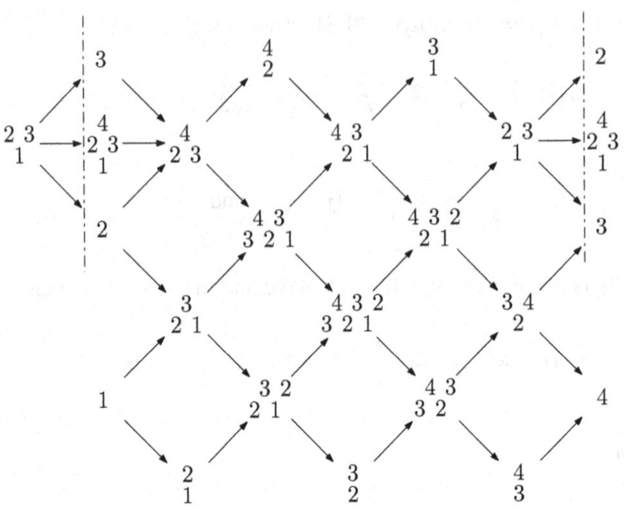

where one has to identify along the vertical dashed lines in such a way that the upper part of $\Gamma(\mathrm{mod}\, A)$ lies on a Möbius strip.

Example IV.1.15. The previous example shows that even if the Auslander–Reiten quiver is not acyclic, it is sometimes possible to use the knitting procedure together perhaps with other ingredients to construct the quiver. We give another example of this type.

Let A be given by the quiver

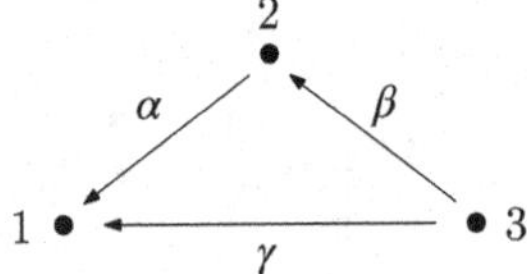

bound by $\beta\alpha = 0$. Here

$$P_1 = 1, \quad P_2 = \begin{smallmatrix} 2 \\ 1 \end{smallmatrix} \quad \text{and} \quad P_3 = \begin{smallmatrix} & 3 & \\ 1 & & 2 \end{smallmatrix}.$$

Although $P_1 = \mathrm{rad}\, P_2$, the radical of P_3 has two indecomposable summands P_1 and $S_2 = 2$, the latter being neither projective nor injective. We compute $\tau^{-1}S_2$. We have a minimal injective copresentation of S_2:

$$0 \longrightarrow S_2 \longrightarrow I_2 \xrightarrow{j} I_3 \longrightarrow 0$$

and, because of Lemma III.1.8, $\tau^{-1}S_2 = \mathrm{Coker}\, \nu^{-1}j$. Applying ν^{-1}, we get

$$P_2 \xrightarrow{\nu^{-1}j} P_3 \longrightarrow \tau^{-1}S_2 \longrightarrow 0.$$

Because $P_2 = \begin{smallmatrix}2\\1\end{smallmatrix}$ and $P_3 = \begin{smallmatrix}3\\2\,1\end{smallmatrix}$, the morphism $\nu^{-1}j$ maps the simple top of P_2 to the isomorphic summand of the socle of P_3. Hence, $\tau^{-1}S_2 = \operatorname{Coker}\nu^{-1}j = \begin{smallmatrix}3\\1\end{smallmatrix}$ and we have an almost split sequence

$$0 \longrightarrow 2 \longrightarrow \begin{smallmatrix}3\\2\,1\end{smallmatrix} \longrightarrow \begin{smallmatrix}3\\1\end{smallmatrix} \longrightarrow 0.$$

Thus, we have the following full subquiver of $\Gamma(\operatorname{mod} A)$

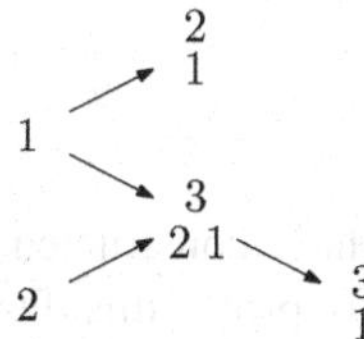

in which all projectives are present. Knitting gives $\Gamma(\operatorname{mod} A)$

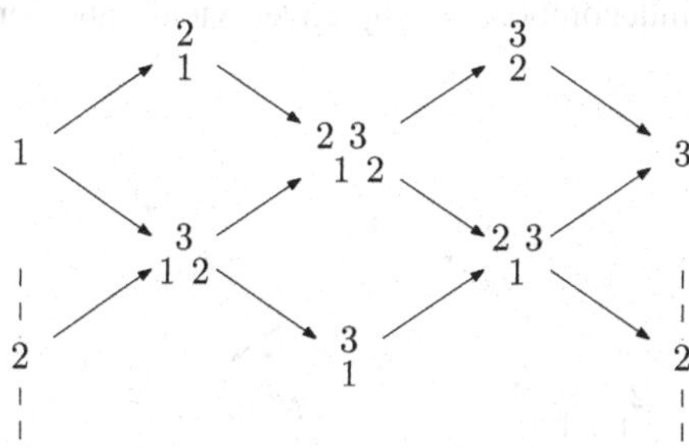

where one identifies along the vertical dotted lines.

For the next example, we recall from Subsection I.1.4 the definition of dimension vector: if $S_1, \ldots, S_n$ form a complete set of representatives of the isoclasses of simple A-modules, and M is an indecomposable A-module, the integer $\mu_i(M)$ denotes the number of composition factors of M that are isomorphic to S_i for each i with $1 \le i \le n$. The dimension vector of M is then defined as $\underline{\dim}\, M = (\mu_1(M), \ldots, \mu_n(M))$.

Example IV.1.16. Let A be the Kronecker algebra with quiver

$$1 \bullet \underset{\beta}{\overset{\alpha}{\leftleftarrows}} \bullet\, 2$$

The knitting procedure, starting from the indecomposable projectives $P_1 = 1$ and $P_2 = \begin{smallmatrix} & 2 & \\ 1 & & 1 \end{smallmatrix}$ gives a connected component of $\Gamma(\operatorname{mod} A)$

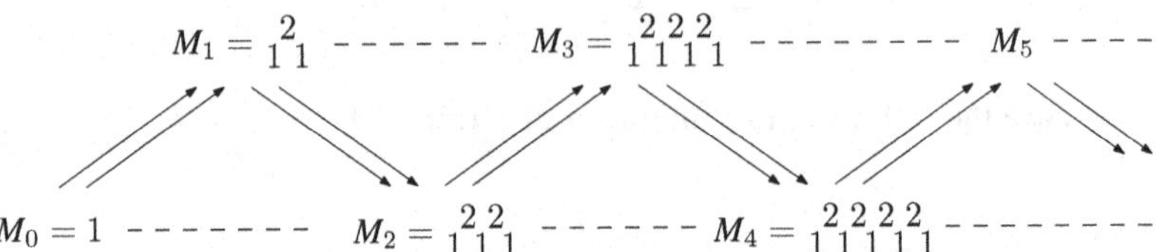

The modules M_i are precisely those constructed in Example II.4.3.

This component is infinite. To prove this fact, it is convenient to use the dimension vectors of the modules occurring in the component.

Indeed, an immediate induction shows that, for every indecomposable module M_t in this component with $t \geq 0$, we have $\underline{\dim}\, M_t = (t+1, t)$. Dually, knitting backwards from the indecomposable injectives yields another infinite connected component

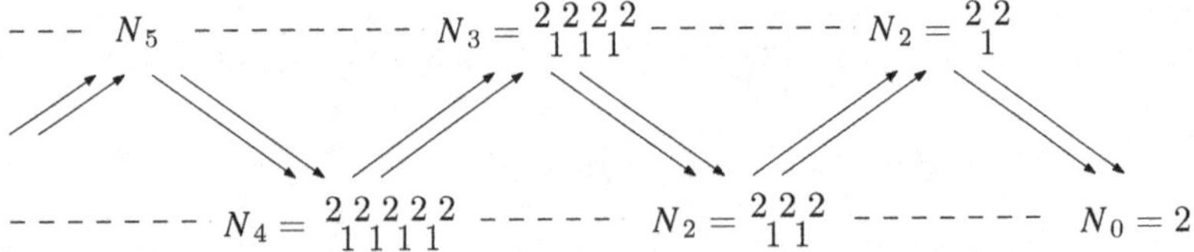

For every indecomposable N_s in this component, we have $\underline{\dim} N_s = (s, s+1)$. In particular, these two components are disjoint.

IV.1.4 The combinatorial structure of the Auslander–Reiten quiver

The particular combinatorial structure of Auslander–Reiten quivers sometimes allows statements to be neatly formulated in graphical terms that may otherwise appear technical. We describe this structure in more detail.

Given a quiver $Q = (Q_0, Q_1)$ and a point $x \in Q_0$, we denote the set of arrows entering x by

$$x^- = \{\alpha \in Q_1 : t(\alpha) = x\}$$

and the set of arrows leaving x by

$$x^+ = \{\alpha \in Q_1 : s(\alpha) = x\}.$$

We have used informally in Subsection IV.1.2 the expression locally finite quiver. We give a formal definition of local finiteness.

Definition IV.1.17. A quiver $Q = (Q_0, Q_1)$ is called **locally finite** if, for each $x \in Q_0$, the sets x^+ and x^- are finite.

In particular, in a locally finite quiver, every point has finitely many neighbours.

For instance, every finite quiver is locally finite. Also, every connected component of an Auslander–Reiten quiver is locally finite, because every almost split sequence has at most finitely many indecomposable middle terms. We now define the notion of translation quiver.

Definition IV.1.18. A **translation quiver** is a pair (Γ, τ) where $\Gamma = (\Gamma_0, \Gamma_1)$ is a locally finite quiver without loops and $\tau : \Gamma_0 \setminus \Gamma_0' \longrightarrow \Gamma_0 \setminus \Gamma_0''$ is a bijection defined between two subsets of Γ_0 such that, for any $x \in \Gamma_0 \setminus \Gamma_0'$, and any direct predecessor y of x, there is a bijection from the set of arrows from y to x to the set of arrows from τx to y.

The partially defined bijection τ is called the **translation**. The set Γ_0' of points on which τ is not defined is called the set of **projective points**, and the set Γ_0'' of points on which τ^{-1} is not defined is called the set of **injective points**. A **full translation subquiver** (Ω, ω) of (Γ, τ) is a pair such that Ω is a full subquiver of Γ and, if $x \in \Omega_0$ is such that $\tau x \in \Omega_0$, then $\omega x = \tau x$. If there is no ambiguity, we denote a translation quiver (Γ, τ) simply as Γ.

Given a nonprojective point x in a translation quiver Γ, the full translation subquiver having as points x, τx and all direct predecessors of x has the following shape:

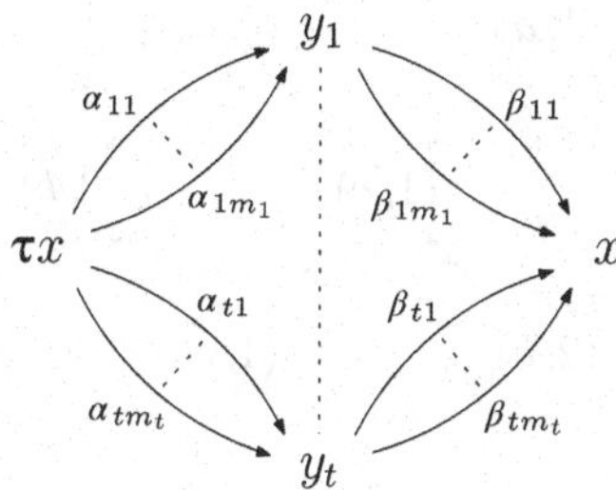

and is called a **mesh**. The following lemma is now obvious.

Lemma IV.1.19. *The Auslander–Reiten quiver of a finite dimensional algebra, equipped with the Auslander–Reiten translation τ has a natural translation quiver structure.* □

Of course, there exist translation quivers that are not Auslander–Reiten quivers. We give examples below.

Example IV.1.20. Let $Q = (Q_0, Q_1)$ be a finite, connected and acyclic quiver. We define its **repetitive quiver** $\mathbb{Z}\,Q$ as follows. The set of points of $\mathbb{Z}\,Q$ is the set

$$(\mathbb{Z}\,Q)_0 = \mathbb{Z} \times Q_0 = \{(n, i) : n \in \mathbb{Z}, i \in Q_0\}.$$

For each arrow $\alpha : i \longrightarrow j$ in Q and each $n \in \mathbb{Z}$, there exist two arrows:

$$(n, \alpha) : (n, i) \longrightarrow (n, j) \quad \text{and} \quad (n, \alpha') : (n + 1, j) \longrightarrow (n, i)$$

in $\mathbb{Z}\,Q$, and all arrows in $\mathbb{Z}\,Q$ are of one of these forms. For a given $n \in \mathbb{Z}$, the full subquiver $\{n\} \times Q$ of $\mathbb{Z}\,Q$ with points $\{n\} \times Q_0 = \{(n, i) : i \in Q_0\}$ is isomorphic to Q and therefore $\mathbb{Z}\,Q$ may be viewed as consisting of an infinity of copies of Q indexed by $n \in \mathbb{Z}$, together with additional arrows (n, α') going from the copy with index $n + 1$ to the copy with index n. The translation is defined for every $(n, i) \in (\mathbb{Z}\,Q)_0$ by $\tau(n, i) = (n + 1, i)$. The translation is thus an automorphism of $\mathbb{Z}\,Q$.

Let, for instance, Q be the quiver

then $\mathbb{Z}\,Q$ is given by

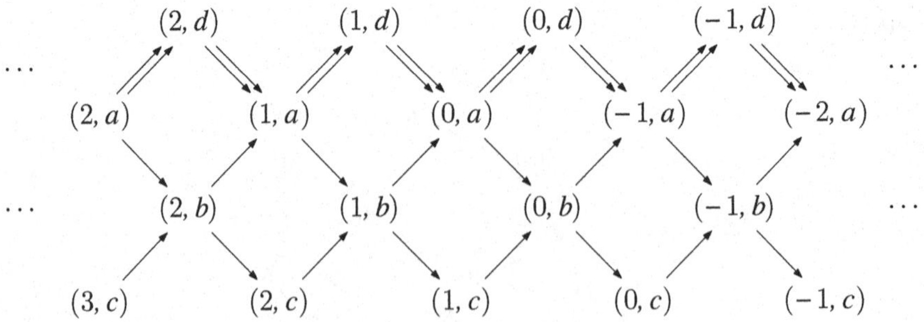

Clearly, repetitive quivers are not Auslander–Reiten quivers: they contain neither projectives nor injectives. However, repetitive quivers may occur as connected components of the Auslander–Reiten quivers of some algebras.

Translation quivers consist of meshes glued together. There are, therefore, two types of paths in a translation quiver, those that factor through a mesh and those that do not factor. This leads to the following definition.

Definition IV.1.21. A path $x_0 \longrightarrow x_1 \longrightarrow \ldots \longrightarrow x_m$ in a translation quiver Γ is called **sectional** if, for every i with $0 \le i \le m-2$, we have $\tau x_{i+2} \ne x_i$.

Sectional paths are easy to read off on a translation quiver. For instance, in the example above, there is a sectional path

$$(2, c) \longrightarrow (1, b) \longrightarrow (0, a) \longrightarrow (0, d),$$

but we may see that no path from $(2, c)$ to $(-1, a)$ is sectional.

If there is a sectional path from x_0 to x_m in Γ, then x_0 is said to be a **sectional predecessor** of x_m, and x_m a **sectional successor** of x_0.

Example IV.1.22. Let Q be the quiver $\mathbb{A}_\infty$

(infinitely many points and infinitely many arrows all oriented to the right). Then $\mathbb{Z}\,Q$ has the following configuration:

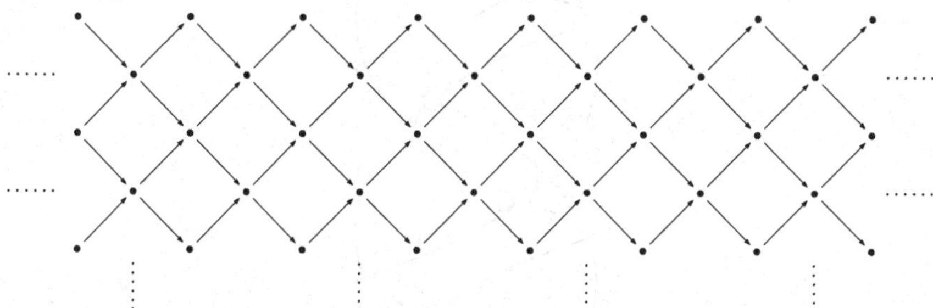

This repetitive quiver has infinite sectional paths: for each $n \in \mathbb{Z}$, one can construct the following infinite path, which is clearly sectional:

$$(n, 1) \longrightarrow (n, 2) \longrightarrow \ldots \longrightarrow (n, i) \longrightarrow (n, i+1) \longrightarrow \ldots$$

Example IV.1.23. Let $\mathbb{Z}\,\mathbb{A}_\infty$ be as constructed above and $m > 0$ an integer. Identify each point $x \in (\mathbb{Z}\,\mathbb{A}_\infty)_0$ with $\tau^m x$ and each arrow $x \longrightarrow y$ with the arrow $\tau^m x \longrightarrow \tau^m y$. This identification gives a translation quiver, which we call **stable tube** (of **rank** m, if we need to be more precise). For $m = 2$, for instance, the quiver looks like

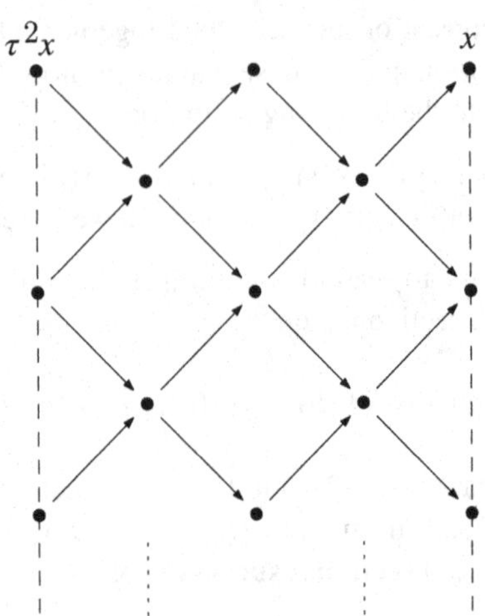

where one has to identify along the vertical dotted lines giving an infinite cylinder, as follows

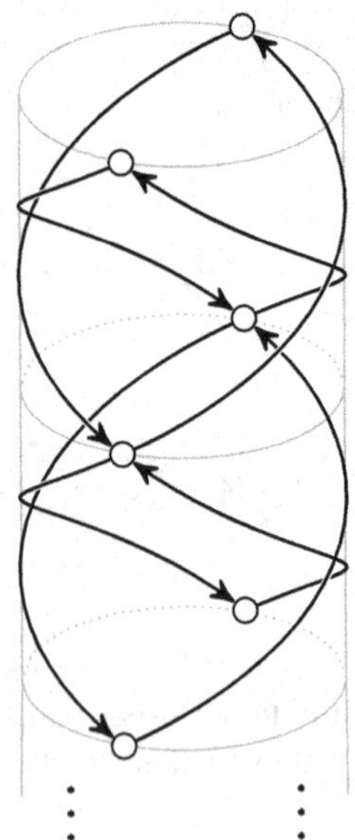

This translation quiver also has infinite sectional paths.

Stable tubes also occur as components of the Auslander–Reiten quiver for some algebras, see, for instance, Section IV.4.

Example IV.1.24. Let Q be a quiver. One can construct both $\mathbb{N}\,Q$ and $(-\mathbb{N})Q$ in the same fashion as we have done for $\mathbb{Z}\,Q$. Let Q be the quiver

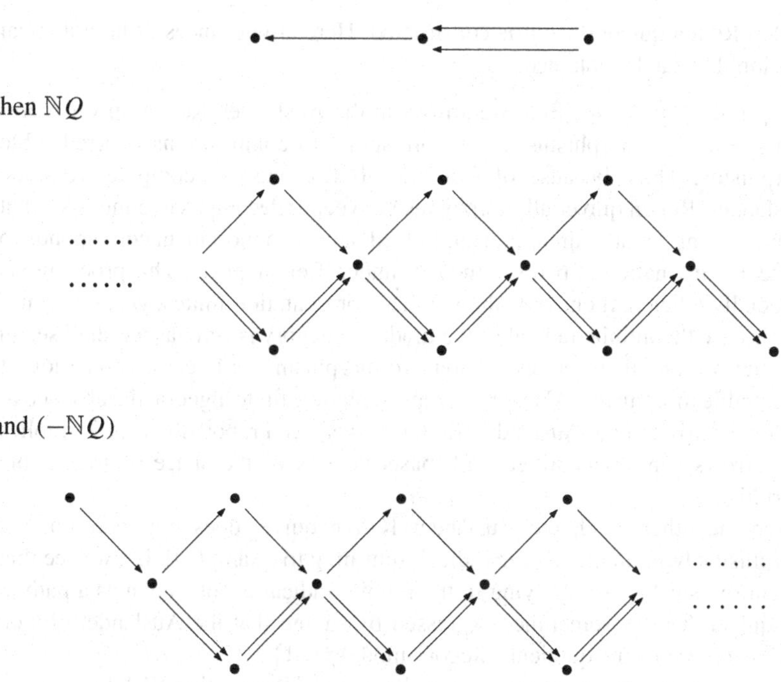

For $\mathbb{N}\,Q$, each point of the form $(0, x)$ for $x \in Q_0$ is an injective point whereas, for $(-\mathbb{N})Q$, such a point is projective.

In Example IV.1.16 above, the connected component of the Auslander–Reiten quiver of the Kronecker algebra containing the projectives is of the form $(-\mathbb{N})Q^{op}$, whereas the component containing the injectives is of the form $\mathbb{N}\,Q^{op}$, where Q is the quiver of the Kronecker algebra.

IV.1.5 The use of Auslander–Reiten quivers

Why does one draw an Auslander–Reiten quiver? Certainly, it is an interesting way to record and visualise the information we have about almost split sequences and the way they fit together. This is already structural knowledge on the module category. But there is more than that: the Auslander–Reiten quiver is also a computational tool. For instance, the knitting procedure explained in Subsection IV.1.3, when applicable, gives a recursive way to construct indecomposable modules and irreducible morphisms. If the connected component of the Auslander–Reiten quiver one is knitting turns out to be finite, then, because of Auslander's theorem, see Remark IV.1.6(e), one gets in this way a complete set of isoclasses of indecomposable modules. In addition, one can extract new information from the

Auslander–Reiten quiver once it is constructed. Here are instances of homological information that can be obtained.

(a) *Computing morphisms.* Because arrows in the Auslander–Reiten quiver represent irreducible morphisms, paths correspond to compositions of irreducible morphisms. Thus, because of Corollary II.4.6, one can compute from the Auslander–Reiten quiver all morphisms between indecomposable modules that do not belong to the infinite radical. Indeed, any such morphism corresponds to a linear combination of paths in the Auslander–Reiten quiver. This procedure is especially efficient if one is dealing with a representation-finite algebra, because in this case the infinite radical of the module category is zero, as we shall see in Chapter VI, and therefore every nonzero morphism is a linear combination of irreducible morphisms. Also, over a representation-finite algebra, the absence of multiple arrows in the Auslander–Reiten quiver, see Proposition IV.1.7, implies that arrows can be identified with basis vectors of the space of irreducible morphisms.

On the other hand, the Auslander–Reiten quiver does not represent the morphisms lying in the infinite radical: returning to Example II.4.3, we see that the morphism $S_1 \longrightarrow M$ lying in the infinite radical is not shown as a path in the quiver. This is sometimes expressed by saying that the Auslander–Reiten quiver represents the quotient category $\operatorname{mod} A/\operatorname{rad}_A^\infty$.

A nice application is the following: because of Proposition III.1.11, one can recognise modules of small homological dimensions by means of morphisms: indeed, for a module M, we have $\operatorname{pd} M \leq 1$ (or $\operatorname{id} M \leq 1$) if and only if $\operatorname{Hom}_A(DA, \tau M) = 0$ (or $\operatorname{Hom}_A(\tau^{-1}M, A) = 0$ respectively).

(b) *Computing extensions.* Because of the Auslander–Reiten formulae, Theorem III.2.4, extensions of the first order can be viewed as morphism sets: indeed, let M, N be A-modules, then

$$\operatorname{Ext}_A^1(M, N) \cong D\underline{\operatorname{Hom}}_A(\tau^{-1}N, M) \cong D\overline{\operatorname{Hom}}_A(N, \tau M).$$

Extensions of higher order can be reduced to extensions of the first order using dimension shifting: let M be a module and $0 \longrightarrow L \longrightarrow P \longrightarrow M \longrightarrow 0$ a short exact sequence with P projective, then, for every $n > 1$, we have $\operatorname{Ext}_A^n(M, -) \cong \operatorname{Ext}_A^{n-1}(L, -)$.

One can also get information on the tensor product and torsion groups between modules. Indeed we recall well-known homological formulae: if M, N are A-modules then we have, for every $n \geq 0$, a functorial isomorphism

$$\operatorname{Tor}_n^A(M, DN) \cong D\operatorname{Ext}_A^n(M, N).$$

In particular,

$$M \otimes_A DN \cong D\operatorname{Hom}_A(M, N).$$

These will be proved in Proposition V.2.17 below.

In a given problem, it may not be necessary to know the whole of the Auslander–Reiten quiver, but a small part of it may suffice to obtain the required information. We illustrate these computations in an example.

Example IV.1.25. Let A be given by the quiver

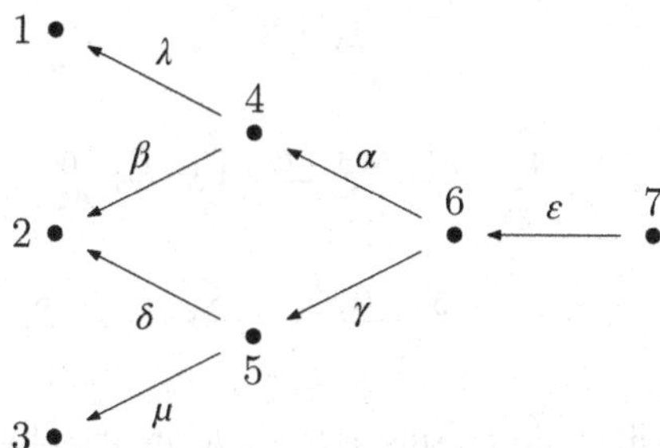

bound by $\epsilon\alpha = 0$, $\alpha\lambda = 0$, $\gamma\mu = 0$, $\alpha\beta = \gamma\delta$. Its Auslander–Reiten quiver $\Gamma(\text{mod}\, A)$ is given by

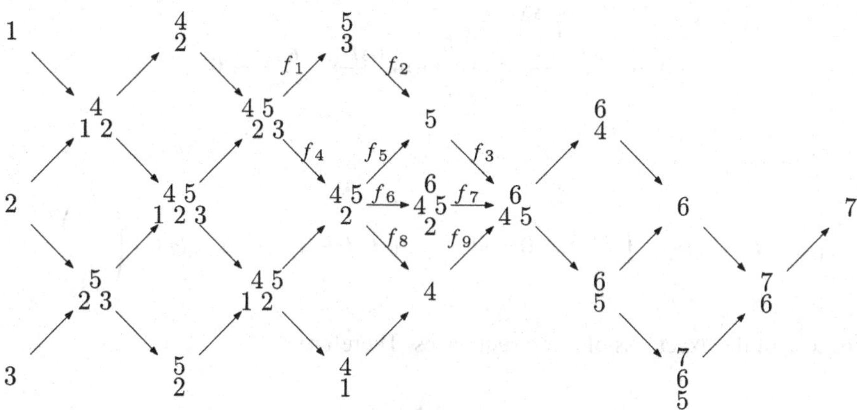

Suppose we wish to find

$$\text{Hom}_A\left(\begin{smallmatrix}4\,5\\2\,3\end{smallmatrix}, \begin{smallmatrix}6\\4\,5\end{smallmatrix}\right).$$

Because A is representation-finite, every nonzero morphism is a sum of compositions of irreducible morphisms (see the previous remarks), and each of these compositions is a path

$$\begin{smallmatrix}4&5\\2&3\end{smallmatrix} \rightsquigarrow \begin{smallmatrix}6\\4\,5\end{smallmatrix}$$

in $\Gamma(\operatorname{mod} A)$. Now we have exactly four paths of this form, namely

$$\begin{smallmatrix}4&5\\2&3\end{smallmatrix} \xrightarrow{f_1} \begin{smallmatrix}5\\3\end{smallmatrix} \xrightarrow{f_2} 5 \xrightarrow{f_3} \begin{smallmatrix}6\\4\,5\end{smallmatrix},$$

$$\begin{smallmatrix}4&5\\2&3\end{smallmatrix} \xrightarrow{f_4} \begin{smallmatrix}4\,5\\2\end{smallmatrix} \xrightarrow{f_5} 5 \xrightarrow{f_3} \begin{smallmatrix}6\\4\,5\end{smallmatrix},$$

$$\begin{smallmatrix}4&5\\2&3\end{smallmatrix} \xrightarrow{f_4} \begin{smallmatrix}4\,5\\2\end{smallmatrix} \xrightarrow{f_6} \begin{smallmatrix}6\\4\,5\\2\end{smallmatrix} \xrightarrow{f_7} \begin{smallmatrix}6\\4\,5\end{smallmatrix}, \text{ and}$$

$$\begin{smallmatrix}4&5\\2&3\end{smallmatrix} \xrightarrow{f_4} \begin{smallmatrix}4\,5\\2\end{smallmatrix} \xrightarrow{f_8} 4 \xrightarrow{f_9} \begin{smallmatrix}6\\4\,5\end{smallmatrix},$$

where the irreducible morphisms $f_1, \ldots, f_9$ are the obvious inclusions and projections. But we also have almost split sequences

$$0 \longrightarrow \begin{smallmatrix}4&5\\2&3\end{smallmatrix} \xrightarrow{\begin{pmatrix}f_1\\f_4\end{pmatrix}} \begin{smallmatrix}5\\3\end{smallmatrix} \oplus \begin{smallmatrix}4\,5\\2\end{smallmatrix} \xrightarrow{(f_2 f_5)} 5 \longrightarrow 0 \qquad \text{and}$$

$$0 \longrightarrow \begin{smallmatrix}4\,5\\2\end{smallmatrix} \xrightarrow{\begin{pmatrix}f_5\\f_6\\f_8\end{pmatrix}} 5 \oplus \begin{smallmatrix}6\\4\,5\\2\end{smallmatrix} \oplus 4 \xrightarrow{(f_3 f_7 f_9)} \begin{smallmatrix}6\\4\,5\end{smallmatrix} \longrightarrow 0.$$

so that

$$f_2 f_1 + f_5 f_4 = (f_2 f_5)\begin{pmatrix}f_1\\f_4\end{pmatrix} = 0 \text{ and } f_3 f_5 + f_7 f_6 + f_9 f_8 = (f_3 f_7 f_9)\begin{pmatrix}f_5\\f_6\\f_8\end{pmatrix} = 0$$

because of the exactness of these sequences. Therefore,

$$\operatorname{Hom}_A\left(\begin{smallmatrix}4&5\\2&3\end{smallmatrix}, \begin{smallmatrix}6\\4\,5\end{smallmatrix}\right)$$

is two-dimensional and one may take, for instance, $\{f_3 f_2 f_1, f_9 f_8 f_4\}$ as a basis of this vector space. Observe that the morphisms $f_3 f_2 f_1$ and $f_9 f_8 f_4$ have as their respective images the simple modules 5 and 4. In addition, the morphism

$$f_7 f_6 f_4 = -f_3 f_5 f_4 - f_9 f_8 f_4 = f_3 f_2 f_1 - f_9 f_8 f_4$$

factors through the projective–injective

$$\begin{smallmatrix} & 6 & \\ 4 & & 5 \\ & 2 & \end{smallmatrix}$$

and thus both

$$\underline{\mathrm{Hom}}_A\left(\begin{smallmatrix}4 & & 5 & \\ & 2 & & 3\end{smallmatrix}, \begin{smallmatrix} & 6 & \\ 4 & & 5\end{smallmatrix}\right) \quad \text{and} \quad \overline{\mathrm{Hom}}_A\left(\begin{smallmatrix}4 & & 5 & \\ & 2 & & 3\end{smallmatrix}, \begin{smallmatrix} & 6 & \\ 4 & & 5\end{smallmatrix}\right)$$

are one-dimensional vector spaces, generated respectively by $\{\underline{f_3 f_2 f_1}\}$ and $\{\overline{f_3 f_2 f_1}\}$.

Assume that we wish to find

$$\mathrm{Ext}_A^1\left(\begin{smallmatrix} & 6 & \\ 4 & & 5\end{smallmatrix}, \begin{smallmatrix} & 4 & \\ 1 & & 2\end{smallmatrix}\right).$$

Applying the Auslander–Reiten formula yields

$$\mathrm{Ext}_A^1\left(\begin{smallmatrix} & 6 & \\ 4 & & 5\end{smallmatrix}, \begin{smallmatrix} & 4 & \\ 1 & & 2\end{smallmatrix}\right) \cong D\overline{\mathrm{Hom}}_A\left(\begin{smallmatrix} & 4 & \\ 1 & & 2\end{smallmatrix}, \tau\left(\begin{smallmatrix} & 6 & \\ 4 & & 5\end{smallmatrix}\right)\right) \cong$$

$$\cong D\overline{\mathrm{Hom}}_A\left(\begin{smallmatrix} & 4 & \\ 1 & & 2\end{smallmatrix}, \begin{smallmatrix} 4 & & 5 \\ & 2 & \end{smallmatrix}\right) = D\,\mathrm{Hom}_A\left(\begin{smallmatrix} & 4 & \\ 1 & & 2\end{smallmatrix}, \begin{smallmatrix} 4 & & 5 \\ & 2 & \end{smallmatrix}\right).$$

Indeed, no morphism

$$\begin{smallmatrix} & 4 & \\ 1 & & 2\end{smallmatrix} \longrightarrow \begin{smallmatrix} 4 & & 5 \\ & 2 & \end{smallmatrix}$$

factors through an injective module, because no injective lies on a path

$$\begin{smallmatrix} & 4 & \\ 1 & & 2\end{smallmatrix} \rightsquigarrow \begin{smallmatrix} 4 & & 5 \\ & 2 & \end{smallmatrix}.$$

Now, it is easily seen that

$$\mathrm{Hom}_A\left(\begin{smallmatrix} & 4 & \\ 1 & & 2\end{smallmatrix}, \begin{smallmatrix} 4 & & 5 \\ & 2 & \end{smallmatrix}\right)$$

is one-dimensional and spanned by a morphism with image $\begin{smallmatrix}4\\2\end{smallmatrix}$. Therefore,

$$\mathrm{Ext}_A^1\left(\begin{smallmatrix} & 6 & \\ 4 & & 5\end{smallmatrix}, \begin{smallmatrix} & 4 & \\ 1 & & 2\end{smallmatrix}\right)$$

is one-dimensional. It is actually easy to see that there is a nonsplit short exact sequence

$$0 \longrightarrow \begin{smallmatrix}4\\1\,2\end{smallmatrix} \longrightarrow \begin{smallmatrix}4\\1\end{smallmatrix} \oplus \begin{smallmatrix}6\\4\,5\\2\end{smallmatrix} \longrightarrow \begin{smallmatrix}6\\4\,5\end{smallmatrix} \longrightarrow 0$$

with the obvious morphisms. The class of this sequence is a basis of the Ext^1-space under consideration. Similarly, we have

$$\operatorname{Ext}_A^1\left(\begin{smallmatrix}7\\6\end{smallmatrix}, \begin{smallmatrix}4\\2\end{smallmatrix}\right) \cong D\overline{\operatorname{Hom}}_A\left(\begin{smallmatrix}4\\2\end{smallmatrix}, \tau\left(\begin{smallmatrix}7\\6\end{smallmatrix}\right)\right) \cong D\overline{\operatorname{Hom}}_A\left(\begin{smallmatrix}4\\2\end{smallmatrix}, \begin{smallmatrix}6\\5\end{smallmatrix}\right) = 0$$

because

$$\operatorname{Hom}_A\left(\begin{smallmatrix}4\\2\end{smallmatrix}, \begin{smallmatrix}6\\5\end{smallmatrix}\right) = 0.$$

Suppose now we want to compute

$$\operatorname{Ext}_A^2\left(\begin{smallmatrix}7\\6\end{smallmatrix}, \begin{smallmatrix}5\\2\,3\end{smallmatrix}\right).$$

Clearly, we have a short exact sequence

$$0 \longrightarrow 5 \longrightarrow \begin{smallmatrix}7\\6\\5\end{smallmatrix} \longrightarrow \begin{smallmatrix}7\\6\end{smallmatrix} \longrightarrow 0$$

with a projective middle term. Dimension shifting and the Auslander–Reiten formulae give

$$\operatorname{Ext}_A^2\left(\begin{smallmatrix}7\\6\end{smallmatrix}, \begin{smallmatrix}5\\2\,3\end{smallmatrix}\right) \cong \operatorname{Ext}_A^1\left(5, \begin{smallmatrix}5\\2\,3\end{smallmatrix}\right) \cong D\overline{\operatorname{Hom}}_A\left(\begin{smallmatrix}5\\2\,3\end{smallmatrix}, \tau(5)\right) \cong$$

$$\cong D\overline{\operatorname{Hom}}_A\left(\begin{smallmatrix}5\\2\,3\end{smallmatrix}, \begin{smallmatrix}4\,5\\2\,3\end{smallmatrix}\right) = D\operatorname{Hom}_A\left(\begin{smallmatrix}5\\2\,3\end{smallmatrix}, \begin{smallmatrix}4\,5\\2\,3\end{smallmatrix}\right)$$

which is one-dimensional. It is easily seen that a basis for this extension space is given by the class of the exact sequence

$$0 \longrightarrow \begin{smallmatrix}5\\2\,3\end{smallmatrix} \longrightarrow \begin{smallmatrix}5\\2\end{smallmatrix} \oplus \begin{smallmatrix}5\\3\end{smallmatrix} \longrightarrow \begin{smallmatrix}7\\6\\5\end{smallmatrix} \longrightarrow \begin{smallmatrix}7\\6\end{smallmatrix} \longrightarrow 0.$$

Looking for the indecomposable modules M such that $\operatorname{Hom}_A(DA, \tau M) = 0$, we find all indecomposable modules M such that $\operatorname{pd} M \leq 1$. These are

$$\left\{ 1, 2, 3, \begin{smallmatrix} 4 \\ 1\,2 \end{smallmatrix}, \begin{smallmatrix} 5 \\ 2\,3 \end{smallmatrix}, \begin{smallmatrix} 6 \\ 4\,5 \\ 2 \end{smallmatrix}, \begin{smallmatrix} 4 \\ 2 \end{smallmatrix}, \begin{smallmatrix} & 4 & & 5 & \\ 1 & & 2 & & 3 \end{smallmatrix}, \begin{smallmatrix} 5 \\ 2 \end{smallmatrix}, \begin{smallmatrix} 4 & & 5 & \\ & 2 & & 3 \end{smallmatrix}, \begin{smallmatrix} & 4 & & 5 \\ 1 & & 2 & \end{smallmatrix}, \begin{smallmatrix} 5 \\ 3 \end{smallmatrix}, \begin{smallmatrix} 4 \\ 1 \end{smallmatrix}, 5, 4, \begin{smallmatrix} 6 \\ 4\,5 \end{smallmatrix}, \begin{smallmatrix} 7 \\ 6 \\ 5 \end{smallmatrix} \right\}.$$

Similarly, all indecomposable modules M such that $\operatorname{id} M \leq 1$ are

$$\left\{ 7, \begin{smallmatrix} 7 \\ 6 \end{smallmatrix}, \begin{smallmatrix} 7 \\ 6 \\ 5 \end{smallmatrix}, \begin{smallmatrix} 6 \\ 4\,5 \\ 2 \end{smallmatrix}, \begin{smallmatrix} 5 \\ 3 \end{smallmatrix}, \begin{smallmatrix} 4 \\ 1 \end{smallmatrix}, 6, \begin{smallmatrix} 6 \\ 4 \end{smallmatrix}, \begin{smallmatrix} 6 \\ 5 \end{smallmatrix}, \begin{smallmatrix} 6 \\ 4\,5 \end{smallmatrix}, 5, \begin{smallmatrix} 4\,5 \\ 1\,2 \end{smallmatrix} \right\}.$$

If one looks for pd 7, one can construct the following minimal projective resolution, which gives $\operatorname{pd} 7 = 3$:

$$0 \longrightarrow 1 \longrightarrow \begin{smallmatrix} 4 \\ 1\,2 \end{smallmatrix} \longrightarrow \begin{smallmatrix} 6 \\ 4\,5 \\ 2 \end{smallmatrix} \longrightarrow \begin{smallmatrix} 7 \\ 6 \\ 5 \end{smallmatrix} \longrightarrow 7 \longrightarrow 0.$$

Doing the same calculation for the other simple modules, one easily gets $\operatorname{gl.dim.} A = 3$.

Exercises for Section IV.1

Exercise IV.1.1. For each of the following bound quivers, compute the corresponding Auslander–Reiten quiver.

(a) $1 \xleftarrow{\varepsilon} 2 \xleftarrow{\delta} 3 \xleftarrow{\gamma} 4 \xleftarrow{\beta} 5 \xleftarrow{\alpha} 6 \qquad \alpha\beta\gamma\delta = 0,\ \gamma\delta\varepsilon = 0$

(b) 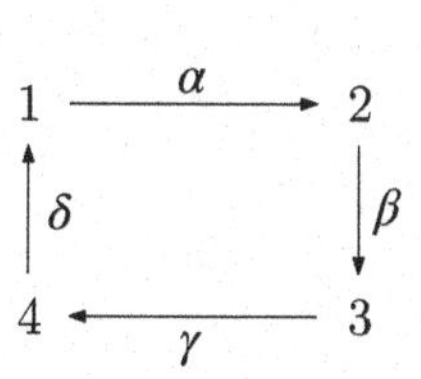 $$\begin{array}{ccc} 1 & \xrightarrow{\alpha} & 2 \\ \delta \uparrow & & \downarrow \beta \\ 4 & \xleftarrow[\gamma]{} & 3 \end{array} \qquad \alpha\beta = 0,\ \beta\gamma = 0$$

(c) 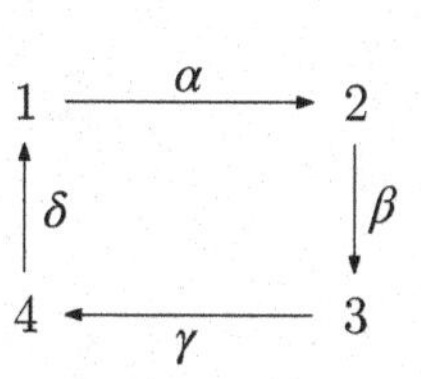 $$\begin{array}{ccc} 1 & \xrightarrow{\alpha} & 2 \\ \delta \uparrow & & \downarrow \beta \\ 4 & \xleftarrow[\gamma]{} & 3 \end{array} \qquad \alpha\beta\gamma = 0$$

(d) Quiver: $1 \xrightarrow{\alpha} 2 \xrightarrow{\beta} 3 \xrightarrow{\gamma} 4 \xrightarrow{\delta} 5 \xrightarrow{\epsilon} 6 \xrightarrow{\lambda} 1$

$\alpha\beta\gamma = 0,\ \beta\gamma\delta = 0$
$\gamma\delta\epsilon = 0,\ \delta\epsilon\lambda = 0$
$\epsilon\lambda\alpha = 0,\ \lambda\alpha\beta = 0$

(e) Quiver: $3 \xrightarrow{\alpha} 1$, $3 \xrightarrow{\beta} 2$, $4 \xrightarrow{\gamma} 3$, $5 \xrightarrow{\delta} 4$, $6 \xrightarrow{\varepsilon} 5$, $6 \xrightarrow{\lambda} 7$

$\delta\gamma\beta = 0$

(f) Quiver: $3 \xrightarrow{\beta} 1$, $3 \xrightarrow{\delta} 2$, $4 \xrightarrow{\alpha} 3$, $5 \xrightarrow{\gamma} 3$

$\alpha\beta = 0$
$\gamma\delta = 0$

(g) Quiver: $3 \xrightarrow{\beta} 1$, $3 \xrightarrow{\delta} 2$, $4 \xrightarrow{\alpha} 3$, $5 \xrightarrow{\gamma} 3$

$\alpha\beta = 0$
$\alpha\delta = 0$
$\gamma\beta = 0$
$\gamma\delta = 0$

(h) Quiver: $3 \xrightarrow{\beta} 1$, $3 \xrightarrow{\delta} 2$, $4 \xrightarrow{\alpha} 3$, $5 \xrightarrow{\gamma} 3$

$\alpha\beta = 0$

(i) Quiver: $1 \xrightarrow{\alpha} 2 \xrightarrow{\beta} 3 \xrightarrow{\delta} 4 \xrightarrow{\varepsilon} 5 \xrightarrow{\lambda} 6$, $7 \xrightarrow{\gamma} 3$

$\gamma\delta\varepsilon\lambda = 0,$
$\beta\delta\varepsilon\lambda = 0$

(j) Quiver: $4 \xrightarrow{\alpha} 2 \xrightarrow{\beta} 1$, $4 \xrightarrow{\gamma} 3 \xrightarrow{\delta} 1$

$\alpha\beta = \gamma\delta$

(k) Quiver: $4 \xrightarrow{\alpha} 2 \xrightarrow{\beta} 1$, $4 \xrightarrow{\gamma} 3 \xrightarrow{\delta} 1$

$\alpha\beta = 0$
$\gamma\delta = 0$

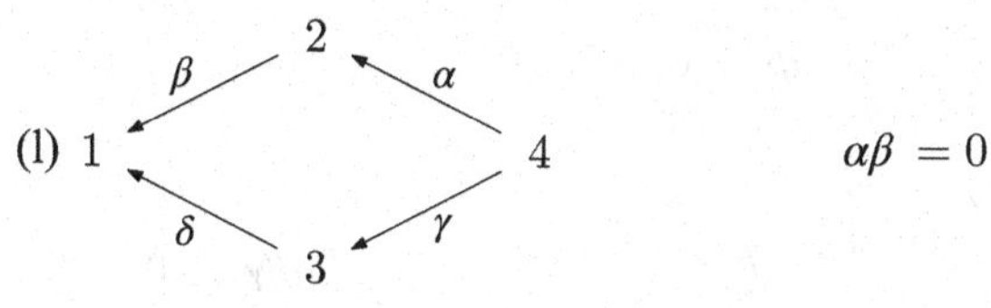

$\alpha\beta = 0$

(m)

$\alpha\beta = 0$
$\gamma\delta = 0$
$\gamma\varepsilon = 0$

(n)

$\beta\gamma = \delta\varepsilon$
$\alpha\beta = 0$
$\gamma\lambda = 0$

(o)

$\alpha\beta = 0$
$\gamma\delta = 0$
$\delta\varepsilon = 0$

(p)

$\gamma\varepsilon = 0$
$\beta\delta = 0$
$\alpha\beta = 0$

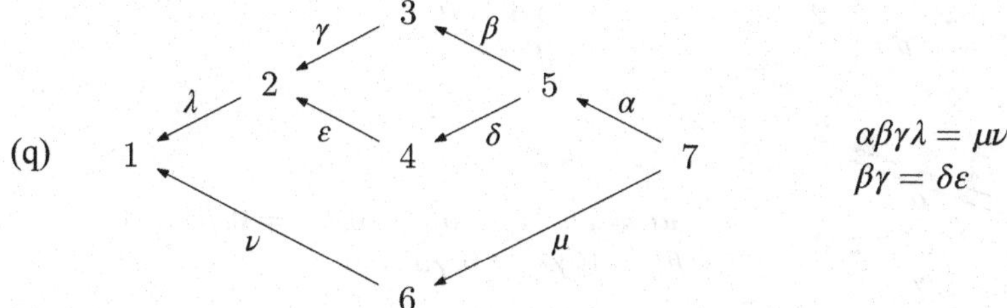

$\alpha\beta\gamma\lambda = \mu\nu$
$\beta\gamma = \delta\varepsilon$

(r)

$\beta\gamma = \nu\varepsilon$
$\alpha\nu = \lambda\mu$

(s) $1 \xleftarrow{\beta} 2 \circlearrowleft \alpha$ $\qquad \alpha\beta = 0,\ \alpha^2 = 0$

(t) $1 \xleftarrow{\varepsilon} 2 \xleftarrow{\delta} 3 \xleftarrow{\gamma} 4 \xleftarrow{\beta} 8 \xleftarrow{\alpha} 9$, $\quad 1 \xleftarrow{\omega} 5 \xleftarrow{\nu} 6 \xleftarrow{\mu} 7 \xleftarrow{\lambda} 8$

$$\alpha\beta\gamma = 0 \qquad \alpha\lambda\mu = 0 \qquad \beta\gamma\delta\varepsilon = \lambda\mu\nu\omega$$

(u) $1 \xleftarrow{\delta} 2 \xleftarrow{\gamma} 3 \underset{\beta}{\overset{\alpha}{\rightleftarrows}} 4$ $\qquad \alpha\beta = 0$

(v) $1 \underset{\beta}{\overset{\alpha}{\leftrightarrows}} 2 \underset{\delta}{\overset{\gamma}{\leftrightarrows}} 3$

$$\alpha\beta = \gamma\delta \qquad \delta\alpha = 0 \qquad \beta\gamma = 0$$

(w) $4 \xrightarrow{\alpha} 2 \xrightarrow{\beta} 1$, $\quad 4 \xrightarrow{\gamma} 3 \xrightarrow{\delta} 1$, $\quad 1 \xrightarrow{\varepsilon} 4$

$$\alpha\beta = \gamma\delta,\ \beta\varepsilon = 0 \qquad \delta\varepsilon = 0,\ \varepsilon\alpha = 0 \qquad \varepsilon\gamma = 0$$

(x) $4 \xrightarrow{\alpha} 2 \xrightarrow{\beta} 1$, $\quad 4 \xrightarrow{\gamma} 3 \xrightarrow{\delta} 1$, $\quad 1 \xrightarrow{\varepsilon} 4$

$$\alpha\beta = \gamma\delta \qquad \delta\varepsilon\alpha = 0 \qquad \beta\varepsilon\gamma = 0$$

(y) $\gamma \circlearrowright 1 \underset{\beta}{\overset{\alpha}{\leftrightarrows}} 2$

$$\alpha\beta = \gamma^2 \qquad \gamma\alpha = 0 \qquad \beta\gamma = 0$$

(z) $1 \underset{\lambda}{\overset{\alpha}{\leftrightarrows}} 2$, $\quad 1 \underset{\mu}{\overset{\beta}{\leftrightarrows}} 3$, $\quad 1 \underset{\nu}{\overset{\gamma}{\leftrightarrows}} 4$

$$\lambda\alpha = \mu\beta = \nu\gamma,\ \alpha\mu = 0,\ \alpha\nu = 0,\ \beta\lambda = 0,$$
$$\beta\nu = 0,\ \gamma\lambda = 0,\ \gamma\mu = 0$$

Exercise IV.1.2. Let A be the path algebra of the quiver

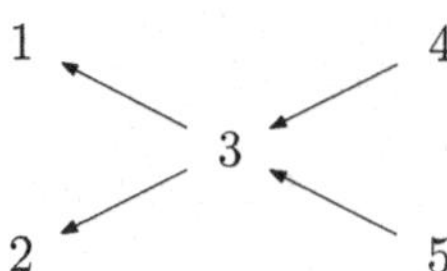

(a) Construct the component of $\Gamma(\mathrm{mod}\, A)$ containing the indecomposable projective modules and show that this component contains no indecomposable injective. Next, show that this component is infinite.
(b) Construct the component of $\Gamma(\mathrm{mod}\, A)$ containing the indecomposable injective modules and show that this component contains no indecomposable projective. Next, show that this component is infinite.
(c) Exhibit an indecomposable A-module belonging to neither of these components, so that $\Gamma(\mathrm{mod}\, A)$ has at least three connected components.

Exercise IV.1.3. Let A be the matrix algebra

$$A = \begin{pmatrix} k & 0 & 0 & 0 \\ k & k & 0 & 0 \\ k & 0 & k & 0 \\ k & 0 & 0 & k \end{pmatrix} = \left\{ \begin{pmatrix} a & 0 & 0 & 0 \\ b & c & 0 & 0 \\ d & 0 & e & 0 \\ f & 0 & 0 & g \end{pmatrix} : a, b, \ldots, g \in \mathbf{k} \right\}$$

with the ordinary matrix addition and multiplication. Let S_1 be the only simple projective A-module. Prove that $\dim(\mathrm{Hom}_A(S_1, \tau^{-1}S_1)) = 2$.

Exercise IV.1.4. Let A be given by the quiver

$$4 \underset{\varepsilon}{\overset{\delta}{\rightleftarrows}} 3 \xleftarrow{\ \gamma\ } 1 \underset{\beta}{\overset{\alpha}{\rightleftarrows}} 2$$

bound by $\beta\alpha = 0$, $\delta\epsilon = 0$ and $\alpha\beta\gamma = \gamma\epsilon\delta$.

(a) Compute the almost split sequences ending in the simple modules S_2 and S_3.
(b) Compute the almost split sequences that have projective–injective middle terms.
(c) Deduce the Auslander–Reiten quiver of A.

Exercise IV.1.5. Let A be a finite dimensional algebra. Prove that there is no arrow in $\Gamma(\mathrm{mod}\, A)$ from an injective to a projective, but there may be arrows from a projective to an injective.

Exercise IV.1.6. Let A_n be the algebra given by the quiver

$$1 \xleftarrow{\alpha_{n-1}} 2 \xleftarrow{\alpha_n} 3 \quad \cdots\cdots \quad n-1 \xleftarrow{\alpha_1} n$$

bound by $\alpha_i\alpha_{i+1} = 0$ for all i. Compute, in terms of n, the extension group $\mathrm{Ext}^j_{A_n}(S_n, S_1)$ for all j such that $1 \leq j < n$.

Exercise IV.1.7. Let Q, Q' be quivers that have the same tree as the underlying graph. Prove that $\mathbb{Z}\,Q$ and $\mathbb{Z}\,Q'$ are isomorphic quivers.

Exercise IV.1.8. Give an example showing that the result of Exercise IV.1.7 above is no longer true if one does not assume that the underlying graph is a tree.

Exercise IV.1.9. Show that the following translation quiver is not an Auslander–Reiten quiver

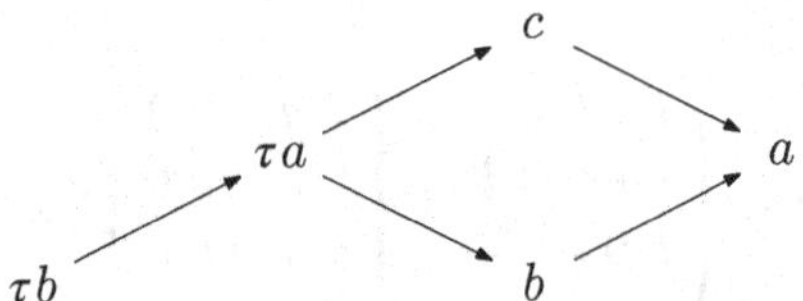

IV.2 Postprojective and preinjective components

IV.2.1 Definitions and characterisations

This section is devoted to the study of two types of connected components that occur frequently in Auslander–Reiten quivers. These are the so-called postprojective and preinjective components, the second being dual to the first. The reason for looking at this type of component is that, if they exist, then they can always be constructed using the knitting algorithm. In particular, postprojective and preinjective components are acyclic. Thus, we start by explaining what we mean by an acyclic component.

Let A be an algebra. We recall from Definition II.4.7 that, if M, N are indecomposable A-modules, then a radical path from M to N in $\mathrm{ind}\,A$ of length t is a sequence

$$M = M_0 \xrightarrow{f_1} M_1 \longrightarrow \ldots \longrightarrow M_{t-1} \xrightarrow{f_t} M_t = N$$

where all M_i are indecomposable and all f_i are radical morphisms. A **cycle** through an indecomposable module M is a radical path from M to itself of length greater than or equal to one.

If $M = M_0 \xrightarrow{f_1} M_1 \longrightarrow \dots \longrightarrow M_{t-1} \xrightarrow{f_t} M_t = M$ is a cycle through M with all f_i irreducible morphisms (that is, a path of irreducible morphisms), then it can be identified to a cycle in the quiver $\Gamma(\operatorname{mod} A)$. In this case, $t \geq 2$, because there are no irreducible morphisms from a module to itself. In addition, all the M_i lie in the same connected component of $\Gamma(\operatorname{mod} A)$, so that one can speak of a cycle in that component. A component Γ of $\Gamma(\operatorname{mod} A)$ is called **acyclic** if it contains no cycle.

For examples of cycles in $\Gamma(\operatorname{mod} A)$, we refer to Examples IV.1.10, IV.1.14 and IV.1.15 above. All other components constructed in Subsection IV.1.3 are acyclic. To motivate the next definition, we refer the reader to Example IV.1.9 above showing the Auslander–Reiten quiver of the algebra given by the quiver

$$\underset{\bullet}{1} \xleftarrow{\ \gamma\ } \underset{\bullet}{2} \xleftarrow{\ \beta\ } \underset{\bullet}{3} \xleftarrow{\ \alpha\ } \underset{\bullet}{4}$$

bound by $\alpha\beta\gamma = 0$. The quiver is acyclic and can be represented as follows:

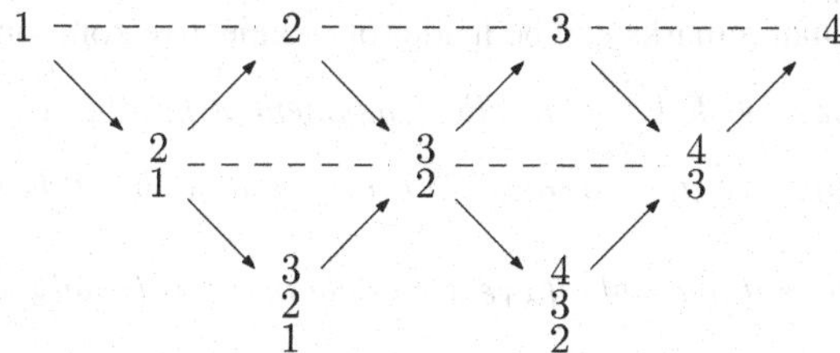

that is, every indecomposable module M can be written in the form $\tau^{-t}P$, where $t \geq 0$ and P is indecomposable projective, or, equivalently, in the form $\tau^s I$, where $s \geq 0$ and I is indecomposable injective. This leads us to the following definition.

Definition IV.2.1. An acyclic component Γ of $\Gamma(\operatorname{mod} A)$ is called **postprojective** (or **preinjective**) provided that every indecomposable M in Γ can be written in the form $\tau^{-t}P$, with $t \geq 0$ and P an indecomposable projective (or in the form $\tau^s I$, with $s \geq 0$ and I an indecomposable injective respectively). An indecomposable module is called **postprojective** (or **preinjective**) if it belongs to a postprojective component (or preinjective component respectively).

These concepts are dual to each other: applying the duality functor D to a postprojective component yields a preinjective component in the module category of the opposite algebra, and conversely.

The example preceding the definition shows that, given a representation-finite algebra with acyclic Auslander–Reiten quiver, then this whole quiver is at the same time a postprojective and a preinjective component.

We observe that $M \cong \tau^{-t}P$ holds if and only if $\tau^t M \cong P$. Therefore, an acyclic component is postprojective if and only if, for every M in that component, there exists $t \geq 0$ such that $\tau^t M$ is indecomposable projective. Another observation is that, if $M \cong \tau^{-t}P$ lies in a postprojective component, where $t \geq 0$ and P is indecomposable projective, then t and P are unique. Indeed, if $M \cong \tau^{-t}P \cong \tau^{-r}P'$, then we may assume without loss of generality that $t \geq r$ and we get that $P \cong \tau^{t-r}P'$ is projective, from which we deduce that $t = r$ and $P \cong P'$. The dual observations hold for preinjective components.

We say that the indecomposable A-modules M and N belong to the same τ-orbit if there exists an integer t such that $M \cong \tau^t N$. Thus, if M, N belong to the same τ-orbit, then they must belong to the same component of $\Gamma(\mathrm{mod}\,A)$. Clearly, belonging to the same τ-orbit is an equivalence relation. The equivalence classes are the sets $\{\tau^t M : t \in \mathbb{Z}, M \text{ in } \mathrm{ind}\,A\}$, called the **$\tau$-orbits** of $\Gamma(\mathrm{mod}\,A)$. Thus, one can define postprojective components to be acyclic components in which every indecomposable lies in the τ-orbit of an indecomposable projective. This implies that the number of τ-orbits in a postprojective component is bounded by the number of isoclasses of indecomposable projective modules. In particular, the number is finite. As we now see, postprojective components of Auslander–Reiten quivers can also be characterised as being those acyclic components in which every module has only finitely many predecessors, or, equivalently, such that every path ending in a module is finite. Dual remarks can be made for preinjective components.

Proposition IV.2.2. *Let Γ be an acyclic component of* $\Gamma(\mathrm{mod}\,A)$. *Then,*

(a) *Γ is postprojective if and only if every module in it has finitely many predecessors in Γ.*
(b) *Γ is preinjective if and only if every module in it has finitely many successors in Γ.*

Proof. We only prove (a) because the proof of (b) is dual.

Assume first that Γ is postprojective and that there exists a module M in Γ that has infinitely many predecessors in Γ. Then, there exists an infinite path of irreducible morphisms:

$$\ldots \longrightarrow M_i \longrightarrow M_{i-1} \longrightarrow \ldots \longrightarrow M_1 \longrightarrow M_0 = M$$

with all M_i in Γ. The acyclicity of Γ implies that $M_i \not\cong M_j$ for $i \neq j$. By hypothesis, for each $j > 0$, there exists $t_j \geq 0$ such that $\tau^{t_j} M_j$ is indecomposable projective. Because the infinitely many M_j distribute among the finitely many τ-orbits of projectives in Γ, there must exist an indecomposable projective module P such that the set $J = \{j : \tau^{t_j} M_j = P\}$ is infinite. Consider the function $t : J \longrightarrow \mathbb{N}$ defined by $j \mapsto t_j$. It cannot be strictly decreasing, because J is an infinite set. Therefore, there exist $i, j \in J$ such that we have both $i < j$ and $t_i < t_j$. But then $\tau^{t_i} M_i \cong P \cong \tau^{t_j} M_j$ yields $M_i \cong \tau^{t_j - t_i} M_j$. This relation implies the existence of a path of irreducible morphisms $M_i \rightsquigarrow M_j$ in Γ of length at least one. In addition, $i < j$ also states that there is a path of irreducible morphisms $M_j \rightsquigarrow M_i$ in Γ of length at least one. Combining these paths, we get a cycle in Γ, a contradiction. This proves necessity.

Conversely, assume that Γ is not postprojective, but every indecomposable module in Γ has only finitely many predecessors in Γ. Because Γ was assumed to be acyclic, but not postprojective, there exists an indecomposable module M in Γ such that, for every $n \geq 0$, $\tau^n M$ is not projective. Then, for each $n \geq 0$, there exists a path of irreducible morphisms:

$$\tau^n M \longrightarrow * \longrightarrow \tau^{n-1} M \longrightarrow * \longrightarrow \ldots \longrightarrow \tau M \longrightarrow * \longrightarrow M$$

in Γ, and this contradicts our hypothesis that M has only finitely many predecessors in Γ. □

For examples of infinite postprojective and preinjective components, we refer the reader to Example IV.1.16, where we constructed these components for the Kronecker algebra.

Other important properties of postprojective and preinjective components are recorded in the following propositions. Observe that, in the same fashion as for the radical in Subsection II.1.3, we can define a subfunctor $\mathrm{rad}_A^\infty(M,-)$ of $\mathrm{Hom}_A(M,-)$ and a subfunctor $\mathrm{rad}_A^\infty(-,M)$ of $\mathrm{Hom}_A(-,M)$.

Proposition IV.2.3. *Let Γ be a component of $\Gamma(\mathrm{mod}\, A)$ and M, N indecomposable modules. Assume that N belongs to Γ.*

(a) *If Γ is postprojective, then*

  (i) $\mathrm{rad}_A^\infty(-,N) = 0$. *In particular, N has only finitely many predecessors in* $\mathrm{ind}\, A$.

  (ii) *If M is an indecomposable module such that $\mathrm{rad}_A(M,N) \neq 0$, then there exists a path of irreducible morphisms $M \rightsquigarrow N$ in Γ.*

(b) *If Γ is preinjective, then*

  (i) $\mathrm{rad}_A^\infty(N,-) = 0$. *In particular, N has only finitely many successors in* $\mathrm{ind}\, A$.

  (ii) *If M is an indecomposable module such that* $\mathrm{rad}_A(N,M) \neq 0$, *then there exists a path of irreducible morphisms $N \rightsquigarrow M$ in Γ.*

Proof. We prove only (a) because the proof of (b) is dual.

(i) Let $f\colon M \longrightarrow N$ be a nonzero morphism in $\mathrm{rad}_A^\infty(M,N)$, where M is an indecomposable module. Because of Proposition II.4.9, there exist for every $i \geq 0$, a path of irreducible morphisms

$$N_i \xrightarrow{g_i} N_{i-1} \longrightarrow \ldots \longrightarrow N_1 \xrightarrow{g_1} N_0 = N$$

and a morphism $f_i \in \mathrm{rad}_A^\infty(M,N_i)$ such that $g_1 \ldots g_i f_i \neq 0$. The acyclicity of Γ guarantees that the N_i are distinct. But this contradicts the fact that, because of Proposition IV.2.2, N has only finitely many predecessors in Γ. Therefore, $\mathrm{rad}_A^\infty(M,N) = 0$. The second statement follows directly.

(ii) This follows from Corollary II.4.8(b) using item (i).

□

Proposition IV.2.4. *Let Γ be a postprojective or preinjective component of $\Gamma(\operatorname{mod} A)$ and M an indecomposable in Γ. Then,* $\operatorname{End} M \cong \mathbf{k}$ *and* $\operatorname{Ext}_A^1(M, M) = 0$.

Proof. Suppose $\operatorname{End} M \ncong \mathbf{k}$. Because $\operatorname{End} M$ is local, we infer that $\operatorname{rad}_A(M, M) = \operatorname{rad}(\operatorname{End} M) \neq 0$. Thus, there exists a nonzero radical morphism $M \longrightarrow M$. Because of Proposition IV.2.3, this means that there exists a cycle through M lying in Γ, a contradiction to its acyclicity.

Assume $\operatorname{Ext}_A^1(M, M) \neq 0$. Because of the Auslander–Reiten formula

$$\operatorname{Ext}_A^1(M, M) = \mathrm{D}\overline{\operatorname{Hom}}_A(M, \tau M)$$

we deduce the existence of a nonzero morphism $M \longrightarrow \tau M$, and hence of a path of irreducible morphisms $M \rightsquigarrow \tau M$ in Γ. Now, this path gives rise to a cycle $M \rightsquigarrow \tau M \longrightarrow * \longrightarrow M$ lying in Γ, and this is again a contradiction. □

IV.2.2 Postprojective and preinjective components for path algebras

Let Q be a finite, connected and acyclic quiver. Because of Proposition I.2.28, the path algebra $\mathbf{k}Q$ is hereditary. We now prove that $\mathbf{k}Q$ always admits exactly one postprojective component containing all indecomposable projectives and one preinjective component containing all indecomposable injectives. Recall that, to each point x in Q, we can assign an indecomposable projective $\mathbf{k}Q$-module P_x, an indecomposable injective $\mathbf{k}Q$-module I_x and a simple $\mathbf{k}Q$-module S_x such that $P_x / \operatorname{rad} P_x \cong S_x \cong \operatorname{soc} I_x$.

We need the following lemma.

Lemma IV.2.5. *If $A = \mathbf{k}Q$, then:*

(a) *The predecessors of points in $\Gamma(\operatorname{mod} A)$ corresponding to indecomposable projective A-modules also correspond to indecomposable projective modules.*
(b) *The successors of points in $\Gamma(\operatorname{mod} A)$ corresponding to indecomposable injective A-modules also correspond to indecomposable injective modules.*

Proof. We prove only (a), because the proof of (b) is dual.

Assume that P is indecomposable projective and that $f\colon M \longrightarrow P$ is irreducible. Because of Theorem II.2.24 and Example II.2.21, f is the inclusion of a direct summand of $\operatorname{rad} P$ into P. But then M is isomorphic to a submodule of P and hence is projective, because $A = \mathbf{k}Q$ is hereditary. This proves that every immediate predecessor of P is projective. The statement follows from an easy induction. □

It follows from the preceding lemma that the full subquiver of $\Gamma(\mathrm{mod}\, A)$ consisting of the indecomposable projectives is connected and so lies in a unique component of $\Gamma(\mathrm{mod}\, A)$. Dually, the full subquiver of $\Gamma(\mathrm{mod}\, A)$ consisting of the indecomposable injectives is connected and so lies in a unique component of $\Gamma(\mathrm{mod}\, A)$.

Lemma IV.2.6. *If $A = \mathbf{k}Q$, then the full subquiver of $\Gamma(\mathrm{mod}\, A)$ consisting of the indecomposable projective (or injective) A-modules is connected and isomorphic to Q^{op}.*

Proof. We give the proof for the projectives, the proof for the injectives being similar.

We know that the full subquiver of $\Gamma(\mathrm{mod}\, A)$ consisting of the projectives is connected. To prove that the subquiver is isomorphic to Q^{op}, it suffices to prove that, for any points x, y in Q, we have

$$\mathrm{Irr}_A(P_y, P_x) \cong e_x \left(\frac{\mathrm{rad}\, A}{\mathrm{rad}^2\, A}\right) e_y.$$

We have $P_x = e_x A$, $P_y = e_y A$, and hence functorial isomorphisms

$$\begin{aligned}\mathrm{rad}_A(P_y, P_x) &= \mathrm{rad}_A(e_y A, e_x A) \cong \mathrm{Hom}_A(e_y A, \mathrm{rad}(e_x A)) \\ &\cong \mathrm{rad}(e_x A) e_y \cong e_x(\mathrm{rad}\, A) e_y.\end{aligned}$$

Passing to the radical square, we have similarly

$$\mathrm{rad}^2_A(P_y, P_x) = \mathrm{rad}(e_y A, \mathrm{rad}(e_x A)).$$

Because $A = \mathbf{k}Q$ is hereditary, $\mathrm{rad}(e_x A)$ is projective (decomposable in general) so $\mathrm{rad}_A(e_y A, \mathrm{rad}(e_x A))$ consists of those morphisms that are not sections, that is, those whose image lies in $\mathrm{rad}(\mathrm{rad}(e_x A)) = \mathrm{rad}^2(e_x A)$. Hence,

$$\begin{aligned}\mathrm{rad}^2_A(P_y, P_x) &\cong \mathrm{Hom}_A(e_y A, \mathrm{rad}^2(e_x A)) \\ &\cong \mathrm{rad}^2(e_x A) e_y \cong e_x(\mathrm{rad}^2\, A) e_y.\end{aligned}$$

Passing to the quotients, we get

$$\mathrm{Irr}_A(P_y, P_x) \cong \frac{\mathrm{rad}_A(P_y, P_x)}{\mathrm{rad}^2_A(P_y, P_x)} \cong \frac{e_x(\mathrm{rad}\, A) e_y}{e_x(\mathrm{rad}^2\, A) e_y} \cong e_x \left(\frac{\mathrm{rad}\, A}{\mathrm{rad}^2\, A}\right) e_y \ ,$$

as required. □

Proposition IV.2.7. *Let Q be a finite, connected and acyclic quiver and $A = \mathbf{k}Q$. Then $\Gamma(\mathrm{mod}\, \mathbf{k}Q)$ admits a unique postprojective component containing all indecomposable projectives and a unique preinjective component containing all indecomposable injectives.*

Proof. Because of Lemma IV.2.6, there is a full connected subquiver of $\Gamma(\operatorname{mod} A)$ isomorphic to Q^{op} and consisting of all indecomposable projectives. Let Γ be the unique connected component of $\Gamma(\operatorname{mod} A)$ containing this subquiver.

We claim that every indecomposable module in Γ is of the form $\tau^{-t} P_x$ with $t \geq 0$ and $x \in Q_0$. If this is not the case, then there exists a module M that is not in the τ-orbit of a projective. Without loss of generality, we may assume that there exist a module $L \cong \tau^{-n} P$ and an irreducible morphism $L \longrightarrow M$ or $M \longrightarrow L$. Assume the former; then, the irreducible morphism $\tau^{-n} P \longrightarrow M$ induces an irreducible morphism $P \longrightarrow \tau^n M$ (by applying τ^n). By hypothesis, M does not lie in the τ-orbit of a projective, and so, in particular, $\tau^n M$ is nonzero, and nonprojective. But then we have an irreducible morphism $\tau^{n+1} M \longrightarrow P$ and Lemma IV.2.5 implies that $\tau^{n+1} M$ is projective, a contradiction to our assumption on M.

It remains to prove that Γ is acyclic. Indeed, suppose that

$$M = M_0 \xrightarrow{f_1} M_1 \longrightarrow \ldots \longrightarrow M_{m-1} \xrightarrow{f_m} M_m = M$$

is a cycle (of irreducible morphisms) in Γ. Because of our claim above, for each i, there exist an $m_i \geq 0$ and an $x_i \in Q_0$ such that $M_i = \tau^{-t_i} P_{x_i}$. Let $t = \min\{t_i : 1 \leq i \leq m\}$. Then we have a cycle

$$\tau^t M = \tau^t M_0 \xrightarrow{f_1} \tau^t M_1 \longrightarrow \ldots \longrightarrow \tau^t M_{m-1} \xrightarrow{f_m} \tau^t M_m = \tau^t M$$

in Γ. In addition, this cycle contains a projective module. Because of Lemma IV.2.5, all the $\tau^t M_i$ are projective; thus, we get a cycle of indecomposable projective modules in Γ. However, the full subquiver of Γ consisting of the indecomposable projectives is isomorphic to Q^{op}, because of Lemma IV.2.6. The acyclicity of Q^{op} then yields a contradiction.

The proof is similar for the preinjective components. □

Example IV.2.8. Let Q be the quiver

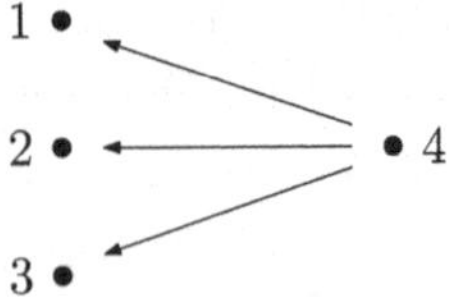

The indecomposable projective $\mathbf{k}Q$-modules are

$$P_1 = 1 \qquad P_2 = 2 \qquad P_3 = 3 \qquad P_4 = \begin{smallmatrix} & 4 & \\ 1 & 2 & 3 \end{smallmatrix}$$

whereas the indecomposable injective $\mathbf{k}Q$-modules are

$$I_1 = \begin{smallmatrix} 4 \\ 1 \end{smallmatrix} \qquad I_2 = \begin{smallmatrix} 4 \\ 2 \end{smallmatrix} \qquad I_3 = \begin{smallmatrix} 4 \\ 3 \end{smallmatrix} \qquad I_4 = 4.$$

The full subquiver containing the indecomposable projectives is

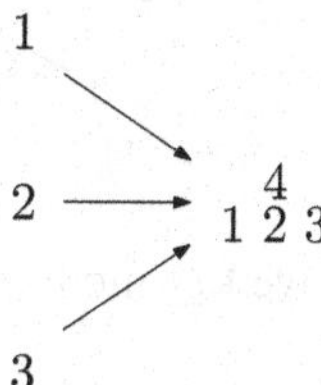

Using the knitting procedure, we construct the postprojective component of $\Gamma(\mathrm{mod}\,\mathbf{k}Q)$

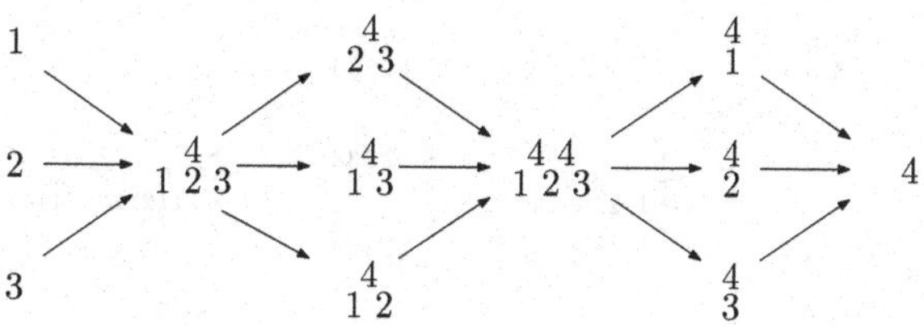

Because we reach the injectives, this is the whole Auslander–Reiten quiver (thus, $\mathbf{k}Q$ is representation-finite). Alternatively, one could construct the quiver starting from the injectives, that is, with the full connected subquiver

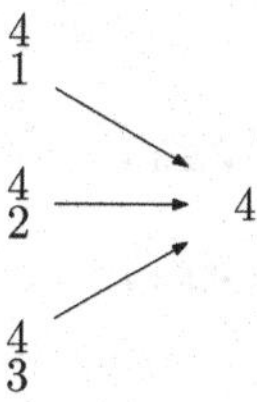

and knit backwards.

Example IV.2.9. Let Q be the quiver

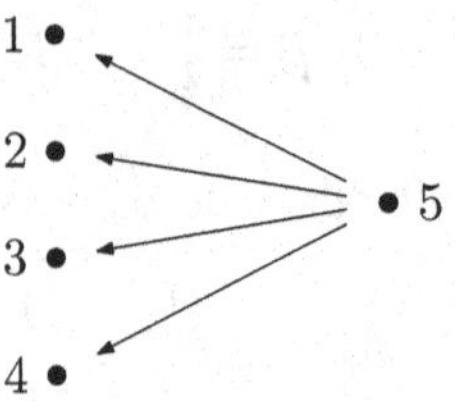

The indecomposable projective $\mathbf{k}Q$-modules are

$$P_1 = 1 \qquad P_2 = 2 \qquad P_3 = 3 \qquad P_4 = 4 \qquad P_5 = {}^{\;\;5}_{1\,2\,3\,4}.$$

Using the knitting procedure one gets the beginning of the postprojective component of $\Gamma(\operatorname{mod} \mathbf{k}Q)$

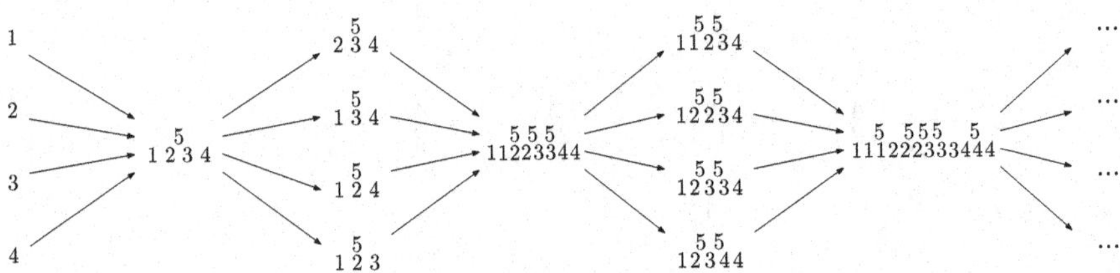

This component is infinite and never reaches an injective. To see this, it is enough to show (by induction, for instance) that $\dim_{\mathbf{k}}(\tau^{-n} P_5) = 5{+}6n$ and $\dim_{\mathbf{k}}(\tau^{-n} P_i) = 1{+}3n$, for each $n \geq 1$ and $i \in \{1, \ldots, 4\}$. Dualising the procedure, we can construct the preinjective component, which is infinite and does not contain any projective.

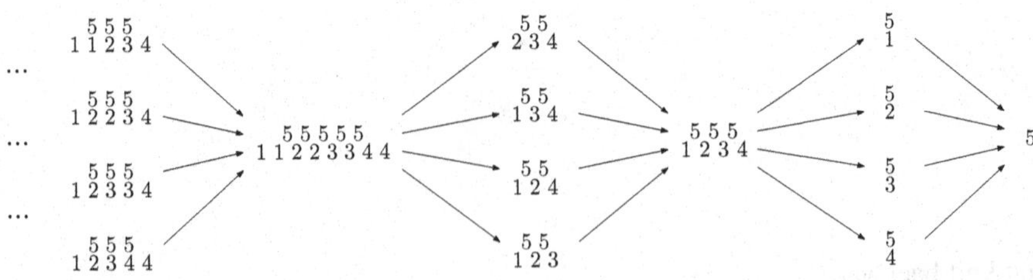

Example IV.2.10. Let A be the Kronecker algebra with quiver

$$1 \bullet \underset{\beta}{\overset{\alpha}{\leftleftarrows}} \bullet\, 2$$

As we have seen, $\Gamma(\text{mod}\,A)$ contains an infinite postprojective component containing all indecomposable projectives, and an infinite preinjective component containing all indecomposable injectives. One could ask whether these two are the only components of $\Gamma(\text{mod}\,A)$. The answer is negative. Indeed, consider the indecomposable module $M = {}^{2}_{1}$ of Examples II.4.3 and III.3.11. It lies in neither the postprojective nor the preinjective component of $\Gamma(\text{mod}\,A)$. Indeed, as seen in Example IV.1.16 above, the dimension of every indecomposable module lying in one of these two components is odd, whereas $\dim_{\mathbf{k}} M = 2$. Now, we have a projective cover morphism $p \colon P_2 \longrightarrow M$. Because P_2 and M belong to different components, it follows from Corollary II.4.5 that $p \in \text{rad}_A^{\infty}(P_2, M)$. In particular, $\text{rad}_A^{\infty}(P_2, M) \neq 0$.

Dually, the epimorphism $q \colon M \longrightarrow I_2$ with kernel S_1 lies in $\text{rad}_A^{\infty}(M, I_2)$. Now the composition $qp \colon P_2 \longrightarrow I_2$ is easily seen to equal the projection of P_2 onto $P_2/\text{rad}\,P_2 \cong S_2 = I_2$. Therefore, $qp \in (\text{rad}_A^{\infty}(P_2, I_2))^2$ is nonzero. This shows that not only do we have $\text{rad}_A^{\infty}(P_2, I_2) \neq 0$ but also $(\text{rad}_A^{\infty}(P_2, I_2))^2 \neq 0$.

IV.2.3 Indecomposables determined by their composition factors

It is a standard question in representation theory to ask which indecomposable modules over a given algebra are uniquely determined by their composition factors. More precisely, let M, N be indecomposable modules over an algebra A. Assume that M and N have exactly the same composition factors: does this condition imply that $M \cong N$? In fact, we defined in Subsection I.1.4 a numerical invariant, called the dimension vector, which counts the composition factors of a module. Our problem may be reformulated as follows: assume that M, N are indecomposable A-modules such that $\underline{\dim}\, M = \underline{\dim}\, N$: do we then have $M \cong N$? This is not true in general, as the following example shows.

Example IV.2.11. Let A be the $\mathbf{k}$-algebra given by the quiver

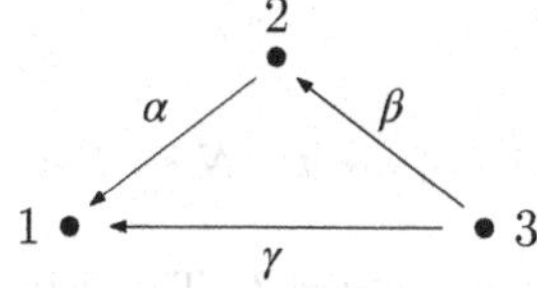

bound by $\beta\alpha = 0$.

The indecomposable projective $P_3 = {}^{\;2}_{1\,3}$ and the indecomposable injective $I_1 = {}^{2\,3}_{\;1}$ have the same composition factors but are not isomorphic.

There are, however, some situations where the answer to our question is positive. This is the case, for instance, when the indecomposable modules lie in a postprojective (or preinjective) component.

Theorem IV.2.12. *Let Γ be a postprojective (or preinjective) component of $\Gamma(\text{mod}\,A)$. If M, N are indecomposable modules in Γ such that $\underline{\dim} M = \underline{\dim} N$, then $M \cong N$.*

Proof. Up to duality, we may assume that Γ is postprojective.

We assume that $M \ncong N$ and let $M \longrightarrow \oplus_{i=1}^{r} E_i$, $N \longrightarrow \oplus_{j=1}^{s} F_j$ be left minimal almost split, where each of the E_i, F_j is indecomposable. We consider the sets $\mathscr{E}$ and $\mathscr{F}$ of predecessors of the E_i and the F_j in Γ, namely

$$\mathscr{E} = \{K \in \Gamma : \text{ there is a path of irreducible morphisms } K \rightsquigarrow E_i, \text{ for some } i\}$$

$$\mathscr{F} = \{L \in \Gamma : \text{ there is a path of irreducible morphisms } L \rightsquigarrow F_j, \text{ for some } j\}$$

Because Γ is a postprojective component, both sets $\mathscr{E}$ and $\mathscr{F}$ are finite, see Proposition IV.2.2. In addition, $\mathscr{E}$ and $\mathscr{F}$ are closed under predecessors, that is, if $K \in \mathscr{E}$ (or $L \in \mathscr{F}$) and there is a path of irreducible morphisms $K' \rightsquigarrow K$ (or $L' \rightsquigarrow L$ respectively), then $K' \in \mathscr{E}$ (or $L' \in \mathscr{F}$ respectively).

We divide our proof into several steps.

We first claim that $\tau^{-1}M \notin \mathscr{E}$ and $\tau^{-1}N \notin \mathscr{F}$. Indeed, assume that $\tau^{-1}M \in \mathscr{E}$. Then there exist i_0 and a path of irreducible morphisms $\tau^{-1}M \rightsquigarrow E_{i_0}$. However, we have an almost split sequence

$$0 \longrightarrow M \longrightarrow \oplus_{i=1}^{r} E_i \longrightarrow \tau^{-1}M \longrightarrow 0.$$

Hence, we get a cycle $\tau^{-1}M \rightsquigarrow E_{i_0} \longrightarrow \tau^{-1}M$ in Γ, a contradiction to its acyclicity. The proof is the same for $\tau^{-1}N \notin \mathscr{F}$. We have established our first claim.

We next claim that either $\tau^{-1}N \notin \mathscr{E}$ or $\tau^{-1}N \notin \mathscr{F}$. For, assume that this is not the case, then both $\tau^{-1}N \in \mathscr{E}$ and $\tau^{-1}N \in \mathscr{F}$. Thus, there exist i, j and paths of irreducible morphisms $\tau^{-1}N \rightsquigarrow E_i$ and $\tau^{-1}M \rightsquigarrow F_j$. Using the previous almost split sequence and the corresponding one for N, we get a cycle of irreducible morphisms

$$\tau^{-1}M \rightsquigarrow F_j \longrightarrow \tau^{-1}N \rightsquigarrow E_i \longrightarrow \tau^{-1}M$$

again a contradiction to the acyclicity of Γ. This establishes our second claim.

Because of this second claim, we may assume without loss of generality that $\tau^{-1}N \notin \mathscr{E}$.

Now we claim that, for every $X \in \mathscr{E}$, we have

$$\dim_{\mathbf{k}}(\text{Hom}(X, M)) = \dim_{\mathbf{k}}(\text{Hom}(X, N)).$$

We prove the statement by induction. We define $h(X)$, for an indecomposable module X in Γ to be the maximal length of a path in Γ from a projective to X.

If $h(X) = 0$, then X is projective and the statement follows from our hypothesis that $\underline{\dim}\, M = \underline{\dim}\, N$ and Lemma I.1.19. Assume $h(X) = n$. From our hypothesis, we may assume that X is not projective. Hence, there exists an almost split sequence:

$$(*) \qquad 0 \longrightarrow \tau X \longrightarrow \oplus_{k=1}^{t} Y_i \longrightarrow X \longrightarrow 0$$

with the Y_i indecomposable. Clearly, $X \in \mathscr{E}$ implies $\tau X, Y_k \in \mathscr{E}$ for all k. In addition, $h(\tau X) < n$ and $h(Y_k) < n$, for all k. Furthermore, τX is isomorphic neither to M nor to N because neither $\tau^{-1}N$ nor $\tau^{-1}M$ belongs to $\mathscr{E}$, see our first claim.

We claim that $\operatorname{Ext}_A^1(X, M) = 0$. Indeed, the Auslander–Reiten formula $\operatorname{Ext}_A^1(X, M) \cong D\underline{\operatorname{Hom}}_A(\tau^{-1}M, X)$ states that if this is not the case, then there exists a nonzero morphism $\tau^{-1}M \longrightarrow X$ that gives rise to a cycle $\tau^{-1}M \rightsquigarrow X \rightsquigarrow E_i \longrightarrow \tau^{-1}M$ in Γ, a contradiction. Therefore, applying $\operatorname{Hom}_A(-, M)$ to the almost split sequence $(*)$ yields

$$0 \longrightarrow \operatorname{Hom}_A(X, M) \longrightarrow \oplus_{k=1}^{t} \operatorname{Hom}_A(Y_k, M) \longrightarrow \operatorname{Hom}_A(\tau X, M) \longrightarrow 0.$$

Similarly, we get a short exact sequence

$$0 \longrightarrow \operatorname{Hom}_A(X, N) \longrightarrow \oplus_{k=1}^{t} \operatorname{Hom}_A(Y_k, N) \longrightarrow \operatorname{Hom}_A(\tau X, N) \longrightarrow 0.$$

Because $h(Y_k) < n$ for all k and $h(\tau X) < n$, the induction hypothesis gives that

$$\begin{aligned} \dim_{\mathbf{k}} \operatorname{Hom}_A(X, M) &= \textstyle\sum_{k=1}^{t} \dim_{\mathbf{k}} \operatorname{Hom}_A(Y_k, M) - \dim_{\mathbf{k}} \operatorname{Hom}_A(\tau X, M) \\ &= \textstyle\sum_{k=1}^{t} \dim_{\mathbf{k}} \operatorname{Hom}_A(Y_k, N) - \dim_{\mathbf{k}} \operatorname{Hom}_A(\tau X, N) \\ &= \dim_{\mathbf{k}} \operatorname{Hom}_A(X, N). \end{aligned}$$

This completes the proof of our claim.

Because $M \in \mathscr{E}$, the previous claim gives that $\operatorname{Hom}_A(M, N)$ contains a nonzero morphism f. Because we assumed $M \ncong N$, the morphism f cannot be an isomorphism and so must factor through the left minimal almost split morphism $M \longrightarrow \oplus_{i=1}^{r} E_i$. Therefore, there exists i such that $\operatorname{Hom}_A(E_i, N) \neq 0$. Applying our last claim again yields $\operatorname{Hom}_A(E_i, M) \neq 0$, leading to a cycle $M \rightsquigarrow E_i \rightsquigarrow M$, a contradiction that completes the proof. □

As an easy application, if A is a representation-finite algebra whose Auslander–Reiten quiver is acyclic, then the indecomposable A-modules are uniquely determined by their dimension vectors.

Exercises for Section IV.2

Exercise IV.2.1. Construct the postprojective and preinjective components for each algebra given by the following bound quivers

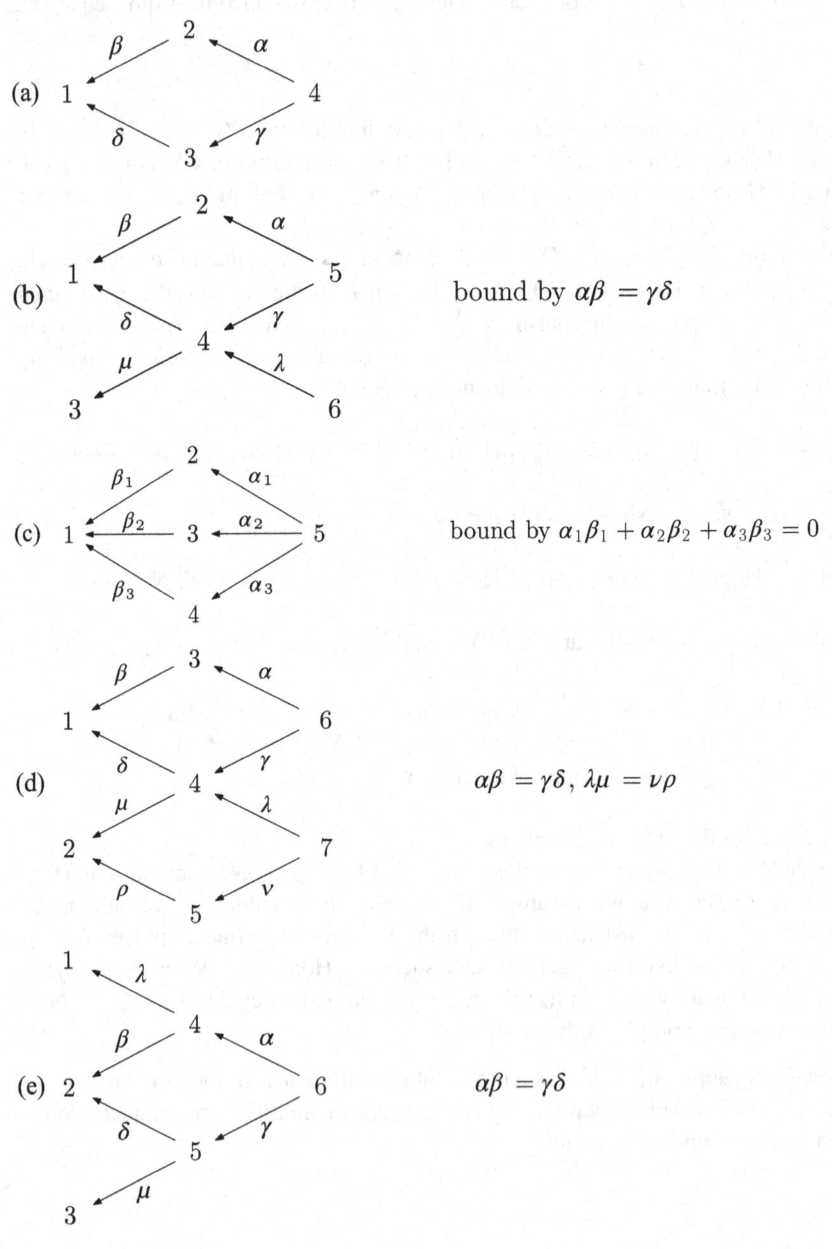

(f)
$$\begin{array}{ccccc} & 2 & \xleftarrow{\ \beta\ } & 5 & \\ & \uparrow{\scriptstyle\delta} & & \uparrow{\scriptstyle\alpha} & \\ 1 \xleftarrow{\ \varepsilon\ } & 3 & \xleftarrow{\ \gamma\ } & 6 & \\ & \downarrow{\scriptstyle\nu} & & \downarrow{\scriptstyle\lambda} & \\ & 4 & \xleftarrow{\ \mu\ } & 7 & \end{array} \qquad \alpha\beta = \gamma\delta, \gamma\nu = \lambda\mu$$

Exercise IV.2.2. Let A be the algebra given by the quiver

$$1 \underset{\delta}{\overset{\gamma}{\leftleftarrows}} 2 \xleftarrow{\ \beta\ } 3 \xleftarrow{\ \alpha\ } 4$$

bound by $\alpha\beta = 0$, $\beta\gamma = 0$. Prove that there exists one indecomposable projective in the preinjective component, and that there exists one indecomposable projective that belongs neither to the postprojective nor to the preinjective component.

Exercise IV.2.3. Let $A = \mathbf{k}Q$, where Q is the quiver

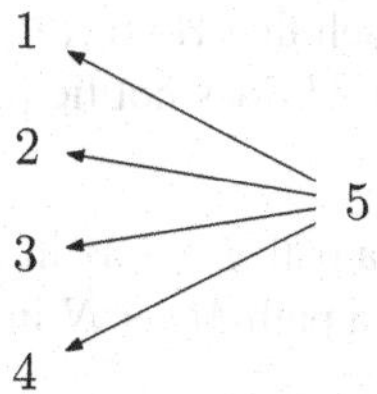

Prove that the indecomposable module $P_5/(S_3 \oplus S_4)$ belongs neither to the postprojective nor to the preinjective component.

Exercise IV.2.4. If Γ is a postprojective or preinjective component, and M lies in Γ, prove that $\mathrm{Ext}^i_A(M, M) = 0$ for all $i \geq 1$.

Exercise IV.2.5. Let $A = \mathbf{k}Q$ be a hereditary algebra. Prove that the following conditions are equivalent:

(a) A is representation-finite.
(b) The postprojective component of $\Gamma(\mathrm{mod}\, A)$ contains injectives.
(c) The preinjective component of $\Gamma(\mathrm{mod}\, A)$ contains projectives.
(d) The postprojective and the preinjective components coincide.

Exercise IV.2.6. Let A be a representation-finite algebra with an acyclic Auslander–Reiten quiver. Prove that $\mathrm{gl.dim.}\, A < \infty$.

Exercise IV.2.7. Let A be a representation-finite algebra and P_x, P_y indecomposable projective modules such that $\underline{\dim}\, P_x = \underline{\dim}\, P_y$. Prove that, if $P_x \not\cong P_y$, then there is a cycle in $\Gamma(\operatorname{mod} A)$ passing through P_x and P_y.

Exercise IV.2.8.

(a) Let Γ be a postprojective component of $\Gamma(\operatorname{mod} A)$. Prove that, for every indecomposable module M in Γ, there exists a path of irreducible morphisms between indecomposable modules

$$P = L_0 \xrightarrow{f_1} L_1 \longrightarrow \ \ldots \ \xrightarrow{f_t} L_t = M$$

with P projective and $f_t \ldots f_1 \neq 0$.

(b) Let Γ be a preinjective component of $\Gamma(\operatorname{mod} A)$. Prove that, for every indecomposable module M in Γ, there exists a path of irreducible morphisms between indecomposable modules

$$M = N_0 \xrightarrow{g_1} N_1 \longrightarrow \ \ldots \ \xrightarrow{g_t} N_t = I$$

with I injective and $g_t \ldots g_1 \neq 0$.

Exercise IV.2.9. Let A be a representation-finite algebra. Assume that there exists an indecomposable module M such that $\operatorname{Hom}_A(P, M) \neq 0$ for all indecomposable projective A-modules P and that M does not lie on any cycle in $\Gamma(\operatorname{mod} A)$. Prove that:

(a) If L is such that there exists a path $L \rightsquigarrow M$ in $\Gamma(\operatorname{mod} A)$, then $\operatorname{pd} L \leq 1$.
(b) If N is such that there exists a path $M \rightsquigarrow N$ in $\Gamma(\operatorname{mod} A)$, then $\operatorname{id} N \leq 1$.

Deduce that $\operatorname{pd} M \leq 1$ and $\operatorname{id} M \leq 1$.

IV.3 The depth of a morphism

IV.3.1 The depth

Let A be an algebra, and M, N indecomposable modules. As seen in Lemma II.2.2, an irreducible morphism $f\colon M \longrightarrow N$ lies in $\operatorname{rad}_A(M, N) \setminus \operatorname{rad}^2_A(M, N)$. We are now interested in the composition of irreducible morphisms in $\operatorname{ind} A$. Indeed, let

$$M = M_0 \xrightarrow{f_1} M_1 \longrightarrow \ldots \xrightarrow{f_t} M_t = N$$

be a path of irreducible morphisms between indecomposable modules, we certainly have $f_t \ldots f_1 \in \operatorname{rad}^t_A(M, N)$. But it is not clear whether $f_t \ldots f_1$ does not also belong to a higher power of the radical, or even whether it is nonzero (actually, if $f_t \dots f_1 = 0$, then $f_t \dots f_1 \in \mathrm{rad}_A^\infty(M, N)$). It is useful to define a numerical invariant expressing to which highest power of the radical a morphism belongs. For this purpose, we recall from Subsection II.4.1 that, given modules M, N, there exists a sequence of vector subspaces:

$$\mathrm{Hom}_A(M, N) = \mathrm{rad}_A^0(M, N) \supseteq \mathrm{rad}_A(M, N) \supseteq \mathrm{rad}_A^2(M, N) \supseteq \dots \supseteq \mathrm{rad}_A^\infty(M, N).$$

Thus, given any morphism $f : M \longrightarrow N$ not lying in the infinite radical, there exists a unique integer $d \geq 0$ such that $f \in \mathrm{rad}_A^d(M, N) \setminus \mathrm{rad}_A^{d+1}(M, N)$.

Definition IV.3.1. Let A be an algebra, M, N modules (not necessarily indecomposable) and $f : M \longrightarrow N$ a morphism. We say that the **depth of** f is infinite if $f \in \mathrm{rad}_A^\infty(M, N)$, and otherwise is the unique natural number d such that $f \in \mathrm{rad}_A^d(M, N) \setminus \mathrm{rad}_A^{d+1}(M, N)$. We denote the depth of f by $\mathrm{dp}(f)$.

Thus, the zero morphism always has infinite depth. Let M, N be indecomposable modules. Then

(a) The morphisms from M to N of depth zero are exactly the isomorphisms.
(b) The morphisms from M to N of depth one are exactly the irreducible.
(c) If M, N lie in distinct components of the Auslander–Reiten quiver of A, then, because of Corollary II.4.8, every morphism from M to N has infinite depth.
(d) If $f : M \longrightarrow N$ has depth d, then, because of Corollary II.4.8, there is a path of irreducible morphisms of length d from M to N.

Example IV.3.2. Let A be given by the quiver

$$\alpha \circlearrowright \overset{1}{\bullet} \xrightarrow{\ \beta\ } \overset{2}{\bullet}$$

bound by $\alpha\beta = 0$ and $\alpha^2 = 0$. Its Auslander–Reiten quiver is as follows:

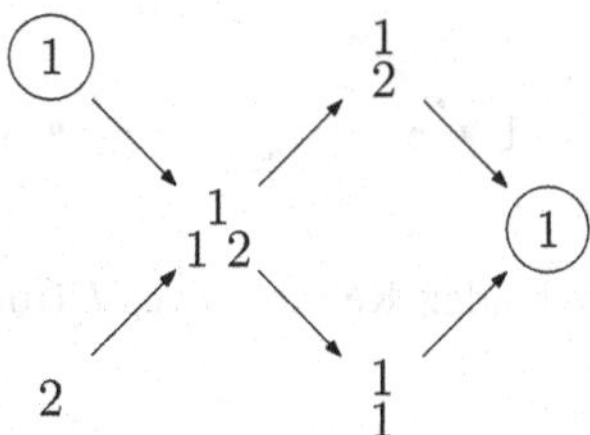

where one identifies the two copies of $S_1 = 1$. Consider the composition

$$f = f_3 f_2 f_1 : P_1 = \begin{smallmatrix} & 1 & \\ 1 & & 2 \end{smallmatrix} \longrightarrow \begin{smallmatrix} 1 \\ 1 \end{smallmatrix} \longrightarrow 1 \longrightarrow \begin{smallmatrix} & 1 & \\ 1 & & 2 \end{smallmatrix}$$

mapping the simple top S_1 of P_1 to the isomorphic direct summand of its socle. We claim that $\mathrm{dp}(f) = 3$. Indeed, f is the composition of three irreducible morphisms and so lies in $\mathrm{rad}_A^3(P_1, P_1)$. On the other hand, there is no nonzero path from P_1 to P_1 of length at least four. Indeed, the almost split sequence

$$0 \longrightarrow \begin{smallmatrix}1\\1\,2\end{smallmatrix} \longrightarrow \begin{smallmatrix}1\\2\end{smallmatrix} \oplus \begin{smallmatrix}1\\1\end{smallmatrix} \longrightarrow 1 \longrightarrow 0$$

gives that any such path is of the form f^m for some $m \geq 2$, then $f^2 = 0$ implies the statement. Thus, $f \in \mathrm{rad}_A^3(P_1, P_1) \setminus \mathrm{rad}^4(P_1, P_1)$ and so $\mathrm{dp}(f) = 3$.

IV.3.2 The depth of a sectional path

As we have done for translation quivers, we define sectional paths for the module category (compare with Definition IV.1.21). We say that a path of irreducible morphisms

$$M = M_0 \xrightarrow{f_1} M_1 \longrightarrow \dots \xrightarrow{f_t} M_t = N$$

between indecomposable modules of length $t \geq 1$ is **sectional** if, for every i with $1 < i \leq t$, we have $\tau M_i \ncong M_{i-2}$ (that is, the path factors through no mesh of the Auslander–Reiten quiver). Our objective in this subsection is to compute the depth of the composition $f_t \dots f_1$ and, in particular, to see whether the composition is nonzero.

We start with an example that illustrates the typical behaviour of such a composition.

Example IV.3.3. Let A be given by the quiver

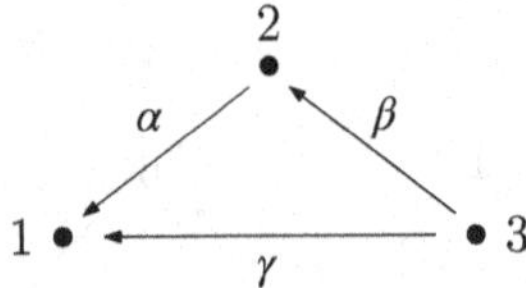

bound by $\beta\alpha = 0$. The Auslander–Reiten quiver $\Gamma(\mathrm{mod}\, A)$ is as follows

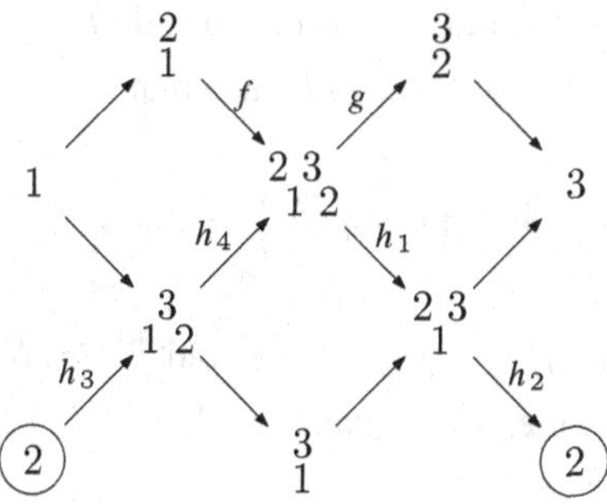

where one identifies the two copies of the simple $S_2 = 2$, see Example IV.1.15. The following path is sectional

$$\begin{smallmatrix}2\\1\end{smallmatrix} \xrightarrow{f} \begin{smallmatrix}2\ 3\\1\ 2\end{smallmatrix} \xrightarrow{h_1} \begin{smallmatrix}2\ 3\\1\end{smallmatrix} \xrightarrow{h_2} 2 \xrightarrow{h_3} \begin{smallmatrix}3\\1\ 2\end{smallmatrix} \xrightarrow{h_4} \begin{smallmatrix}2\ 3\\1\ 2\end{smallmatrix} \xrightarrow{g} \begin{smallmatrix}3\\2\end{smallmatrix}$$

and is nonzero, with image S_2. It is the composition of six irreducible morphisms; therefore, $\mathrm{dp}(gh_4h_3h_2h_1f) \geq 6$.

Lemma IV.3.4. *Let*

$$M = M_0 \xrightarrow{f_1} M_1 \longrightarrow \dots \longrightarrow M_{t-1} \xrightarrow{f_t} M_t = N$$

be a sectional path. If there exist morphisms $g\colon M \longrightarrow L$ and $f_t'\colon L \longrightarrow N$ such that $(f_t, f_t')\colon M_{t-1} \oplus L \longrightarrow N$ is right minimal almost split, then $f_t'g + f_t \dots f_1 \notin \mathrm{rad}_A^{t+1}(M, N)$.

Proof. The proof is done by induction on $t \geq 1$. Assume $t = 1$ and that there exist $f_1\colon M \longrightarrow N$ irreducible, $g\colon M \longrightarrow L$ and $f_1'\colon L \longrightarrow N$ such that $(f_1, f_1')\colon M \oplus L \longrightarrow N$ is right minimal almost split. The morphism $\begin{pmatrix}1\\g\end{pmatrix}\colon M \longrightarrow M \oplus L$ is a section, because $(1\ 0)\begin{pmatrix}1\\g\end{pmatrix} = 1$. Then, $f_1'g + f_1 = (f_1\ f_1')\begin{pmatrix}1\\g\end{pmatrix}$ is the composition of the right minimal almost split, and therefore irreducible, morphism $(f_1\ f_1')$ with a section. Thus, it is irreducible because of Corollary II.2.25. In particular, $f_1'g + f_1 \notin \mathrm{rad}_A^2(M, N)$.

Assume that $t \geq 2$ and the statement holds true for all sectional paths of length $t - 1$. If the statement does not hold true for the path

$$M = M_0 \xrightarrow{f_1} M_1 \longrightarrow \dots \longrightarrow M_{t-1} \xrightarrow{f_t} M_t = N,$$

then there exist morphisms $g\colon M \longrightarrow L$ and $f_t'\colon L \longrightarrow N$ such that $(f_t\ f_t')\colon M_{t-1} \oplus L \longrightarrow N$ is right minimal almost split and also $f_t'g + f_t \dots f_1$ belongs to $\mathrm{rad}_A^{t+1}(M, N)$.

Because $f'_t g + f_t \dots f_1$ is a radical morphism and $(f_t\ f'_t)$ is right minimal almost split, there exists $\begin{pmatrix} h \\ h' \end{pmatrix} : M \longrightarrow M_{t-1} \oplus L$ such that

$$f_t h + f'_t h' = \begin{pmatrix} f_t & f'_t \end{pmatrix} \begin{pmatrix} h \\ h' \end{pmatrix} = f'_t g + f_t \dots f_1 .$$

In addition, the fact that $f'_t g + f_t \dots f_1 \in \operatorname{rad}_A^{t+1}(M, N)$ implies that $\begin{pmatrix} h \\ h' \end{pmatrix} \in \operatorname{rad}_A^t(M, M_{t-1} \oplus L)$. The previous equality reads

$$f_t(h - f_{t-1} \dots f_1) + f'_t(h' - g) = 0$$

and therefore the morphism

$$\begin{pmatrix} h - f_{t-1} \dots f_1 \\ h' - g \end{pmatrix} : M \longrightarrow M_{t-1} \oplus L$$

factors through the kernel K of $(f_t\ f'_t)$ that is, there exists $k\colon M \longrightarrow K$ such that the following diagram with an exact row is commutative.

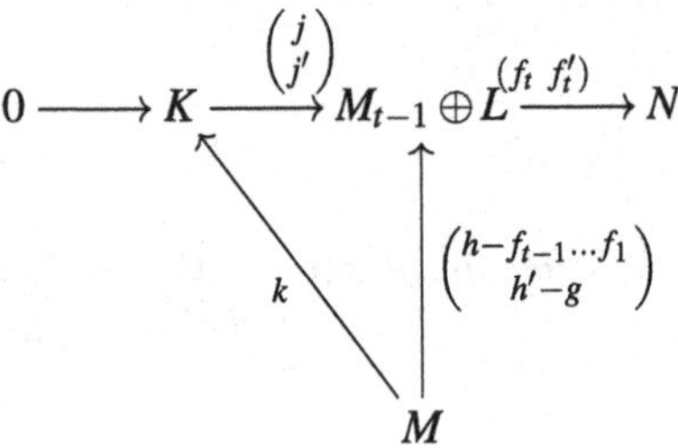

Thus, $h - f_{t-1} \dots f_1 = jk$ or, equivalently, $h = jk + f_{t-1} \dots f_1$. Now, the morphism $(j\ f_{t-1})\colon K \oplus M_{t-2} \longrightarrow M_{t-1}$ is irreducible. Indeed, we have two cases to consider: if N is projective, then $K = 0$ and clearly $(0\ f_{t-1})\colon 0 \oplus M_{t-1} \longrightarrow M_{t-1}$ is irreducible, whereas, if N is not projective, then the above left exact sequence is almost split because $(f_t\ f'_t)$ is right minimal almost split. Hence, j is irreducible and so is f_{t-1}; therefore, $(j\ f_{t-1})$ is irreducible because of Theorem II.2.24.

Applying Theorem II.2.24 again, there exist a right minimal almost split morphism $(j\ l\ f_{t-1})\colon K \oplus K' \oplus M_{t-2} \longrightarrow M_{t-1}$ and a morphism

$$\begin{pmatrix} k \\ 0 \\ f_{t-2} \dots f_1 \end{pmatrix} : M \longrightarrow K \oplus K' \oplus M_{t-2}$$

(if $t = 2$, then $M_{t-2} = M_0 = M$ and we take $f_{t-2} \dots f_1$ to be the identity 1_M) such that

$$(j\ l\ f_{t-1}) \begin{pmatrix} k \\ 0 \\ f_{t-2} \dots f_1 \end{pmatrix} = jk + f_{t-1} f_{t-2} \dots f_1 = h,$$

which lies in $\mathrm{rad}_A^t(M, M_{t-1})$. Then the induction hypothesis gives a contradiction. This completes the proof. □

This lemma allows us to prove that the composition of morphisms lying on a sectional path is always nonzero. In addition, its depth equals the length of the path.

Theorem IV.3.5. *Let* $M = M_0 \xrightarrow{f_1} M_1 \longrightarrow \dots \longrightarrow M_{t-1} \xrightarrow{f_t} M_t = N$ *be a sectional path, then* $\mathrm{dp}(f_t \dots f_1) = t$. *In particular,* $f_t \dots f_1 \neq 0$.

Proof. Indeed, it is easy to see that $f_t \dots f_1 \in \mathrm{rad}_A^t(M, N)$. Suppose that $f_t \dots f_1 \in \mathrm{rad}_A^{t+1}(M, N)$. Because f_t is irreducible, Theorem II.2.24 gives $f_t' \colon L \longrightarrow N$ such that $(f_t, f_t') \colon M_{t-1} \oplus L \longrightarrow N$ is right minimal almost split. Taking $g \colon M \longrightarrow L$ equal to zero, we get a contradiction to the previous lemma and the result is established.

In addition, if $f_t \dots f_1 = 0$, then the composition belongs to the infinite radical and so it would have infinite depth, a contradiction. □

The next result is due to Bautista and Smalø.

Theorem IV.3.6. *Let* $M_0 \xrightarrow{f_1} M_1 \longrightarrow \dots \longrightarrow M_{t-1} \xrightarrow{f_t} M_t$ *be a path of irreducible morphisms between indecomposable modules. If* $f_t = f_1$, *then this path is not sectional.*

Proof. Indeed, assume that the path is sectional and consider the cyclic subpath

$$M = M_0 \xrightarrow{f_1} M_1 \longrightarrow \dots \xrightarrow{f_{t-1}} M_{t-1} = M.$$

If we compose this cycle with itself m times, then we still get a sectional path, because $f_1 \colon M \longrightarrow M_1$ and $f_t \colon M_{t-1} \longrightarrow M_t$ are the same. Writing $f = f_{t-1} \dots f_1$, Theorem IV.3.5 above says that $f^m \neq 0$ for all m. However, $f \in \mathrm{rad}_A(M, M) = \mathrm{rad}(\mathrm{End}\, M)$; hence, f is nilpotent, a contradiction. □

The previous theorem is sometimes expressed by saying that "there are no sectional cycles". But one has to be careful: what the theorem really says is that there are no sectional "cycles" where the first morphism and the last coincide. We show this in an example.

Example IV.3.7. Let A be the algebra of Example IV.3.3 and consider the following sectional path

$$\begin{smallmatrix} 2\,3 \\ 1\,2 \end{smallmatrix} \xrightarrow{h_1} \begin{smallmatrix} 2\,3 \\ 1 \end{smallmatrix} \xrightarrow{h_2} 2 \xrightarrow{h_3} \begin{smallmatrix} 3 \\ 1\,2 \end{smallmatrix} \xrightarrow{h_4} \begin{smallmatrix} 2\,3 \\ 1\,2 \end{smallmatrix}.$$

The path starts and ends at the same module; therefore, it is a cycle in the module category. However, if we add, as in the previous theorem, the morphism h_4 either at the beginning or h_1 at the end, then the new path is no longer sectional.

IV.3.3 Composition of two irreducible morphisms

It is not true in general that the depth of a composition of irreducible morphisms equals the length of the corresponding path. As an example, we study here the shortest nontrivial paths of irreducible morphisms, namely those of length two. Assume that

$$L \xrightarrow{f} M \xrightarrow{g} N$$

is a path of irreducible morphisms between indecomposable modules. Because f, g are irreducible, we have $gf \in \mathrm{rad}^2(L, N)$. Because of Theorem IV.3.5, if $\tau N \not\cong L$ then $gf \neq 0$ and $\mathrm{dp}(gf) = 2$. In particular, $gf \notin \mathrm{rad}_A^3(L, N)$. If, however, the path is not sectional, then it is reasonable to ask when do we have $gf = 0$ and, even if $gf \neq 0$, do we have $\mathrm{dp}(gf) \geq 3$? We start with an example showing that this situation may occur.

Example IV.3.8. Let A be the algebra of Example IV.1.15, that is the one given by the quiver

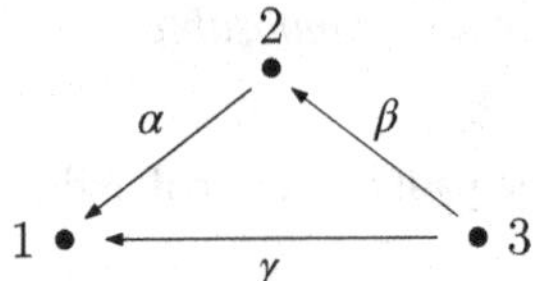

bound by $\beta\alpha = 0$. Because

$$0 \longrightarrow {\scriptstyle\begin{smallmatrix}2\\1\end{smallmatrix}} \xrightarrow{f} {\scriptstyle\begin{smallmatrix}2&&3\\&1&&2\end{smallmatrix}} \xrightarrow{g} {\scriptstyle\begin{smallmatrix}3\\2\end{smallmatrix}} \longrightarrow 0$$

is an almost split sequence, we have $gf = 0$. Consider the morphism $h = h_4h_3h_2h_1$ of Example IV.3.7. As seen before, $h \neq 0$. Therefore, define

$$f' = f + hf \colon {\scriptstyle\begin{smallmatrix}2\\1\end{smallmatrix}} \longrightarrow {\scriptstyle\begin{smallmatrix}2&&3\\&1&&2\end{smallmatrix}}.$$

We claim that f' is irreducible. Indeed, it is easy to see that $h^2 = 0$ and therefore

$$(1 + h)(1 - h) = 1 = (1 - h)(1 + h),$$

that is, $(1 + h)$ is invertible. Applying Corollary II.2.25, we deduce that $f' = (1 + h)f$ is irreducible, because f is. In addition, $gf = 0$ implies that

$$gf' = g(f + hf) = gf + ghf = ghf.$$

But we have seen in Example IV.3.3 that ghf is the composition of six irreducible morphisms lying on a sectional path. Therefore, $\mathrm{dp}(ghf) = 6$, because of Theorem IV.3.5. Then, gf' also has depth 6. In particular, $gf' \in \mathrm{rad}_A^3\left(\begin{smallmatrix}2\\1\end{smallmatrix}, \begin{smallmatrix}3\\2\end{smallmatrix}\right)$.

Lemma IV.3.9. *Let $f: L \longrightarrow M$ and $g: M \longrightarrow N$ be irreducible morphisms between indecomposables. If $\mathrm{dp}(gf) \geq 3$, then there exist almost split sequences:*

$$0 \longrightarrow L \xrightarrow{h} M \xrightarrow{g} N \longrightarrow 0 \quad \text{and} \quad 0 \longrightarrow L \xrightarrow{f} M \xrightarrow{h'} N \longrightarrow 0.$$

Proof. Because $gf \in \mathrm{rad}_A^3(L, N)$, we infer from Theorem IV.3.5 that the path $L \xrightarrow{f} M \xrightarrow{g} N$ is not sectional and so $\tau N \cong L$.

We claim that g is right minimal almost split. If this is not the case, then, because of Theorem II.2.24, there exists an irreducible morphism $g': M' \longrightarrow N$ such that $(g\ g'): M \oplus M' \longrightarrow N$ is right minimal almost split.

Because of Theorem II.2.31, there exists an almost split sequence:

$$0 \longrightarrow L \xrightarrow{\left(\begin{smallmatrix}k\\k'\end{smallmatrix}\right)} M \oplus M' \xrightarrow{(g\ g')} N \longrightarrow 0.$$

Because $gf \in \mathrm{rad}_A^3(L, N)$, there exists a factorisation $gf = vw$ where $w \in \mathrm{rad}_A^2(L, X)$ and $v \in \mathrm{rad}_A(X, N)$ for some module X. It follows from the definition of almost split sequences and the fact that v is radical that there exists $\left(\begin{smallmatrix}u\\u'\end{smallmatrix}\right): X \longrightarrow M \oplus M'$ such that

$$v = (g\ g')\left(\begin{smallmatrix}u\\u'\end{smallmatrix}\right) = gu + gu'.$$

Now,

$$(g\ g')\left(\begin{smallmatrix}uw-f\\u'w\end{smallmatrix}\right) = guw - gf + g'u'w = (gu + g'u')w - gf = vw - gf = 0.$$

Hence, $\left(\begin{smallmatrix}uw-f\\u'w\end{smallmatrix}\right)$ factors through $\left(\begin{smallmatrix}k\\k'\end{smallmatrix}\right)$, that is, there exists $l: L \longrightarrow L$ such that

$$\left(\begin{smallmatrix}uw-f\\u'w\end{smallmatrix}\right) = \left(\begin{smallmatrix}k\\k'\end{smallmatrix}\right) l$$

or equivalently

$$\left(\begin{smallmatrix}uw\\u'w\end{smallmatrix}\right) = \left(\begin{smallmatrix}f+kl\\k'l\end{smallmatrix}\right).$$

Because $w \in \mathrm{rad}_A^2(L, X)$ we have $\mathrm{dp}(uw) \geq 2$ so that $\mathrm{dp}(f + kl) \geq 2$ that is, $f + kl \in \mathrm{rad}_A^2(L, M)$. Because f and k are irreducible, this implies that l is not a radical morphism. Therefore, $l : L \longrightarrow L$ is an isomorphism. Then, $k'l$ is irreducible, again because of Corollary II.2.25. But $k'l = u'w$ and $\mathrm{dp}(u'w) \geq 2$ so we get a contradiction. This establishes our claim.

Because of Theorem II.2.31, we deduce the existence of an almost split sequence $0 \longrightarrow L \xrightarrow{h} M \xrightarrow{g} N \longrightarrow 0$. Dualising the argument, one gets the other almost split sequence. □

We deduce necessary and sufficient conditions for the composition of two irreducible morphisms to be nonzero and lie in the radical cube of the module category (that is, have depth at least three).

Theorem IV.3.10. *Let L, M, N be indecomposable modules. The following conditions are equivalent:*

(a) *There exist irreducible morphisms $h\colon L \longrightarrow M$ and $h'\colon M \longrightarrow N$ such that the composite $h'h$ is nonzero and lies in $\mathrm{rad}_A^3(L, N)$;*
(b) *There exist an almost split sequence $0 \longrightarrow L \xrightarrow{f} M \xrightarrow{g} N \longrightarrow 0$ and a morphism $\varphi \in \mathrm{rad}_A^2(M, M)$ such that $g\varphi f \neq 0$;*
(c) *There exists an almost split sequence $0 \longrightarrow L \xrightarrow{f} M \xrightarrow{g} N \longrightarrow 0$ and $\mathrm{rad}_A(L, N) \neq 0$.*

Proof. (a) implies (b). Because of the above lemma, there exists an almost split sequence:

$$0 \longrightarrow L \xrightarrow{f} M \xrightarrow{g} N \longrightarrow 0.$$

Using that $\mathrm{rad}^3(L, N) \neq 0$, we have a path

$$L \xrightarrow{f} M \xrightarrow{l} X \xrightarrow{k} N$$

where f and l are irreducible and $k \in \mathrm{rad}_A(X, N)$ is such that $klf \neq 0$. Clearly, $M \not\cong X$ because there exists no irreducible morphism from M to M. Now, $k\colon X \longrightarrow N$ is a radical morphism; hence, there exists $v\colon X \longrightarrow M$, making the following diagram commute

$$\begin{array}{ccccccccc} & & & & & & X & & \\ & & & & & \swarrow^{v} & \downarrow{\scriptstyle k} & & \\ 0 & \longrightarrow & L & \xrightarrow{f} & M & \xrightarrow{g} & N & \longrightarrow & 0. \end{array}$$

Because $M \not\cong X$, the morphism v is also radical. The morphism l being irreducible, we have $\varphi = vl \in \mathrm{rad}_A^2(M, M)$ and also

$$g\varphi f = gvlf = klf \neq 0.$$

This completes the proof of this implication.

(b) implies (c). Indeed, g, f irreducible and $\varphi \in \mathrm{rad}_A^2(M, M)$ imply $g\varphi f \in \mathrm{rad}_A^4(L, N) \subseteq \mathrm{rad}_A(L, N)$. Then, $g\varphi f \neq 0$ gives $\mathrm{rad}_A(L, N) \neq 0$.

(c) implies (a). Let $u \colon L \longrightarrow N$ be a nonzero radical morphism. Then, there exists $v \colon L \longrightarrow M$ such that $u = gv$:

$$\begin{array}{ccccccccc} & & & & & & L & & \\ & & & & & \swarrow^{v} & \downarrow^{u} & & \\ 0 & \longrightarrow & L & \xrightarrow{f} & M & \xrightarrow{g} & N & \longrightarrow & 0. \end{array}$$

Assume first that v is irreducible. Because of Proposition IV.1.2, the residual class $f + \mathrm{rad}_A^2(L, M)$ is a $\mathbf{k}$-basis for

$$\mathrm{Irr}_A(L, M) = \frac{\mathrm{rad}_A(L, M)}{\mathrm{rad}_A^2(L, M)}.$$

Thus, there exist $\lambda \in \mathbf{k}$ and $v' \in \mathrm{rad}^2(L, M)$ such that $v = \lambda f + v'$. Then, we have

$$0 \neq u = gv = g(\lambda f + v') = gv'.$$

Because $gv' \in \mathrm{rad}_A^3(L, N)$, we get $\mathrm{rad}_A^3(L, N) \neq 0$, as required.

Assume now that v is not irreducible. Because v is radical, there exists $w \colon M \longrightarrow M$ such that $wf = v$

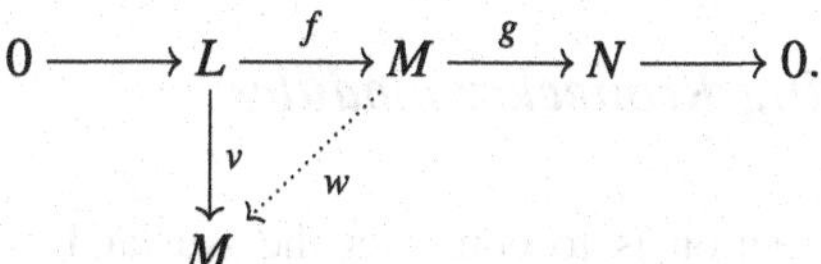

The morphism $w \in \mathrm{End}\, M$ is not an isomorphism, because otherwise v would be irreducible. Consider $h = f + wf \colon L \longrightarrow M$. Because w is a nonisomorphism, it is nilpotent. Let $m \geq 0$ be such that $w^m = 0$. Then

$$(1 + w)(1 - w + w^2 - \ldots + (-1)^{m-1} w^{m-1}) = 1$$

says that $1 + w$ is invertible, so $h = (1 + w)f$ is irreducible, because of Corollary II.2.25. Setting $h' = g$ and using that $gf = 0$, we get

$$h'h = g(f + wf) = gf + gwf = gv = u \neq 0.$$

This completes the proof. $\square$

Exercises for Section IV.3

Exercise IV.3.1. Prove that the equivalent conditions of Theorem IV.3.10 are equivalent to the further conditions:

(a) There exist an almost split sequence

$$0 \longrightarrow L \xrightarrow{f} M \xrightarrow{g} N \longrightarrow 0$$

and nonisomorphisms $\varphi_1 : L \longrightarrow X$ and $\varphi_2 : X \longrightarrow N$ such that X is an indecomposable module not isomorphic to M and $\varphi_2\varphi_1 \neq 0$.

(b) There exist an almost split sequence

$$0 \longrightarrow L \xrightarrow{f} M \xrightarrow{g} N \longrightarrow 0$$

and $\mathrm{rad}_A^4(L, N) \neq 0$.

Exercise IV.3.2. Compute the depth of each nontrivial path in the module category of the algebra of:

(a) Example IV.3.2.
(b) Example IV.3.3.

IV.4 Modules over the Kronecker algebra

IV.4.1 Representing Kronecker modules

The objective of this section is to construct the Auslander–Reiten quiver of the Kronecker algebra, namely the path algebra of the Kronecker quiver K_2

$$1 \bullet \underset{\beta}{\overset{\alpha}{\leftleftarrows}} \bullet 2$$

see I.2.6. Throughout this section, we set $A = \mathbf{k}K_2$. As seen in Example IV.1.16, the quiver $\Gamma(\mathrm{mod}\,A)$ has an infinite postprojective component $\mathscr{P}$ of the form

$$\begin{array}{ccccccccccc}
 & & M_1 = {}^{\;\;2}_{1\,1} & - - - & & & M_3 = {}^{\;2\,2\,2}_{1\,1\,1\,1} & - - - & & & M_5 \; - - - \\
 & \nearrow\!\!\nearrow & & \searrow\!\!\searrow & & \nearrow\!\!\nearrow & & \searrow\!\!\searrow & & \nearrow\!\!\nearrow & & \searrow\!\!\searrow \\
M_0 = 1 & - - - & & & M_2 = {}^{\;2\,2}_{1\,1\,1} & - - - & & & M_4 = {}^{\;2\,2\,2\,2}_{1\,1\,1\,1\,1} & - - - &
\end{array}$$

and an infinite preinjective component $\mathscr{Q}$ of the form

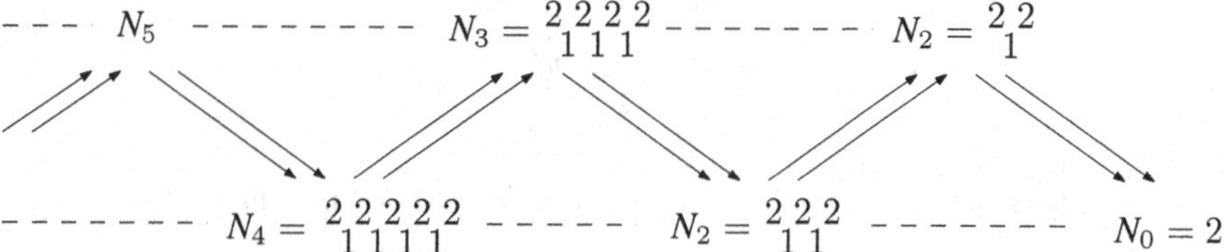

To study representation-infinite algebras, one needs new techniques not discussed so far. Indeed, if we want to compute all the indecomposable modules over a representation-finite algebra, it suffices (though this is not always easy) to construct a finite component of the Auslander–Reiten quiver and then apply Auslander's theorem, see Remark IV.1.6(e). For representation-infinite algebras, we must thus devise techniques that allow us to check whether or not we have obtained all isoclasses of indecomposable modules. These techniques are beyond the scope of this book. For the Kronecker algebra, we use an ad hoc method with elementary, but tedious, linear algebra. In particular, we prove that indecomposable A-modules lying neither in $\mathscr{P}$ nor in $\mathscr{Q}$ must belong to tubes in the sense of Example IV.1.23.

Our first step is to define a category that we call category of representations of K_2 and we denote as rep K_2. To motivate the introduction of this category, consider the indecomposable projective module $P_2 = M_1 = \begin{smallmatrix}2\\1\,1\end{smallmatrix}$. The top of the module is the one-dimensional vector space $P_2e_2 = S_2$ and its radical is $P_2e_1 = S_1 \oplus S_1$, which is two-dimensional. Because of Lemma I.2.15, P_2e_1 admits as a basis $\{\alpha, \beta\}$ and $\alpha = e_2\alpha e_1, \beta = e_2\beta e_1$ belong to the two indecomposable summands of P_2e_1 respectively. This amounts to saying that P_2 consists of two vector spaces P_2e_2 and P_2e_1 and two $\mathbf{k}$-linear maps from P_2e_2 to P_2e_1 explaining how the basis of P_2e_2 maps into P_2e_1.

We let the objects of rep K_2 be quadruples (E_2, E_1, f, g), where E_2, E_1 are $\mathbf{k}$-vector spaces and $f, g : E_2 \longrightarrow E_1$ are $\mathbf{k}$-linear maps. Such a quadruple is called a **representation** of K_2 and is depicted as follows:

$$\left(E_2 \overset{f}{\underset{g}{\rightrightarrows}} E_1 \right)$$

The maps f, g are called the **structural maps** of the representation.

A morphism from $\left(E_2 \overset{f}{\underset{g}{\rightrightarrows}} E_1 \right)$ to $\left(E_2' \overset{f'}{\underset{g'}{\rightrightarrows}} E_1' \right)$ is a pair (u, v) of $\mathbf{k}$-linear maps $u : E_2 \longrightarrow E_2', v : E_1 \longrightarrow E_1'$ such that $vf = f'u$ and $vg = g'u$, that is, u, v are compatible with the structural maps.

$$\begin{array}{ccc} E_2 & \overset{f}{\underset{g}{\rightrightarrows}} & E_1 \\ {\scriptstyle u}\downarrow & & \downarrow{\scriptstyle v} \\ E_2' & \overset{f'}{\underset{g'}{\rightrightarrows}} & E_1' \end{array}$$

The composition of morphisms in rep K_2 is defined componentwise: if (u, v) : $\left(E_2 \overset{f}{\underset{g}{\rightrightarrows}} E_1 \right) \longrightarrow \left(E_2' \overset{f'}{\underset{g'}{\rightrightarrows}} E_1' \right)$ and $(u', v') : \left(E_2' \overset{f'}{\underset{g'}{\rightrightarrows}} E_1' \right) \longrightarrow \left(E_2'' \overset{f''}{\underset{g''}{\rightrightarrows}} E_1'' \right)$, then $(u', v')(u, v) = (u'u, v'v)$.

We prove that the category of representations of the Kronecker quiver is equivalent to the module category of the quiver's path algebra. This is not specific to the quiver K_2. One can define, in the same way, representations of an arbitrary acyclic quiver and this category of representations is equivalent to the module category over the path algebra, see Exercises I.2.22 and I.2.20. In this book, we only need this result in the context of the Kronecker quiver.

Lemma IV.4.1. *We have an equivalence of categories* mod $A \cong$ rep K_2.

Proof. We first construct a **k**-functor F : mod $A \longrightarrow$ rep K_2. Let M be an A-module. Recall from Subsection I.2.6 that

$$A = \left\{ \begin{pmatrix} a & 0 \\ (b, c) & d \end{pmatrix} : a, b, c, d \in \mathbf{k} \right\}$$

has $e_1 = \begin{pmatrix} 1 & 0 \\ 0 & 0 \end{pmatrix}$ and $e_2 = \begin{pmatrix} 0 & 0 \\ 0 & 1 \end{pmatrix}$ forming a complete set of primitive orthogonal idempotents. Set $E_1 = Me_1$, $E_2 = Me_2$: these are finite dimensional vector spaces because so is M. The structural maps f, g are defined by

$$f(x) = x \begin{pmatrix} 0 & 0 \\ (1, 0) & 0 \end{pmatrix} \quad \text{and} \quad g(x) = x \begin{pmatrix} 0 & 0 \\ (0, 1) & 0 \end{pmatrix}$$

for $x \in E_2$. The matrix equalities

$$\begin{pmatrix} 0 & 0 \\ (1, 0) & 0 \end{pmatrix} \begin{pmatrix} 1 & 0 \\ 0 & 0 \end{pmatrix} = \begin{pmatrix} 0 & 0 \\ (1, 0) & 0 \end{pmatrix} \quad \text{and} \quad \begin{pmatrix} 0 & 0 \\ (0, 1) & 0 \end{pmatrix} \begin{pmatrix} 0 & 0 \\ 0 & 1 \end{pmatrix} = \begin{pmatrix} 0 & 0 \\ (0, 1) & 0 \end{pmatrix}$$

show that $f(x), g(x) \in E_1$. Clearly, f, g are **k**-linear. Let $\varphi : M \longrightarrow N$ be a morphism of A-modules. Then set $F(\varphi) = (u, v)$ where $u : Me_1 \longrightarrow Ne_1$, v :

$Me_2 \longrightarrow Ne_2$ are the restrictions of φ to Me_1 and Me_2 respectively. It is easily checked that F is indeed a $\mathbf{k}$-functor.

We next construct $G : \operatorname{rep} K_2 \longrightarrow \operatorname{mod} A$. Let $(E_2 \overset{f}{\underset{g}{\rightrightarrows}} E_1)$ be a representation of K_2. Then the vector space $M = E_2 \oplus E_1$ becomes a right A-module if one defines multiplication as follows:

$$\begin{pmatrix} x \\ y \end{pmatrix} \begin{pmatrix} a & 0 \\ (b, c) & d \end{pmatrix} = \begin{pmatrix} xa \\ f(x)b + g(x)c + yd \end{pmatrix}$$

for $x \in E_2$, $y \in E_1$, $a, b, c, d \in \mathbf{k}$. Let $G(E_2 \overset{f}{\underset{g}{\rightrightarrows}} E_1) = M$. A morphism $(u, v) : (E_2 \overset{f}{\underset{g}{\rightrightarrows}} E_1) \longrightarrow (E_2' \overset{f'}{\underset{g'}{\rightrightarrows}} E_1')$ in $\operatorname{rep} K_2$ induces a $\mathbf{k}$-linear map $\left(\begin{smallmatrix} u & 0 \\ 0 & v \end{smallmatrix}\right) : E_2 \oplus E_1 \longrightarrow E_2' \oplus E_1'$. It is A-linear because

$$\begin{aligned}
\left[\begin{pmatrix} u & 0 \\ 0 & v \end{pmatrix} \begin{pmatrix} x \\ y \end{pmatrix}\right] \begin{pmatrix} a & 0 \\ (b, c) & d \end{pmatrix} &= \begin{pmatrix} u(x) \\ v(y) \end{pmatrix} \begin{pmatrix} a & 0 \\ (b, c) & d \end{pmatrix} \\
&= \begin{pmatrix} u(xa) \\ f'u(x)b + g'u(x)c + v(y)d \end{pmatrix} \\
&= \begin{pmatrix} u(xa) \\ vf(x)b + vg(x)c + v(y)d \end{pmatrix} \\
&= \begin{pmatrix} u & 0 \\ 0 & v \end{pmatrix} \left[\begin{pmatrix} x \\ y \end{pmatrix} \begin{pmatrix} a & 0 \\ (b, c) & d \end{pmatrix}\right].
\end{aligned}$$

We set $G(u, v) = \begin{pmatrix} u & 0 \\ 0 & v \end{pmatrix}$. The rest of the proof is the verification that G is a $\mathbf{k}$-functor, quasi-inverse to F. □

Because of this lemma, the problem of studying the module category of the Kronecker algebra reduces to that of studying the category $\operatorname{rep} K_2$. It is useful to reformulate some statements in this latter context. For instance, if $(u, v) : (E_2 \rightrightarrows E_1) \longrightarrow (E_2' \rightrightarrows E_1')$ and $(u', v') : (E_2' \rightrightarrows E_1') \longrightarrow (E_2'' \rightrightarrows E_1'')$ are morphisms, then the sequence of representations

$$0 \longrightarrow (E_2 \rightrightarrows E_1) \xrightarrow{(u,v)} (E_2' \rightrightarrows E_1') \xrightarrow{(u',v')} (E_2'' \rightrightarrows E_1'') \longrightarrow 0$$

is exact if and only if the sequences of vector spaces $0 \longrightarrow E_2 \xrightarrow{u} E_2' \xrightarrow{u'} E_2'' \longrightarrow 0$ and $0 \longrightarrow E_1 \xrightarrow{v} E_1' \xrightarrow{v'} E_1'' \longrightarrow 0$ are both exact. In particular, the structural maps of $(\ E_2 \rightrightarrows E_1 \ )$ are (isomorphic to) the restrictions of the structural maps of $(\ E_2' \rightrightarrows E_1' \ )$. Direct sums of representations occur in the usual way:

$$(\ E_2 \underset{g}{\overset{f}{\rightrightarrows}} E_1 \ ) \oplus (\ E_2' \underset{g'}{\overset{f'}{\rightrightarrows}} E_1' \ ) = (\ E_2 \oplus E_2' \underset{\begin{pmatrix} g & 0 \\ 0 & g' \end{pmatrix}}{\overset{\begin{pmatrix} f & 0 \\ 0 & f' \end{pmatrix}}{\rightrightarrows}} E_1 \oplus E_1' \ )$$

In the sequel, we consider this equivalence $\operatorname{mod} A \cong \operatorname{rep} K_2$ as an identification. We show how to view indecomposable postprojective modules as representations. Let $P_1 = 1$, $P_2 = \begin{smallmatrix} & 2 & \\ 1 & & 1 \end{smallmatrix}$ be the indecomposable projective A-modules. Applying the formula of Lemma IV.4.1 for F yields

$$F(P_1) = (\ 0 \underset{0}{\overset{0}{\rightrightarrows}} \mathbf{k} \ ) \quad \text{and} \quad F(P_2) = (\ \mathbf{k} \underset{\binom{0}{1}}{\overset{\binom{1}{0}}{\rightrightarrows}} \mathbf{k}^2 \ ).$$

In general, $M_n = \begin{smallmatrix} 2 \ 2 \ \cdots 2 \\ 1 \ 1 \ \cdots 1 \ 1 \end{smallmatrix}$ is such that $\dim_{\mathbf{k}} M_n e_2 = n$, $\dim_{\mathbf{k}} M_n e_1 = n + 1$. Let $\{u_1, \dots, u_n\}$ be a basis for $M_n e_2$, and let $\{v_1, \dots, v_{n+1}\}$ be a basis for $M_n e_1$. Then, $f(u_i) = v_i$, $g(u_i) = v_{i+1}$, for each i such that $1 \le i \le n$, owing to the definitions of f, g in Lemma IV.4.1 above. Diagrammatically, the actions of f, g are depicted as

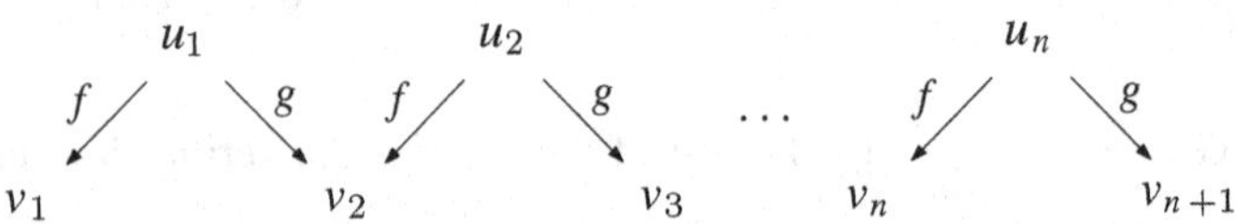

In other words, we recover the picture for $M_n = \begin{smallmatrix} 2 \ 2 \ \cdots 2 \\ 1 \ 1 \ \cdots 1 \ 1 \end{smallmatrix}$ obtained in Example II.4.3.

For instance, taking $M_2 = \begin{smallmatrix} 2 \ 2 \\ 1 \ 1 \ 1 \end{smallmatrix}$, and letting $\{u_1, u_2\}$, $\{v_1, v_2, v_3\}$ be the canonical bases in $M_2 e_2$, $M_2 e_1$ respectively, we get that f, g are respectively given by the matrices

$$\begin{pmatrix} 1 & 0 \\ 0 & 1 \\ 0 & 0 \end{pmatrix} \quad \text{and} \quad \begin{pmatrix} 0 & 0 \\ 1 & 0 \\ 0 & 1 \end{pmatrix}.$$

Both f, g are injective: each of them maps a basis of E_2 to a linearly independent set in E_1.

The situation with preinjective modules is dual and we leave it to the reader as an exercise.

We now construct an infinite family of representations of K_2 that are neither postprojective nor preinjective. Let $n \geq 1$, $\lambda \in \mathbf{k}$, and $J_n(\lambda)$ be the $n \times n$-Jordan block

$$J_n(\lambda) = \begin{pmatrix} \lambda & 0 & 0 & \cdots & 0 \\ 1 & \lambda & 0 & \cdots & 0 \\ 0 & 1 & \lambda & \cdots & 0 \\ \vdots & \vdots & \vdots & \ddots & \vdots \\ 0 & 0 & 0 & \cdots & \lambda \end{pmatrix}.$$

Denoting by I_n the $n \times n$-identity matrix, we set

$$H_\lambda^n = (\mathbf{k}^n \overset{I_n}{\underset{J_n(\lambda)}{\rightrightarrows}} \mathbf{k}^n).$$

Thus, if $n = 1$, we have $H_\lambda^1 = (\mathbf{k} \overset{1}{\underset{\lambda}{\rightrightarrows}} \mathbf{k})$. In general, if we denote by $\{u_1, \ldots, u_n\}$ the canonical basis for both coordinate spaces $\mathbf{k}^n$, we get $f(u_i) = u_i$ for all i, whereas $g(u_i) = \lambda u_i + u_{i+1}$ for all $i \neq n$, and $g(u_n) = \lambda u_n$.

This corresponds to the picture

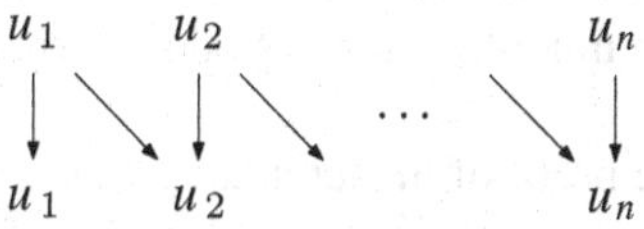

In particular, the module H_0^2 is the module H of Example III.3.11.

Lemma IV.4.2. *For every n, λ, the representation H_λ^n is indecomposable.*

Proof. Because representations are modules, as seen in Lemma IV.4.1, it suffices to prove that End H_λ^n is a local algebra. An endomorphism of H_λ^n is a pair (U, V) of $n \times n$-matrices compatible with the structural matrices I_n, $J_n(\lambda)$. Compatibility with I_n gives $U = V$, whereas compatibility with $J_n(\lambda)$ is expressed by the equality $UJ_n(\lambda) = J_n(\lambda)U$. Let $U = [a_{ij}]_{i,j}$ where $a_{ij} \in \mathbf{k}$ for all i, j. Comparing the products $UJ_n(\lambda)$ and $J_n(\lambda)U$ yields $a_{ij} = 0$ if $i < j$, $a_{ii} = a_{jj}$ and $a_{ij} = a_{i+1,j+1}$ if $i > j$. Thus, End H_λ^n is isomorphic to the matrix algebra

$$R = \left\{ \begin{pmatrix} a_1 & 0 & 0 & \cdots & 0 \\ a_2 & a_1 & 0 & \cdots & 0 \\ a_3 & a_2 & a_1 & \cdots & 0 \\ \vdots & \vdots & \vdots & \ddots & \vdots \\ a_n & \cdots & \cdots & \cdots & a_1 \end{pmatrix} \mid a_i \in \mathbf{k} \text{ for } 1 \le i \le n \right\}$$

equipped with the ordinary matrix operations. Now, let

$$I = \left\{ \begin{pmatrix} 0 & 0 & 0 & \cdots & 0 \\ a_2 & 0 & 0 & \cdots & 0 \\ a_3 & a_2 & 0 & \cdots & 0 \\ \vdots & \vdots & \vdots & \ddots & \vdots \\ a_n & \cdots & \cdots & \cdots & 0 \end{pmatrix} \mid a_i \in \mathbf{k} \text{ for } 2 \le i \le n \right\}.$$

Then, I is an ideal of R, we have $I^n = 0$ and $R/I \cong \mathbf{k}$. Because of Theorem I.1.7, $I = \operatorname{rad} R$. In addition, I is a maximal two-sided ideal in R. Therefore, R is local and H_λ^n is indecomposable. □

Corollary IV.4.3.

(a) *The indecomposable representations H_λ^n are neither postprojective nor preinjective ;*
(b) End $H_\lambda^1 = \mathbf{k}$;
(c) $H_\lambda^n \cong H_\mu^m$ *if and only if* $n = m$ *and* $\lambda = \mu$.

Proof.

(a) As seen in Example IV.1.16, the $\mathbf{k}$-dimension of an indecomposable postprojective or preinjective module is always odd, whereas $\dim_{\mathbf{k}} H_\lambda^n = 2n$ for every λ.
(b) This follows from the proof of the lemma.
(c) An isomorphism $H_\lambda^n \longrightarrow H_\mu^m$ is given by a pair of invertible matrices (U, V) compatible with the structural matrices. Then, $n = m$ and compatibility with the identity matrix gives $U = V$. Compatibility with the Jordan blocks yields $UJ_n(\lambda) = J_n(\mu)U$. Because U is invertible, $J_n(\lambda) = U^{-1}J_n(\mu)U$. Uniqueness of the Jordan form gives $\lambda = \mu$.

□

Dually, the representations

$$K_\lambda^n = (\mathbf{k}^n \underset{I_n}{\overset{J_n(\lambda)}{\rightrightarrows}} \mathbf{k}^n)$$

are indecomposable for every n, λ and neither postprojective nor preinjective. We shall see later that, except for one particular case, K_λ^n can be reduced to H_λ^n.

The representations H_λ^n and K_λ^n and the corresponding modules are called **regular**.

IV.4.2 Modules over the Kronecker algebra

The objective of this subsection is to prove that every indecomposable module over the Kronecker algebra is isomorphic to a postprojective, or a preinjective or a regular module. We follow here the proof of Burgermeister, which uses only elementary linear algebra.

As a first reduction, we observe that a representation $\left(E_2 \overset{f}{\underset{g}{\rightrightarrows}} E_1 \right)$ is indecomposable if and only if its dual $\left(\mathrm{D}E_1 \overset{\mathrm{D}f}{\underset{\mathrm{D}g}{\rightrightarrows}} \mathrm{D}E_2 \right)$ is as well. Thus, we may, without loss of generality, assume that $\dim_{\mathbf{k}} E_2 \leq \dim_{\mathbf{k}} E_1$.

From now on, we denote by M a representation $\left(E_2 \overset{f}{\underset{g}{\rightrightarrows}} E_1 \right)$ of K_2 such that, additionally, $\dim_{\mathbf{k}} E_2 \leq \dim_{\mathbf{k}} E_1$.

Lemma IV.4.4. *Assume that* $M = \left(E_2 \overset{f}{\underset{g}{\rightrightarrows}} E_1 \right)$ *is indecomposable and not simple. Then:*

(a) $\operatorname{Ker} f \cap \operatorname{Ker} g = 0$;
(b) $\operatorname{Im} f + \operatorname{Im} g = E_1$.

Proof.

(a) Assume that $F = \operatorname{Ker} f \cap \operatorname{Ker} g$ is nonzero. Then, M has a direct summand of the form $\left(F \rightrightarrows 0 \right)$. Because M is indecomposable, we have $M = \left(F \rightrightarrows 0 \right)$ and $\dim_{\mathbf{k}} F = 1$. But then $M \cong \left(\mathbf{k} \rightrightarrows 0 \right) \cong S_2$, a contradiction.
(b) Assume that $\operatorname{Im} f + \operatorname{Im} g \subsetneqq E_1$ and F are such that $E_1 = F \oplus (\operatorname{Im} f + \operatorname{Im} g)$. Then $\left(0 \rightrightarrows F \right)$ is a direct summand of M. Again, we get $M \cong \left(0 \rightrightarrows \mathbf{k} \right) \cong S_1$, another contradiction.

□

Lemma IV.4.5. *Let* $n = \dim_{\mathbf{k}} E_2$, $n + m = \dim_{\mathbf{k}} E_1$, *for* $m \geq 0$, $d = \dim_{\mathbf{k}} \operatorname{Ker} f$, $d' = \dim_{\mathbf{k}} \operatorname{Ker} g$, *and set* $W = \operatorname{Im} f \cap \operatorname{Im} g$, $V = f^{-1}(W) \cap g^{-1}(W)$. *Then:*

(a) $\dim_{\mathbf{k}} W = n - d - d' - m$;
(b) $\dim_{\mathbf{k}} V \geq n - d - d' - 2m$;
(c) $(V \rightrightarrows W)$ *is a subrepresentation of* M.

Proof.

(a) Because $\operatorname{Im} f + \operatorname{Im} g = E_1$, we have

$$\begin{aligned}\dim_{\mathbf{k}} W &= \dim_{\mathbf{k}}(\operatorname{Im} f) + \dim_{\mathbf{k}}(\operatorname{Im} g) - \dim_{\mathbf{k}} E_1 \\ &= \dim_{\mathbf{k}} E_2 - \dim_{\mathbf{k}}(\operatorname{Ker} f) + \dim_{\mathbf{k}} E_2 - \dim_{\mathbf{k}}(\operatorname{Ker} g) - \dim_{\mathbf{k}} E_1 \\ &= n - d + n - d' - n - m \\ &= n - d - d' - m.\end{aligned}$$

(b) $\dim_{\mathbf{k}} f^{-1}(W) = \dim_{\mathbf{k}} W + \dim_{\mathbf{k}} \operatorname{Ker} f = n - d' - m$. Similarly, $\dim_{\mathbf{k}} g^{-1}(W) = n - d - m$. Therefore,

$$\begin{aligned}\dim_{\mathbf{k}} V &= \dim_{\mathbf{k}} f^{-1}(W) + \dim_{\mathbf{k}} g^{-1}(W) - \dim_{\mathbf{k}}(f^{-1}(W) + g^{-1}(W)) \\ &\geq (2n - d - d' - 2m) - n \\ &= n - d - d' - 2m.\end{aligned}$$

(c) This is obvious.

□

In view of Lemma IV.4.5 above, let $r \geq 0$ be such that $\dim_{\mathbf{k}} V = n - d - d' - 2m + r$. We need more notation. Let $K = \operatorname{Ker} f \cap V$, $K' = \operatorname{Ker} g \cap V$ and L, L' be such that $\operatorname{Ker} f = L \oplus K$, $\operatorname{Ker} g = L' \oplus K'$. Finally, set $k = \dim_{\mathbf{k}} K$, $k' = \dim_{\mathbf{k}} K'$.

Lemma IV.4.6.

(a) *The sum* $V + L$ *is direct. If* H *is such that* $f^{-1}(W) = V \oplus L \oplus H$, *then* $\dim_{\mathbf{k}} H = k + m - r$;
(b) *The sum* $V + L'$ *is direct. If* H' *is such that* $g^{-1}(W) = V \oplus L' \oplus H'$, *then* $\dim_{\mathbf{k}} H' = k' + m - r$;
(c) *The sum* $(V \oplus L \oplus H) + (L' \oplus H')$ *is direct and its dimension is* $n - r$.

Proof.

(a) Assume $x \in V \cap L$, then $x \in V$ and $x \in L \subseteq \operatorname{Ker} f$ so that $x \in K$. Because $L \cap K = 0$, we get $x = 0$. Therefore, $V \cap L = 0$ and the sum $V + L$ is direct. Let H be such that $f^{-1}(W) = V \oplus L \oplus H$. Then

$$\begin{aligned}\dim_{\mathbf{k}} H &= \dim_{\mathbf{k}} f^{-1}(W) - \dim_{\mathbf{k}} L - \dim_{\mathbf{k}} V \\ &= (n - d' - m) - (d - k) - (n - d - d' - 2m + r) \\ &= k + m - r.\end{aligned}$$

(b) This is similar to (a).

(c) Let $x \in (V \oplus L \oplus H) \cap (L' \oplus H')$. Then, $x \in f^{-1}(W) \cap g^{-1}(W) = V$. But $V \cap (L' \oplus H') = 0$. Hence, $x = 0$ and the sum is direct. Its dimension is $(n-d-d'-2m+r)+(d-k)+(d'-k')+(k+m-r)+(k'+m-r) = n-r$.

□

Lemma IV.4.7. *Let X be such that $E_2 = V \oplus L \oplus H \oplus L' \oplus H' \oplus X$ and $Y = f(X) + g(X) \subseteq E_1$. Then we have*

(a) $\operatorname{Im} f = W + f(L') + f(H') + f(X)$;
(b) $\operatorname{Im} g = W + g(L) + g(H) + g(X)$;
(c) *We have isomorphisms $L' \cong f(L')$, $H' \cong f(H')$, $L \cong g(L)$ and $H \cong g(H)$. In addition, $E_1 = W \oplus g(L) \oplus g(H) \oplus f(L') \oplus f(H') \oplus Y$.*

Proof.

(a) Because $f(L) = 0$, we have $f(V) + f(H) = ff^{-1}(W) \subseteq W$. Therefore,

$$\begin{aligned}\operatorname{Im} f &= f(V) + f(H) + f(L') + f(H') + f(X)\\ &\subseteq W + f(L') + f(H') + f(X)\\ &\subseteq \operatorname{Im} f\end{aligned}$$

so that $\operatorname{Im} f = W + f(L') + f(H') + f(X)$.

(b) This is similar to (a).

(c) Because of Lemma IV.4.4(b), we have

$$E_1 = \operatorname{Im} f + \operatorname{Im} g = W + f(L') + f(H') + g(L) + g(H) + Y.$$

The dimension of E_1 does not exceed the sum s of the dimensions of the subspaces on the right-hand side. Because of Lemma IV.4.6(c), we have $\dim_{\mathbf{k}} X = r$; therefore, $\dim_{\mathbf{k}} f(X) \leq \dim_{\mathbf{k}} X = r$ and similarly $\dim_{\mathbf{k}} g(X) \leq r$. Thus, we have

$$\begin{aligned}n + m &\leq s\\ &\leq (n-d-d'-m) + (d-k)+(d-k')+(k+m-r)+(k'+m-r)+2r\\ &= n+m\end{aligned}$$

where we have used Lemma IV.4.6(a) and (b). This implies that $s = n + m$ and we have the stated isomorphisms $L' \cong f(L')$, $H' \cong f(H')$, $L \cong g(L)$, $H \cong g(H)$. In addition, $E = W \oplus g(L) \oplus g(H) \oplus f(L') \oplus f(H') \oplus Y$.

□

Corollary IV.4.8. *M is isomorphic to one of the four subrepresentations*

(a) $M = (\; L \rightrightarrows g(L) \;) \cong K_0^1 = (\; \mathbf{k} \underset{1}{\overset{0}{\rightrightarrows}} \mathbf{k} \;)$;

(b) $M = (\ L' \rightrightarrows f(L')\ ) \cong H_0^1 = (\ \mathbf{k} \underset{0}{\overset{1}{\rightrightarrows}} \mathbf{k}\ )$;

(c) $M = (\ X \rightrightarrows Y\ ) \cong M_2 = (\ \mathbf{k} \underset{\binom{0}{1}}{\overset{\binom{1}{0}}{\rightrightarrows}} \mathbf{k}^2\ )$;

(d) $M = (\ V \oplus H \oplus H' \rightrightarrows W \oplus f(H') \oplus g(H)\ )$.

Proof. Because M was assumed to be indecomposable, it follows from the direct decompositions of E_2, E_1 and the isomorphisms of Lemma IV.4.7 that M is isomorphic to one of the four subrepresentations $(\ L \rightrightarrows g(L)\ )$, $(\ L' \rightrightarrows f(L')\ )$, $(\ X \rightrightarrows Y\ )$ and $(\ V \oplus H \oplus H' \rightrightarrows W \oplus f(H') \oplus g(H)\ )$ and the others vanish. In addition:

(a) If $M = (\ L \rightrightarrows g(L)\ )$ then, because $f(L) = 0$, then M is indecomposable if and only if $\dim_{\mathbf{k}} L = 1$. The statement follows.
(b) This is similar to (a).
(c) Because $\dim_{\mathbf{k}} Y = 2r$ (see proof of Lemma IV.4.7(c)), M is indecomposable if and only if $\dim_{\mathbf{k}} X = 1$, $\dim_{\mathbf{k}} Y = 2$ and $Y = f(X) \oplus g(X)$. It is then isomorphic to M_2.
(d) Follows from the previous arguments.

□

We already know that K_0^1, H_0^1 and M_2 are indecomposable. Thus, it remains to consider case (d). In this case, we have $X = Y = L = L' = 0$; therefore, $d = k$, $d' = k'$, $r = 0$ and

$$\begin{aligned} \dim_{\mathbf{k}} V &= n - d - d' - 2m \\ \dim_{\mathbf{k}} W &= n - d - d' - m \\ \dim_{\mathbf{k}} H &= k + m = d + m \\ \dim_{\mathbf{k}} H' &= k' + m = d' + m. \end{aligned}$$

Also, $K \subseteq f^{-1}(W) = V \oplus H$ and $K \cap H = 0$ imply that $K \subseteq V$. Similarly, $K' \subseteq V$. The following picture shows how the maps f and g act

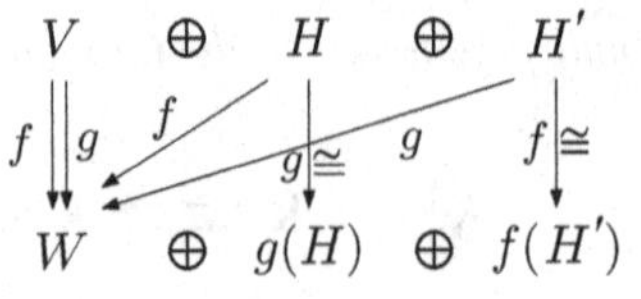

$$\begin{array}{ccccc} V & \oplus & H & \oplus & H' \\ f \Downarrow g & \swarrow f & g \downarrow \cong \quad g \swarrow & & f \downarrow \cong \\ W & \oplus & g(H) & \oplus & f(H') \end{array}$$

Lemma IV.4.9. *The subrepresentation* $(V \rightrightarrows W)$ *is indecomposable.*

Proof. Assume that $(V \rightrightarrows W) \cong (V_1 \rightrightarrows W_1) \oplus (V_2 \rightrightarrows W_2)$ is a nontrivial direct sum decomposition. Let W_1', W_2' be such that $W_1 = f(V_1) \oplus W_1'$, $W_1 = f(V_2) \oplus W_2'$. Because $f^{-1}(W_1') \cap f^{-1}(W_2') \cap H \subseteq f^{-1}(W_1) \cap f^{-1}(W_2) = f^{-1}(W_1 \cap W_2) = 0$, we may choose a direct sum decomposition $H = U_1 \oplus U_2$ so that $f(U_1) = W_1' \subseteq W_1$ and $f(U_2) = W_2' \subseteq W_2$. Similarly, we choose a decomposition $H' = U_1' \oplus U_2'$ such that $g(U_1') \subseteq W_1$ and $g(U_2') \subseteq W_2$. We then get a nontrivial direct sum decomposition $M \cong (V_1 \oplus U_1 \oplus U_1' \rightrightarrows W_1 \oplus g(U_1) \oplus f(U_1')) \oplus (V_2 \oplus U_2 \oplus U_2' \rightrightarrows W_2 \oplus g(U_2) \oplus f(U_2'))$. This contradicts the indecomposability of M. □

We are ready to complete the proof of the classification theorem, saying that every indecomposable module over the Kronecker algebra is postprojective, preinjective or regular.

Theorem IV.4.10. *Let* $M = (E_2 \rightrightarrows E_1)$ *be an indecomposable representation of* K_2 *and* $n = \dim_{\mathbf{k}} E_2$.

(a) *If* $\dim_{\mathbf{k}} E_1 = n$, *then* $M \cong H_\lambda^n$ *for some* $\lambda \in \mathbf{k}$ *or* $M \cong K_0^n$;
(b) *If* $\dim_{\mathbf{k}} E_1 > n$, *then* $M \cong M_n$;
(c) *If* $\dim_{\mathbf{k}} E_1 < n$, *then* $M \cong N_n$.

Proof.

(a) If $\dim_{\mathbf{k}} E_1 = n = \dim_{\mathbf{k}} E_2$, then $m = 0$. We claim that one of f, g is an isomorphism. This is proved by induction on n. If $n = 1$, this is trivial so let $n > 1$. If $d = d' = 0$, then f, g are injective, and hence isomorphisms because $\dim_{\mathbf{k}} E_1 = \dim_{\mathbf{k}} E_2$. If one of d, d' is positive, then we have $\dim_{\mathbf{k}} V = \dim_{\mathbf{k}} W < n$. The induction hypothesis says that, in the indecomposable subrepresentation $(V \rightrightarrows W)$ of Lemma IV.4.9, the restriction of one of f and g, say f, to V is an isomorphism. But then $f^{-1}(W) = V$ implies $H = 0$. Therefore, $f : V \oplus H' \longrightarrow W \oplus f(H')$ is an isomorphism. This establishes our claim.

As a consequence, M is isomorphic to a representation of the form $(E \overset{1}{\underset{g}{\rightrightarrows}} E)$ or $(E \overset{f}{\underset{1}{\rightrightarrows}} E)$, say the former. Consider E as a module over the polynomial algebra $\mathbf{k}[t]$ by setting $t \cdot x = g(x)$ for $x \in E$.

The indecomposability of the representation $(E \overset{1}{\underset{g}{\rightrightarrows}} E)$ implies the indecomposability of the $\mathbf{k}[t]$-module E. The structure theorem of modules over a principal ideal domain implies that there exist an irreducible polynomial p and $s \geq 0$ such that $E \cong \mathbf{k}[t]/ < p^s >$. Because $\mathbf{k}$ is algebraically closed, g can be represented in some basis by a Jordan block $J_n(\lambda)$. Thus, $M \cong H^n_\lambda$. Similarly, $M \cong (E \overset{f}{\underset{1}{\rightrightarrows}} E)$ implies $M \cong K^n_\lambda$ for some λ. Now, if $\lambda \neq 0$, then $J_n(\lambda)$ is invertible and so is f. Therefore, $M \cong (E \overset{1}{\underset{f^{-1}}{\rightrightarrows}} E)$ and we are reduced to the case before.

(b) Assume $m \geq 1$. We prove that if $m \geq 2$, then M is decomposable. This is done by induction on n. If $n = 1$, this is trivial. If $n > 1$, consider $(V \rightrightarrows W)$. We have $\dim_{\mathbf{k}} W = n - d - d' - m = m + \dim_{\mathbf{k}} V$. Because of the induction hypothesis, $(V \rightrightarrows W)$ is decomposable, a contradiction to Lemma IV.4.9.

This implies $m = 1$. We prove by induction on n that $M \cong M_n$. If $n = 1$, then $M \cong M_1 = P_2$. Assume $n = 2$, then $\dim_{\mathbf{k}} V = n - d - d' - 2 \geq 0$ implies $d = d' = 0$ and $V = 0$. Therefore, $\dim_{\mathbf{k}} W = n - d - d' - 1 = 1$, $\dim_{\mathbf{k}} H = d + 1 = 1$ and $\dim_{\mathbf{k}} H' = d' + 1 = 1$. In this case, M can be written in the form

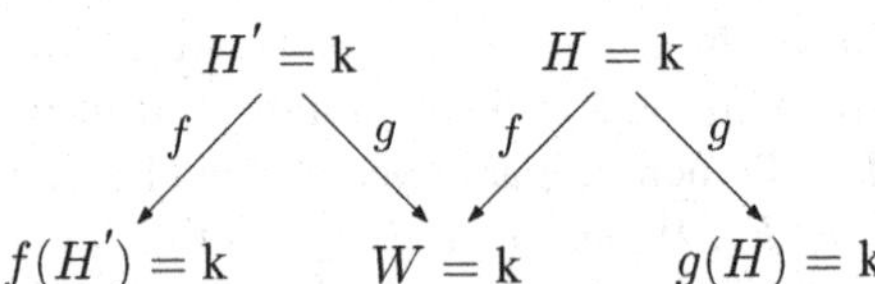

that is, $M \cong M_2$.

Suppose $n > 2$. We know that $(V \rightrightarrows W)$ is indecomposable. We have $t = \dim_{\mathbf{k}} V = n - d - d' - 2 < n$ and $\dim_{\mathbf{k}} W = n - d - d' - 1 = 1 + \dim_{\mathbf{k}} V$. The induction hypothesis applied to $(V \rightrightarrows W)$ yields that the latter is isomorphic to M_t. In particular, f and g are injective (see the remark after Lemma IV.4.1). Therefore, $d = 0$, $d' = 0$ and $t = n - 2$. But then M can be written in the form

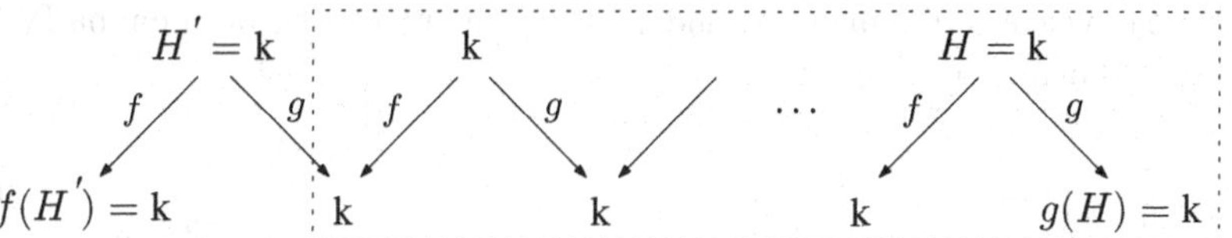

where the part between the dotted lines is the subrepresentation $(V \rightrightarrows W) \cong M_t$. This shows that $M \cong M_n$.

(c) This is similar to (b).

□

This proof is not only long and tedious, but also very particular to the case of the Kronecker algebra. There exist more general techniques, but, as mentioned at the beginning of this section, we shall not cover them in this book.

IV.4.3 The Auslander–Reiten quiver of the Kronecker algebra

We already know that the Auslander–Reiten quiver of the Kronecker algebra admits a postprojective component containing all indecomposables of the form M_n and a preinjective component containing all indecomposables of the form N_n. But there remain modules of the form K_0^n or H_λ^n for some $\lambda \in \mathbf{k}$. To treat both cases as one, we set $K_0^n = H_\infty^n$; thus, our modules are of the form H_λ^n with $\lambda \in \mathbf{k} \cup \{\infty\}$. Alternatively, we may think of λ as ranging over the projective line over $\mathbf{k}$. We recall that, because of Proposition I.2.28, the Kronecker algebra is hereditary.

Lemma IV.4.11. *For each $n \geq 2$, there exist a monomorphism $j_n : H_\lambda^{n-1} \longrightarrow H_\lambda^n$ and an epimorphism $p_n : H_\lambda^n \longrightarrow H_\lambda^{n-1}$ such that we have short exact sequences*

$$0 \longrightarrow H_\lambda^{n-1} \xrightarrow{j_n} H_\lambda^n \xrightarrow{p'_n} H_\lambda^1 \longrightarrow 0$$

$$0 \longrightarrow H_\lambda^1 \xrightarrow{j'_n} H_\lambda^n \xrightarrow{p_n} H_\lambda^{n-1} \longrightarrow 0$$

where $j'_n = j_n \dots j_2$ and $p'_n = p_2 \dots p_n$.

Proof. Indeed, let $j_n = \left(\begin{pmatrix} 0 \\ I_{n-1} \end{pmatrix}, \begin{pmatrix} 0 \\ I_{n-1} \end{pmatrix}\right)$ where each of the coordinate $n \times (n-1)$-matrices has a first row consisting of zeros and I_{n-1} is the identity $(n-1) \times (n-1)$-matrix. It is easily seen that $j_n : H_\lambda^{n-1} \longrightarrow H_\lambda^n$ is a morphism of representations and actually a monomorphism. Similarly, $p_n = ((I_{n-1}, 0), (I_{n-1}, 0)) : H_\lambda^n \longrightarrow H_\lambda^{n-1}$ is an epimorphism of representations. The rest is a straightforward calculation. □

One can visualise the maps p_n and j_n, using the picture before Lemma IV.4.2. Indeed, letting $n = 3$, we have

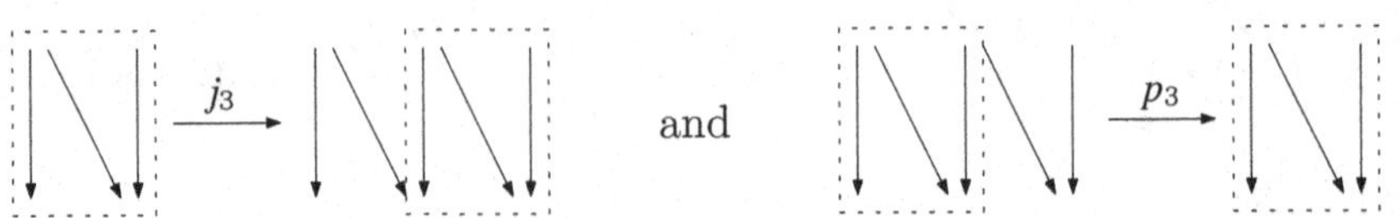

On the left, the module H_λ^2 embeds as a submodule of H_λ^3 by means of j_3, and the quotient is isomorphic to H_λ^1, whereas on the right, we see that p_3 maps H_λ^3 epimorphically onto H_λ^2, with its kernel isomorphic to H_λ^1.

Lemma IV.4.12. *If* $\mathrm{Hom}_A(H_\mu^m, H_\lambda^n) \neq 0$*, then* $\lambda = \mu$*. In particular, if* H_μ^m *and* H_λ^n *belong to the same component of the Auslander–Reiten quiver, then* $\lambda = \mu$*.*

Proof. Assume $\mathrm{Hom}_A(H_\mu^m, H_\lambda^n) \neq 0$ with $\lambda \neq \mu$. We first claim that we may assume $m = n$. If $m < n$ then, because of Lemma IV.4.11, we have epimorphisms

$$H_\mu^n \xrightarrow{p_n} \cdots \xrightarrow{p_{m+1}} H_\mu^m$$

so $p_{m+1} \cdots p_n$ is an epimorphism. Hence, if $f : H_\mu^m \longrightarrow H_\lambda^n$ is nonzero, neither is $f p_{m+1} \cdots p_n : H_\mu^n \longrightarrow H_\lambda^n$. We treat the case $m > n$ similarly, using the j_n instead of the p_n. This establishes our claim.

Assume that $f : H_\mu^n \longrightarrow H_\lambda^n$ is nonzero and let f be represented by the matrices U, V which are compatible with the structural matrices. Compatibility with the identity matrices yields $U = V$. Therefore, $U = [a_{ij}]$ satisfies $U J_n(\mu) = J_n(\lambda) U$. Equating the last columns, we get

$$\mu \begin{pmatrix} a_{1n} \\ a_{2n} \\ \vdots \\ a_{nn} \end{pmatrix} = \begin{pmatrix} 0 \\ a_{1n} \\ \vdots \\ a_{n-1,n} \end{pmatrix} + \lambda \begin{pmatrix} a_{1n} \\ a_{2n} \\ \vdots \\ a_{nn} \end{pmatrix}.$$

Using that $\lambda \neq \mu$, we get $a_{in} = 0$ for all i. Equating the $(n-1)$-st columns and using that $a_{in} = 0$ for all i, we get similarly $a_{i,n-1} = 0$ for all i. Descending induction gives $a_{ij} = 0$ for all i, j, that is, $U = 0$, a contradiction. This proves the first statement.

If H_λ^n and H_μ^m belong to the same component, then there is a sequence of the form $L_0 = H_\lambda^n, L_1, \ldots, L_t = H_\mu^m$ where the L_i are indecomposable and for each i, we have an irreducible morphism $L_i \longrightarrow L_{i+1}$ or $L_{i+1} \longrightarrow L_i$. Applying repeatedly the first statement, we get $\lambda = \mu$. □

This implies that the indecomposable regular modules occur in a family of connected components $(\mathscr{T}_\lambda)_\lambda$ of the Auslander–Reiten quiver, where each $\mathscr{T}_\lambda$ contains all H_λ^n, with $n \geq 1$ (and no H_μ^m for $\mu \neq \lambda$). We now compute the almost split sequences inside each $\mathscr{T}_\lambda$.

Theorem IV.4.13. *For every $n \geq 1$ and $\lambda \in \mathbf{k} \cup \{\infty\}$, there is an almost split sequence of the form*

$$0 \longrightarrow H_\lambda^n \xrightarrow{\binom{j_{n+1}}{p_n}} H_\lambda^{n+1} \oplus H_\lambda^{n-1} \xrightarrow{(-p_{n+1}\ j_n)} H_\lambda^n \longrightarrow 0$$

(where we agree that $H_\lambda^0 = 0$), and also all H_λ^n with $n \geq 1$ belong to a tube of rank 1 in $\Gamma(\mathrm{mod}\, A)$.

Proof. We use induction on n. If $n = 1$, then, because of Lemma IV.4.11, there is a short exact sequence

$$0 \longrightarrow H_\lambda^1 \xrightarrow{j_2} H_\lambda^2 \xrightarrow{p_2} H_\lambda^1 \longrightarrow 0.$$

It is not split because H_λ^2 is indecomposable. Now, the short exact sequences $0 \longrightarrow P_1 \longrightarrow P_2 \longrightarrow H_\lambda^1 \longrightarrow 0$ and $0 \longrightarrow H_\lambda^1 \longrightarrow I_1 \longrightarrow I_2 \longrightarrow 0$ imply that $\tau H_\lambda^1 = H_\lambda^1$, because $\nu P_1 = I_1$, $\nu P_2 = I_2$. Because of Corollary IV.4.3(b), $\mathrm{End}\, H_\lambda^1 \cong \mathbf{k}$. Because of Exercise III.2.6, the sequence is almost split.

Assume $n > 1$. The induction hypothesis says that there is an almost split sequence

$$0 \longrightarrow H_\lambda^{n-1} \xrightarrow{\binom{j_n}{p_{n-1}}} H_\lambda^n \oplus H_\lambda^{n-2} \xrightarrow{(-p_n\ j_{n-1})} H_\lambda^{n-1} \longrightarrow 0.$$

A straightforward computation involving the explicit forms of the maps j, p, given in the proof of Lemma IV.4.11 shows that the square

$$\begin{array}{ccc} H_\lambda^n & \xrightarrow{p_n} & H_\lambda^{n-1} \\ {\scriptstyle j_{n+1}}\downarrow & & \downarrow{\scriptstyle j_n} \\ H_\lambda^{n+1} & \xrightarrow{p_{n+1}} & H_\lambda^n \end{array}$$

commutes. Hence, we have a short exact sequence

$$0 \longrightarrow H_\lambda^n \xrightarrow{\binom{j_{n+1}}{p_n}} H_\lambda^{n+1} \oplus H_\lambda^{n-1} \xrightarrow{(-p_{n+1}\ j_n)} H_\lambda^n \longrightarrow 0.$$

We claim that the sequence is almost split.

First, the short exact sequences $0 \longrightarrow P_1^n \longrightarrow P_2^n \longrightarrow H_\lambda^n \longrightarrow 0$ and $0 \longrightarrow \tau H_\lambda^n \longrightarrow I_1^n \longrightarrow I_2^n \longrightarrow 0$ show that the dimension vector $\underline{\dim}\,(\tau H_\lambda^n) = n\, \underline{\dim}\, I_1 - n\, \underline{\dim}\, I_2 = n\, \underline{\dim}\, P_2 - n\, \underline{\dim}\, P_1 = \underline{\dim}\, H_\lambda^n = (n, n)$. Because of the classification theorem IV.4.10, there exists $\mu \in \mathbf{k} \cup \{\infty\}$ such that $\tau H_\lambda^n \cong H_\mu^n$. On the other hand, τH_λ^n lies in the same Auslander–Reiten component as H_λ^n. Applying Lemma IV.4.12, we get $\mu = \lambda$. This proves that $\tau H_\lambda^n \cong H_\lambda^n$.

Because of the induction hypothesis, p_n and j_n are irreducible. We claim that p_{n+1}, j_{n+1} are irreducible too. If not, then, because of Lemma II.2.2(b), one of them belongs to the radical square. But then $p_{n+1} j_{n+1} \in \mathrm{rad}_A^3(H_\lambda^n, H_\lambda^n)$. The commutativity of the above square gives $j_n p_n \in \mathrm{rad}_a^3(H_\lambda^n, H_\lambda^n)$. Because of Lemma IV.3.9, j_n is a proper epimorphism and p_n a proper monomorphism, a contradiction. This shows that p_{n+1}, j_{n+1} are irreducible. Therefore, the last short exact sequence is almost split.

Knitting together all these almost split sequences, we get a component of the form

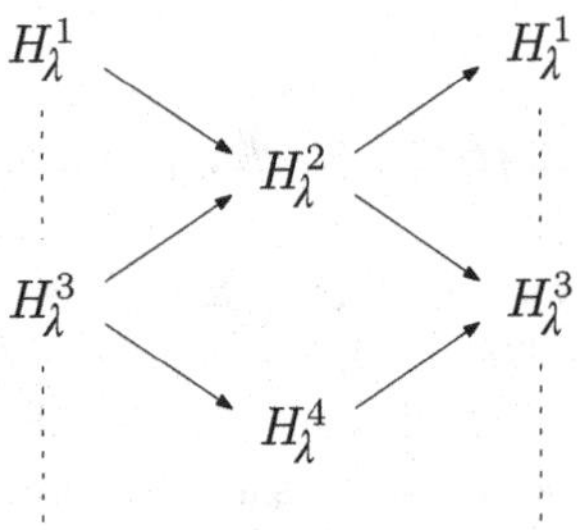

where we identify along the vertical dotted lines to form a cylinder. This is indeed a tube of rank 1. □

We are now ready to describe the Auslander–Reiten quiver of the Kronecker algebra. We need a notation. Let $\mathscr{C}, \mathscr{D}$ be connected components of $\Gamma(\mathrm{mod}\, A)$. We write $\mathrm{Hom}_A(\mathscr{C}, \mathscr{D}) = 0$ if for all M in $\mathscr{C}$ and N in $\mathscr{D}$, we have $\mathrm{Hom}_A(M, N) = 0$.

Corollary IV.4.14. *The Auslander–Reiten quiver* $\Gamma(\mathrm{mod}\, A)$ *consists of a unique postprojective component* $\mathscr{P}$, *a unique preinjective component* $\mathscr{Q}$ *and an infinite family* $(\mathscr{T}_\lambda)_{\lambda \in \mathbf{k} \cup \{\infty\}}$ *of tubes of rank one. In addition:*

(a) $\mathrm{Hom}_A(\mathscr{T}_\lambda, \mathscr{T}_\mu) = 0$ *whenever* $\lambda \neq \mu$*;*
(b) $\mathrm{Hom}_A(\mathscr{T}_\lambda, \mathscr{P}) = 0$ *and* $\mathrm{Hom}_A(\mathscr{Q}, \mathscr{T}_\lambda) = 0$ *for every* λ*;*
(c) *For every* $\lambda \in \mathbf{k} \cup \{\infty\}$, H *in* $\mathscr{T}_\lambda$, M_i *in* $\mathscr{P}$ *and* N_j *in* $\mathscr{Q}$, *we have* $\mathrm{Hom}_A(M_i, H) \neq 0$ *and* $\mathrm{Hom}_A(H, N_j) \neq 0$.

Proof.

(a) This follows from Lemma IV.4.12.
(b) Assume that H in $\mathscr{T}_\lambda$ and M_i in $\mathscr{P}$ are such that $\mathrm{Hom}_A(H, M_i) \neq 0$. Let $f : H \longrightarrow M_i$ be a nonzero morphism. Because H, M_i lie in distinct components, f is a radical morphism. If $i > 1$, there is a right minimal almost split morphism $M_{i-1}^2 \longrightarrow M_i$ through which f factors. Hence, $\mathrm{Hom}_A(H, M_{i-1}) \neq 0$. Descending induction yields $\mathrm{Hom}_A(H, M_1) \neq 0$. But M_1 is simple projective, and we get a contradiction. Therefore, $\mathrm{Hom}_A(\mathscr{T}_\lambda, \mathscr{P}) = 0$. Similarly, $\mathrm{Hom}_A(\mathscr{Q}, \mathscr{T}_\lambda) = 0$.

(c) Let H lie in $\mathscr{T}_\lambda$ and M_i in $\mathscr{P}$. Then, there exist $l \geq 0$ and $k \in \{1, 2\}$ such that $M_i \cong \tau^{-l} P_k$. Because H lies in $\mathscr{T}_\lambda$, we have $\tau H \cong H$; hence, $\tau^{-l} H \cong H$. Therefore, $\mathrm{Hom}_A(M_i, H) \cong \mathrm{Hom}_A(\tau^{-l} P_k, \tau^{-l} H) \cong \mathrm{Hom}_A(P_k, H)$, because of Exercise III.2.3(e). The latter equals $He_k = E_k$ and is seen to be nonzero for every H in $\mathscr{T}_\lambda$ and $k \in \{1, 2\}$. The last statement is proved similarly. □

One may visualise $\Gamma(\mathrm{mod}\, A)$ as follows

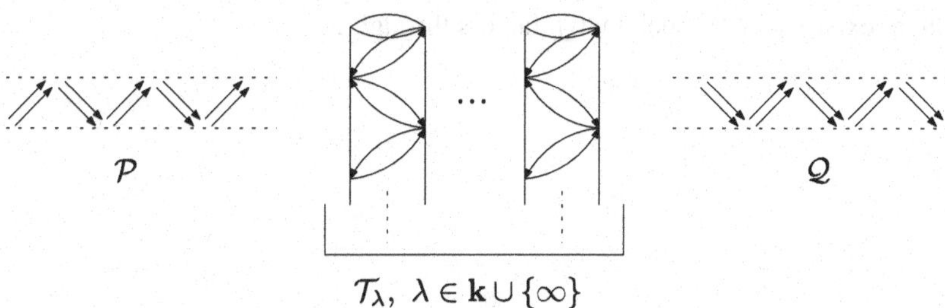

where maps go globally from left to right, taking into account that there is no map from one tube to another, because of Lemma IV.4.12.

Exercises for Section IV.4

Exercise IV.4.1. Prove directly that rep K_2 is an abelian category, without using Lemma IV.4.1.

Exercise IV.4.2. Prove that the finite dimensional **k**-algebra

$$R = \left\{ \begin{pmatrix} a_1 & 0 & \cdots & 0 \\ a_2 & a_1 & \ddots & \vdots \\ \vdots & \ddots & \ddots & 0 \\ a_n & \cdots & a_2 & a_1 \end{pmatrix} \mid a_i \in \mathbf{k} \right\}$$

with the ordinary matrix operations, is isomorphic to $\mathbf{k}[t]/\langle t^n \rangle$.

Exercise IV.4.3. Let A be the Kronecker algebra, S_1 its unique simple projective, S_2 its unique simple injective and $\lambda \in \mathbf{k} \cup \{\infty\}$.

(a) Let M be any indecomposable postprojective module. Prove that there exist E in $\mathscr{T}_\lambda$ and a nonsplit short exact sequence:

$$0 \longrightarrow M \longrightarrow E \longrightarrow S_2 \longrightarrow 0.$$

(b) Let N be any indecomposable preinjective module. Prove that there exist F in $\mathscr{T}_\lambda$ and a nonsplit short exact sequence:

$$0 \longrightarrow S_1 \longrightarrow F \longrightarrow N \longrightarrow 0.$$

Exercise IV.4.4. Let A be the Kronecker algebra. Prove that every morphism from the postprojective to the preinjective component factors through any tube, that is, if M, N are respectively an indecomposable postprojective and an indecomposable preinjective module, $f : M \longrightarrow N$ is a nonzero morphism and $\lambda \in \mathbf{k} \cup \{\infty\}$, prove that there exists X in $\mathscr{T}_\lambda$ such that f factors through X.

Chapter V
Endomorphism algebras

One of the constant lines of thinking in representation theory is the comparison between the module categories of a given algebra and the endomorphism algebra of some "well-chosen" module. For instance, the classical Morita theorem asserts that, given a progenerator P of the module category of an algebra A, that is, a projective module P that is also a generator of mod A, the categories mod A and mod(End P) are equivalent. This implies that, from the point of view of representation theory, we may assume that the algebras we deal with are basic, something we have done consistently. If one takes the endomorphism algebra of a module that is not a progenerator, then these modules categories are not equivalent, but nevertheless several features from one may pass to the other. This approach, initiated with the projectivisation procedure, much used by Auslander and his school, culminated in the now very important tilting theory. The aim of this chapter is to present these topics.

V.1 Projectivisation

V.1.1 The evaluation functor

Let A be an algebra. In Chapter II, we have seen how to translate certain statements about modules into functorial language, that is, to pass from mod A to the category Fun A of contravariant functors from mod A to mod $\mathbf{k}$. As we saw, this has the advantage of reducing several problems about arbitrary A-modules to problems about projective functors, which are easier to handle. Now, if instead of projective functors, we have projective modules (over a different algebra), then the problem may become even simpler.

Accordingly, in this subsection, we start from an algebra A, an A-module T, and set $B = \operatorname{End}_A(T)$. We recall that add T denotes the $\mathbf{k}$-linear full subcategory of

I. Assem, F. U. Coelho, *Basic Representation Theory of Algebras*, Graduate Texts in Mathematics 283, https://doi.org/10.1007/978-3-030-35118-2_5

mod A consisting of the direct sums of summands of T. As we shall see, the functor $\mathrm{Hom}_A(T, -)$ maps modules in add T to projective B-modules. Thus, summands of T become, via this functor, projective modules. For this reason, the technique we are going to see is called "projectivisation": it reduces some questions about T to questions about projective B-modules.

It is convenient to start with the functor category Fun A of contravariant functors from mod A to mod $\mathbf{k}$. As in Subsection II.3.1, we denote, for objects F, G of Fun A, by $\mathrm{Hom}(F, G)$, the space of functorial morphisms from F to G. We define the **evaluation functor** $\mathscr{E}\colon$ Fun $A \longrightarrow$ mod B as follows: for every object F in Fun A, we set

$$\mathscr{E}(F) = F(T)$$

and, for every morphism $\varphi\colon F \longrightarrow G$ in Fun A,

$$\mathscr{E}(\varphi) = \varphi_T\colon F(T) \longrightarrow G(T).$$

That is, functors and functorial morphisms are evaluated on the fixed object T. To prove that the evaluation $\mathscr{E}$ maps Fun A to mod B, we must endow the vector space $F(T)$ with a B-module structure. Let $x \in F(T)$ and $b \in B$. Thus, b is an endomorphism of T, which implies that $F(b)$ is an endomorphism of $F(T)$. We set

$$xb = F(b)(x).$$

We show that this is a B-module structure. Let $b, b' \in B$. Then $F(bb') = F(b')F(b)$ because F is contravariant. Therefore, for every $x \in F(T)$,

$$x(bb') = F(bb')(x) = F(b')F(b)(x) = F(b')(xb) = (xb)b'.$$

Thus, $\mathscr{E}(F)$ is indeed a B-module. In addition, for every functorial morphism $\varphi\colon F \longrightarrow G$, the morphism $\mathscr{E}(\varphi)$ is B-linear, because

$$\mathscr{E}(\varphi)(xb) = \varphi_T(xb) = \varphi_T F(b)(x) = G(b)\varphi_T(x) = \varphi_T(x)b = \mathscr{E}(\varphi)(x)b$$

for every $x \in F(T), b \in B$. This shows that the evaluation is a well-defined functor from Fun A to mod B.

We prove that $\mathscr{E}$ induces an equivalence between mod B and a certain full subcategory of Fun A, which is yet to be determined. To do this, we consider the full subcategory proj B of mod B consisting of the projective B-modules. The following lemma says that proj B is equivalent to the full subcategory $\mathscr{P}(T)$ of Fun A consisting of all functors of the form $\mathrm{Hom}_A(-, T_0)$, with T_0 in add T.

We notice that, if T_0 lies in add T, then $\mathrm{Hom}_A(T, T_0) = \mathscr{E}\,\mathrm{Hom}_A(-, T_0)$ lies in add $\mathrm{Hom}_A(T, T) =$ add B_B, that is, is a projective B-module. Thus, $\mathscr{E}$ maps $\mathscr{P}(T)$ into proj B.

Lemma V.1.1. *Let A be an algebra, T an A-module and $B = \mathrm{End}_A(T)$. Then,*

(a) *For every T_0 in* $\mathrm{add}\, T$, *and every F in* $\mathrm{Fun}\, A$, *we have an isomorphism* $\mathrm{Hom}(\mathrm{Hom}_A(-, T_0), F) \cong \mathrm{Hom}_B(\mathscr{E}\, \mathrm{Hom}_A(-, T_0), \mathscr{E} F)$.
(b) *$\mathscr{E}$ induces an equivalence between $\mathscr{P}(T)$ and* $\mathrm{proj}\, B$.

Proof.

(a) Because we deal with **k**-linear functors, it suffices to prove the statement when $T_0 = T$. Because of Yoneda's lemma II.3.1, we have $\mathrm{Hom}(\mathrm{Hom}_A(-, T), F) \cong F(T)$ as B-modules. On the other hand,

$$\begin{aligned} \mathrm{Hom}_B(\mathscr{E}\, \mathrm{Hom}_A(-, T), \mathscr{E} F) &\cong \mathrm{Hom}_B(\mathrm{Hom}_A(T, T), F(T)) \\ &= \mathrm{Hom}_B(B, F(T)) \\ &\cong F(T)_B \end{aligned} .$$

This establishes the claim.
(b) Let P_0 be an indecomposable projective B-module. There exists a primitive idempotent e_0 in B such that $P_0 = e_0 B = e_0\, \mathrm{Hom}_A(T, T) \cong \mathrm{Hom}_A(T, e_0 T)$. That is, there exists an indecomposable summand $T_0 = e_0 T$ of T_A such that $P_0 = \mathrm{Hom}_A(T, T_0)$. This shows that $\mathscr{E} \colon \mathscr{P}(T) \longrightarrow \mathrm{proj}\, B$ is a well-defined and dense functor. Because of (a), it is also full and faithful. The proof is complete.

□

Because (a) in the above lemma offers only a restricted version of full faithfulness, we cannot expect that the equivalence $\mathscr{E} \colon \mathscr{P}(T) \longrightarrow \mathrm{proj}\, B$ would extend to an equivalence from the whole of $\mathrm{Fun}\, A$ to $\mathrm{mod}\, B$. We may however expect to find a full subcategory of $\mathrm{Fun}\, A$ that would be equivalent to $\mathrm{mod}\, B$. Let $\mathrm{pres}\, \mathscr{P}(T)$ denote the full subcategory of $\mathrm{Fun}\, A$ consisting of all functors F that admit a projective presentation of the form

$$\mathrm{Hom}_A(-, T_1) \longrightarrow \mathrm{Hom}_A(-, T_0) \longrightarrow F \longrightarrow 0$$

with T_0, T_1 lying in $\mathrm{add}\, T$, that is, $\mathrm{Hom}_A(-, T_0), \mathrm{Hom}_A(-, T_1)$ lying in $\mathscr{P}(T)$, such functors F are said to be **$\mathscr{P}(T)$-presented**. We prove that the full subcategory $\mathrm{pres}\, \mathscr{P}(T)$ of $\mathrm{Fun}\, A$ is equivalent to $\mathrm{mod}\, B$.

Theorem V.1.2 (Projectivisation Theorem). *Let A be an algebra, T a module and $B = \mathrm{End}_A(T)$. Then the evaluation functor induces an equivalence of categories $\mathscr{E} \colon \mathrm{pres}\, \mathscr{P}(T) \longrightarrow \mathrm{mod}\, B$.*

Proof. Clearly, $\mathscr{E} \colon \mathrm{pres}\, \mathscr{P}(T) \longrightarrow \mathrm{mod}\, B$ is a well-defined functor, because it is the composition of the inclusion functor $\mathrm{pres}\, \mathscr{P}(T) \hookrightarrow \mathrm{Fun}\, A$ with the evaluation $\mathscr{E} \colon \mathrm{Fun}\, A \longrightarrow \mathrm{mod}\, B$.

We first prove that $\mathscr{E}$ is dense. Let X be a B-module and consider a projective presentation

$$P_1 \xrightarrow{p} P_0 \longrightarrow X \longrightarrow 0$$

in mod B. Because of Lemma V.1.1(b), there exists a morphism $f\colon T_1 \longrightarrow T_0$ with T_0, T_1 in add T such that $\operatorname{Hom}_A(T, f) = \mathscr{E}(f) = p$. Let $F = \operatorname{Coker}\operatorname{Hom}_A(-, f)$. We have an exact sequence

$$\operatorname{Hom}_A(-, T_1) \xrightarrow{\operatorname{Hom}_A(-,f)} \operatorname{Hom}_A(-, T_0) \longrightarrow F \longrightarrow 0$$

in Fun A. Because $T_0, T_1 \in \operatorname{add} T$, we infer that F lies in pres $\mathscr{P}(T)$. Evaluating the previous sequence on T and comparing with the original projective presentation yield a commutative diagram in mod B with exact rows

$$\begin{array}{ccccccc}
\operatorname{Hom}_A(T,T_1) & \xrightarrow{\operatorname{Hom}_A(T,f)} & \operatorname{Hom}_A(T,T_0) & \longrightarrow & F(T) & \longrightarrow & 0 \\
\downarrow{\cong} & & \downarrow{\cong} & & & & \\
P_1 & \xrightarrow{p} & P_0 & \longrightarrow & X & \longrightarrow & 0
\end{array}$$

Consequently, $X \cong F(T) = \mathscr{E}(F)$. This proves density.

To show that $\mathscr{E}$ is full and faithful, let F, G be objects in pres $\mathscr{P}(T)$. We have an exact sequence in Fun A

$$\operatorname{Hom}_A(-, T_1) \longrightarrow \operatorname{Hom}_A(-, T_0) \longrightarrow F \longrightarrow 0.$$

with T_0, T_1 in add M. Applying to this sequence the functor $\operatorname{Hom}(-, G)$ yields an exact sequence

$$0 \longrightarrow \operatorname{Hom}(F, G) \longrightarrow \operatorname{Hom}(\operatorname{Hom}_A(-, T_0), G) \longrightarrow \operatorname{Hom}(\operatorname{Hom}_A(-, T_1), G).$$

Evaluating functors on T, and applying Lemma V.1.1(a), we get a commutative diagram with exact rows

$$\begin{array}{ccccccc}
0 \to & \operatorname{Hom}_B(F(T), G(T)) & \to & \operatorname{Hom}_B(\operatorname{Hom}_A(T,T_0), G(T)) & \to & \operatorname{Hom}_B(\operatorname{Hom}_A(T,T_1), G(T)) \\
 & & & \downarrow{\cong} & & \downarrow{\cong} \\
0 \longrightarrow & \operatorname{Hom}(F,G) & \longrightarrow & \operatorname{Hom}(\operatorname{Hom}_A(-,T_0),G) & \longrightarrow & \operatorname{Hom}(\operatorname{Hom}_A(-,T_1),G)
\end{array}$$

The statement follows. □

In the proof of full faithfulness above, we used that F lies in pres $\mathscr{P}(T)$, but not that G lies in it. This is thanks to the statement of Lemma V.1.1(a).

We have proved that the evaluation functor $\mathscr{E}$ induces horizontal equivalences in the commutative diagram

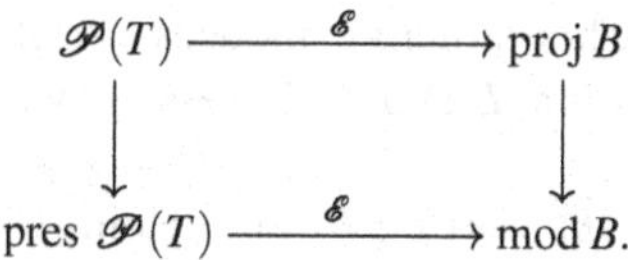

where the vertical arrows represent the inclusion functors.

Clearly, a similar statement holds if one uses covariant functors instead of contravariant ones.

Another question arises: to which functors in pres $\mathscr{P}(T)$ do injective B-modules correspond? Let inj B denote the full subcategory of mod B consisting of the injective B-modules and W the full subcategory of Fun A consisting of all functors of the form $\mathrm{D}\,\mathrm{Hom}_A(T_0, -)$ with T_0 in add M.

Corollary V.1.3. *The equivalence* $\mathscr{E}\colon$ pres $\mathscr{P}(T) \longrightarrow$ mod B *restricts to an equivalence* $W \cong$ inj B.

Proof. Let I_0 be an indecomposable injective B-module and e_0 the corresponding primitive idempotent in B. Then,

$$I_0 = \mathrm{D}(Be_0) = \mathrm{D}(\mathrm{Hom}_A(T, T)e_0) \cong \mathrm{D}\,\mathrm{Hom}_A(e_0T, T).$$

Thus, setting $T_0 = e_0T$, we get $I_0 \cong \mathscr{E}\mathrm{D}\,\mathrm{Hom}_A(T_0, -)$. Let $\mathscr{G}\colon \mathrm{mod}\,B \longrightarrow$ pres $\mathscr{P}(T)$ be a quasi-inverse of $\mathscr{E}$, whose existence is granted by Theorem V.1.2. Then we have $\mathscr{G}I_0 \cong \mathrm{D}\,\mathrm{Hom}_A(T_0, -)$. In particular, $\mathrm{D}\,\mathrm{Hom}_A(T_0, -)$ belongs to pres $\mathscr{P}(T)$. This shows that the evaluation functor $\mathscr{E}\colon W \longrightarrow$ inj B is well-defined and dense.

It is full and faithful because of Yoneda's lemma II.3.1 and Lemma V.1.1(a) (in both its contravariant and covariant versions), which imply that

$$\begin{aligned}
&\mathrm{Hom}_B(\mathscr{E}\mathrm{D}\,\mathrm{Hom}_A(T_0, -), \mathscr{E}\mathrm{D}\,\mathrm{Hom}_A(T_1, -))\\
&\qquad \cong \mathrm{Hom}_B(\mathrm{D}\,\mathrm{Hom}_A(T_0, T), \mathrm{D}\,\mathrm{Hom}_A(T_1, T))\\
&\qquad \cong \mathrm{Hom}_B(\mathrm{Hom}_A(T_1, T), \mathrm{Hom}_A(T_0, T))\\
&\qquad \cong \mathrm{Hom}(\mathrm{Hom}_A(T_1, -), \mathrm{Hom}_A(T_0, -))\\
&\qquad \cong \mathrm{Hom}(\mathrm{D}\,\mathrm{Hom}_A(T_0, -), \mathrm{D}\,\mathrm{Hom}_A(T_1, -))
\end{aligned}$$

where T_1, T_0 lie in add T. This completes the proof. □

So far, we have been dealing with functors. It is time to give a module theoretic interpretation of the projectivisation procedure. For this purpose, we observe what happens when one evaluates a Hom-functor $\mathrm{Hom}_A(-, U)\colon \mathrm{mod}\,A \longrightarrow \mathrm{mod}\,\mathbf{k}$, where U is any A-module. Then,

$$\mathscr{E}\operatorname{Hom}_A(-, U) = \operatorname{Hom}_A(T, U) = \operatorname{Hom}_A(T, -)(U).$$

Now, T has a canonical left B-module structure given by $bt = b(t)$ for $t \in T$ and $b \in B = \operatorname{End}(T)$. This structure induces a right B-module structure on $\operatorname{Hom}_A(T, U)$ as follows: for $b \in B$ and $f : T \longrightarrow U$, we have

$$(fb)(t) = f(bt) = (f \circ b)(t) = \operatorname{Hom}_A(b, T)(f)(t)$$

for each $t \in T$. This is exactly the B-module structure on $\mathscr{E}\operatorname{Hom}_A(-, U)$ defined at the beginning of the subsection. We may thus specialise Lemma V.1.1 and Theorem V.1.2 to this context. We define $\operatorname{pres} T$ to be the full subcategory of $\operatorname{mod} A$ consisting of all modules U_A such that there exists an exact sequence

$$T_1 \longrightarrow T_0 \longrightarrow U \longrightarrow 0$$

with T_0, T_1 in $\operatorname{add} T$. Such modules are called T-**presented** and the sequence is a T-**presentation**.

Corollary V.1.4. *Let A be an algebra, T a module and $B = \operatorname{End}_A(T)$. Then*

(a) *For every T_0 in $\operatorname{add} T$ and U in $\operatorname{mod} A$, there is an isomorphism $\operatorname{Hom}_A(T_0, U) \cong \operatorname{Hom}_B(\operatorname{Hom}_A(T, T_0), \operatorname{Hom}_A(T, U))$ given by $f \mapsto \operatorname{Hom}_A(T, f)$.*
(b) *The functor $\operatorname{Hom}_A(T, -)$ induces an equivalence between $\operatorname{add} T$ and $\operatorname{proj} B$.*
(c) *The functor $\operatorname{D}\operatorname{Hom}_A(-, T)$ induces an equivalence between the full subcategory of $\operatorname{pres} T$ consisting of modules of the form $\operatorname{D}\operatorname{Hom}_A(T_0, T)$ with T_0 in $\operatorname{add} T$, and $\operatorname{inj} B$.*

Proof.

(a) Because of Yoneda's lemma II.3.1 and Lemma V.1.1(a), we have

$$\begin{aligned}\operatorname{Hom}_A(T_0, U) &\cong \operatorname{Hom}(\operatorname{Hom}_A(-, T_0), \operatorname{Hom}_A(-, U))\\ &\cong \operatorname{Hom}_B(\mathscr{E}\operatorname{Hom}_A(-, T_0), \mathscr{E}\operatorname{Hom}_A(-, U))\\ &\cong \operatorname{Hom}_B(\operatorname{Hom}_A(T, T_0), \operatorname{Hom}_A(T, U))\end{aligned}$$

the isomorphism being given by

$$f \mapsto \operatorname{Hom}_A(-, f) \mapsto \mathscr{E}\operatorname{Hom}_A(-, f) = \operatorname{Hom}_A(T, f).$$

(b) As seen in the proof of Lemma V.1.1(b), every indecomposable projective B-module is of the form $P_0 = \operatorname{Hom}_A(T, T_0)$, where T_0 is an indecomposable summand of T. This proves that $\mathscr{E}\colon \operatorname{add} T \longrightarrow \operatorname{proj} B$ is well-defined and dense. Full faithfulness follows from (a).
(c) The proof is similar to that of (b) and left as an exercise.

□

Summing up, the functor $\mathrm{Hom}_A(T, -) : \mathrm{mod}\, A \longrightarrow \mathrm{mod}\, B$ sends objects in $\mathrm{add}\, T$ to projective B-modules whereas the functor $\mathrm{D}\,\mathrm{Hom}_A(-, T)$ sends the objects in $\mathrm{add}\, T$ to injective B-modules.

V.1.2 Projectivising projectives

Let, as in Subsection V.1.1, A be an algebra, T an A-module and $B = \mathrm{End}\, T$. We recall that T admits a left B-module structure defined by setting $bt = b(t)$, for $b \in B, t \in T$. As a result, T becomes a $B - A$-bimodule: indeed, every $b \in B$ is a morphism in $\mathrm{mod}\, A$ from T to itself and so

$$b(ta) = b(t)a = (bt)a$$

for $t \in T$, $a \in A$. We have considered in Subsection V.1.1 the functor $\mathrm{Hom}_A(T, -) : \mathrm{mod}\, A \longrightarrow \mathrm{mod}\, B$. It is reasonable to consider the tensor functor $- \otimes_B T : \mathrm{mod}\, B \longrightarrow \mathrm{mod}\, A$ too, sending each right B-module X to $X \otimes_B T$. The latter has a right A-module structure defined by

$$(x \otimes t)a \ = \ x \otimes (ta)$$

for $x \in X, t \in T, a \in A$. We prove that $X \otimes_B T$ is T-presented.

Lemma V.1.5. *The image of the tensor functor* $- \otimes_B T : \ \mathrm{mod}\, B \longrightarrow \mathrm{mod}\, A$ *lies in* $\mathrm{pres}\, T$.

Proof. Indeed, an arbitrary B-module X admits a projective presentation of the form

$$P_1 \longrightarrow P_0 \longrightarrow X \longrightarrow 0$$

with P_0, P_1 projective B-modules. Applying the right exact functor $- \otimes_B T$ yields an exact sequence in $\mathrm{mod}\, A$

$$P_1 \otimes_B T \longrightarrow P_0 \otimes_B T \longrightarrow X \otimes_B T \longrightarrow 0.$$

Because $B \otimes_B T_A \cong T_A$ and P_1, P_0 lie in $\mathrm{add}\, T$, we have that $P_1 \otimes_B T$ and $P_0 \otimes_B T$ lie in $\mathrm{add}\, T$. Therefore, $X \otimes_B T$ is T-presented. □

This lemma shows the existence of a functor $- \otimes_B T : \mathrm{mod}\, B \longrightarrow \mathrm{pres}\, T$. We ask under which conditions this functor is a quasi-inverse to the restriction to $\mathrm{pres}\, T$ of the functor $\mathrm{Hom}_A(T, -) : \mathrm{mod}\, A \longrightarrow \mathrm{mod}\, B$. For this purpose, we recall the well-known adjunction isomorphism

$$\mathrm{Hom}_A(X \otimes_B T, M) \cong \mathrm{Hom}_B(X, \mathrm{Hom}_A(T, M))$$

bifunctorial in the B-module X and the A-module M. The existence of this isomorphism entails the existence of the functorial morphisms

$$\begin{aligned} \epsilon_M : \mathrm{Hom}_A(T, M) \otimes_B T &\longrightarrow M \\ f \otimes t &\mapsto f(t) \end{aligned}$$

for $t \in T$, $f: T \longrightarrow M$ and

$$\begin{aligned} \delta_X : X &\longrightarrow \mathrm{Hom}_A(T, X \otimes_B T) \\ x &\mapsto (t \mapsto x \otimes t) \end{aligned}$$

for $t \in T$, $x \in X$. These morphisms are respectively called the **counit** and the **unit** of the adjunction. Using these morphisms, we can reprove Corollary V.1.4(b) by showing that $- \otimes_B T$ restricted to proj B is a quasi-inverse of the restriction of $\mathrm{Hom}_A(T, -)$ to add T.

Lemma V.1.6.

(a) *Let T_0 be in* add T*, then ϵ_{T_0} is an isomorphism.*
(b) *Let P_0 be in* proj B*, then δ_{P_0} is an isomorphism.*

Proof. We only prove (a), because the proof of (b) is similar.

Because we deal with **k**-linear functors, it suffices to prove the statement when $T_0 = T$. But in this case, the counit

$$\epsilon_T : \mathrm{Hom}_A(T, T) \otimes_B T = B \otimes_B T \longrightarrow T$$

is the morphism $b \otimes t \mapsto b(t) = bt$ (for $b \in B$, $t \in T$), which defines the left B-module structure of T. Therefore, it is an isomorphism of A-modules. □

We now assume that T is a projective A-module. In this case, there exists an idempotent $e \in A$ such that $T = eA$. Then, $B = \mathrm{End}\, T \cong eAe$.

Proposition V.1.7. *Let $e \in A$ be an idempotent, $T = eA$ and $B = eAe$. Then the restriction to* pres T *of the functor* $\mathrm{Hom}_A(T, -)$ *and the functor* $- \otimes_B T$ *induce quasi-inverse equivalences between* pres T *and* mod B.

Proof. It suffices to prove that, for each T-presented A-module M and each B-module X, the morphisms ϵ_M and δ_X are isomorphisms. Let M be in pres T. Then, there exist T_0, T_1 in add T and an exact sequence

$$T_1 \longrightarrow T_0 \longrightarrow M \longrightarrow 0.$$

Because T is projective, $\mathrm{Hom}_A(T, -)$ is exact. Thus, we get an exact sequence

$$\mathrm{Hom}_A(T, T_1) \longrightarrow \mathrm{Hom}_A(T, T_0) \longrightarrow \mathrm{Hom}_A(T, M) \longrightarrow 0.$$

Applying the right exact functor $- \otimes_B T$ and comparing with the first sequence above yields a commutative diagram with exact rows

$$\begin{array}{ccccccc}
\mathrm{Hom}_A(T,T_1)\otimes_B T & \longrightarrow & \mathrm{Hom}_A(T,T_0)\otimes_B T & \longrightarrow & \mathrm{Hom}_A(T,M)\otimes_B T & \longrightarrow & 0 \\
{\scriptstyle \varepsilon_{T_1}}\downarrow & & {\scriptstyle \varepsilon_{T_0}}\downarrow & & \downarrow{\scriptstyle \varepsilon_M} & & \\
T_1 & \longrightarrow & T_0 & \longrightarrow & M & \longrightarrow & 0
\end{array}$$

Because of Lemma V.1.6, ε_{T_1} and ε_{T_0} are isomorphisms. Hence, so is ε_M. Similarly, if X is a B-module, then there exists a projective presentation

$$P_1 \longrightarrow P_0 \longrightarrow X \longrightarrow 0$$

in mod B. Applying first the right exact functor $- \otimes_B T$ and then the exact functor $\mathrm{Hom}_A(T, -)$ yields a commutative diagram with exact rows

$$\begin{array}{ccccccc}
P_1 & \longrightarrow & P_0 & \longrightarrow & X & \longrightarrow & 0 \\
{\scriptstyle \delta_{P_1}}\downarrow & & {\scriptstyle \delta_{P_0}}\downarrow & & \downarrow{\scriptstyle \delta_X} & & \\
\mathrm{Hom}_A(T, P_1 \otimes_B T) & \longrightarrow & \mathrm{Hom}_A(T, P_0 \otimes_B T) & \longrightarrow & \mathrm{Hom}_A(T, X \otimes_B T) & \longrightarrow & 0
\end{array}$$

Again, Lemma V.1.6 says that $\delta_{P_0}, \delta_{P_1}$ are isomorphisms. Hence, so is δ_X. □

Assuming that T is projective is certainly a strong assumption, but if one reads carefully the proof of the proposition, one sees that what is really needed is that T satisfies the following condition: for any A-module M such that there is an exact sequence $T_1 \longrightarrow T_0 \longrightarrow M \longrightarrow 0$ with T_1, T_0 in add T, the induced sequence

$$\mathrm{Hom}_A(T, T_1) \longrightarrow \mathrm{Hom}_A(T, T_0) \longrightarrow \mathrm{Hom}_A(T, M) \longrightarrow 0$$

in mod B is also exact. This remark will be used in the coming developments.

Example V.1.8. Let A be given by the quiver

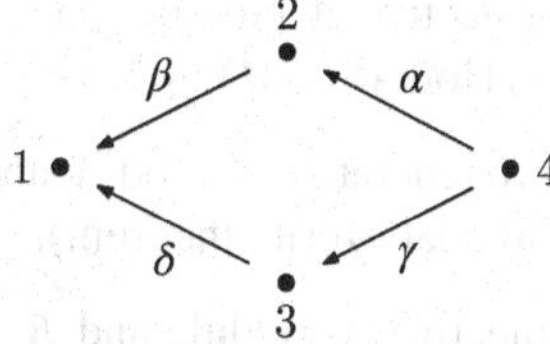

bound by $\alpha\beta = \gamma\delta$. Let $e = e_1 + e_2 + e_4$. Then $T = P_1 \oplus P_2 \oplus P_4$ while $B = eAe$ is the hereditary algebra given by the quiver

$$\overset{1}{\bullet} \longleftarrow \overset{2}{\bullet} \longleftarrow \overset{4}{\bullet}$$

where the point i corresponds to the indecomposable projective A-module P_i, for $i \in \{1, 2, 4\}$. Then, $\operatorname{pres} T$ contains exactly as indecomposable objects the indecomposable A-modules whose minimal projective presentation involves only the projectives P_1, P_2 and P_4. Looking at the Auslander–Reiten quiver of A:

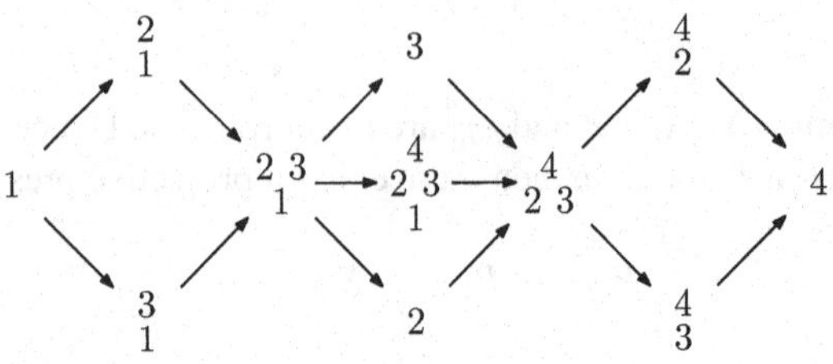

one sees immediately that

$$\operatorname{pres} T = \{1, \begin{smallmatrix}2\\1\end{smallmatrix}, \begin{smallmatrix}&4&\\2&&3\\&1&\end{smallmatrix}, 2, \begin{smallmatrix}&4&\\2&&3\end{smallmatrix}, \begin{smallmatrix}4\\3\end{smallmatrix}\}.$$

Rearranging these modules, one gets a quiver isomorphic to the Auslander–Reiten quiver of $\operatorname{mod} B$

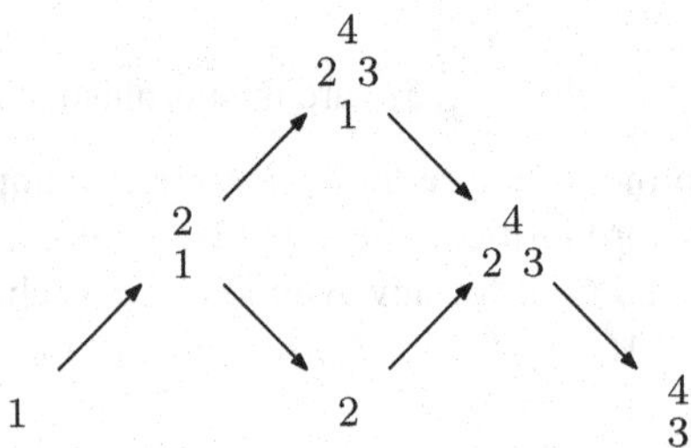

Exercises for Section V.1

Exercise V.1.1. Let T be a projective A-module, and M an A-module in $\operatorname{pres} T$. Prove that $\operatorname{pd} M \leq 1$ implies $\operatorname{pd} \operatorname{Hom}_A(T, M) \leq 1$.

Exercise V.1.2. Let T be a progenerator of $\operatorname{mod} A$ and $B = \operatorname{End} T$. Prove that $\operatorname{mod} A \cong \operatorname{mod} B$ (this is the classical Morita theorem).

Exercise V.1.3. Let I be an injective A-module and $B = \operatorname{End} I$. Prove that $\operatorname{mod} B$ is equivalent to the full subcategory $\operatorname{copres} I$ of $\operatorname{mod} A$ consisting of all A-modules M having an injective copresentation

$$0 \longrightarrow M \longrightarrow I_0 \longrightarrow I_1$$

with I_0, I_1 in $\operatorname{add} I$.

Exercise V.1.4. Let A be an algebra and T an A-module. Given an A-module M, a morphism $f_M : T_M \longrightarrow M$ with T_M in $\operatorname{add} T$ is called a **right** $\operatorname{add} T$**-approximation** if, whenever $f_0 : T_0 \longrightarrow M$ is a morphism with T_0 in $\operatorname{add} T$, there exists $g : T_0 \longrightarrow T_M$ such that $f_0 = f_M g$.

(a) Prove that f_M is a right $\operatorname{add} T$-approximation if and only if the functorial morphism $\operatorname{Hom}_A(-, f_M) : \operatorname{Hom}_A(-, T_M)|_{\operatorname{add} T} \longrightarrow \operatorname{Hom}_A(-, M)|_{\operatorname{add} T}$ is an epimorphism.
(b) Let $\{f_1, \ldots, f_d\}$ be a generating set of the $\operatorname{End} T$-module $\operatorname{Hom}_A(T, M)$ and $f = [f_1 \ldots f_d] : T^d \longrightarrow M$. Prove that f is a right $\operatorname{add} T$-approximation.
(c) With the notation of (b), prove that f is an epimorphism if and only if M is generated by T.
(d) Prove that, for every module M, there always exists a right $\operatorname{add} T$-approximation, which is also right minimal.
(e) Let $f_M : T_M \longrightarrow M$, $f'_M : T'_M \longrightarrow M$ be right $\operatorname{add} T$-approximations that are right minimal. Prove that there exists an isomorphism $g : T'_M \longrightarrow T_M$ such that $f'_M = f_M g$.

Exercise V.1.5. Let A be an algebra and T an A-module. One defines the **left** $\operatorname{add} T$**-approximation** of an A-module M dually to right $\operatorname{add} T$-approximation. State and prove the results corresponding to those of the previous exercise.

Exercise V.1.6. Let A be an algebra, $\mathscr{C}$ a full abelian subcategory of $\operatorname{mod} A$ and T a generator of $\mathscr{C}$, which is projective in $\mathscr{C}$. Prove that $\operatorname{Hom}_A(T, -) \colon \mathscr{C} \longrightarrow \operatorname{mod}(\operatorname{End} T)$ is an equivalence of categories.

Exercise V.1.7. Let A be an algebra, T an A-module and $B = \operatorname{End} T_A$. Prove that, for every projective B-module P and every B-module X, the morphism $g \longmapsto g \otimes T$ induces an isomorphism $\operatorname{Hom}_B(X, P) \cong \operatorname{Hom}_A(X \otimes_B T, P \otimes_B T)$.

Exercise V.1.8. Let A be an algebra, T an A-module and $B = \operatorname{End} T_A$. Prove that, for every module M, the A-module $\operatorname{Hom}_A(T, M) \otimes_B T$ is generated by T_A. Deduce that the morphism $\varepsilon_M \colon \operatorname{Hom}_A(T, M) \otimes_B T \longrightarrow M$ is surjective if and only if M is generated by T.

Exercise V.1.9. Let A be an algebra, T an A-module and $B = \operatorname{End} T_A$. Prove that the morphism $x \longmapsto (f \mapsto f(x))$ from M to $\operatorname{Hom}_{B^{op}}(\operatorname{Hom}_A(M, T), T)$ is injective if and only if M is cogenerated by T.

Exercise V.1.10. For each of the bound quiver algebras A below and each of the idempotents e indicated, compute eAe and show explicitly the equivalence between $\operatorname{pres}(eA)$ and $\operatorname{mod}(eAe)$.

(a) $1 \xleftarrow{\gamma} 2 \xleftarrow{\beta} 3 \xleftarrow{\alpha} 4$ $\qquad$ $\alpha\beta\gamma = 0$; $e = e_1 + e_2 + e_4$.

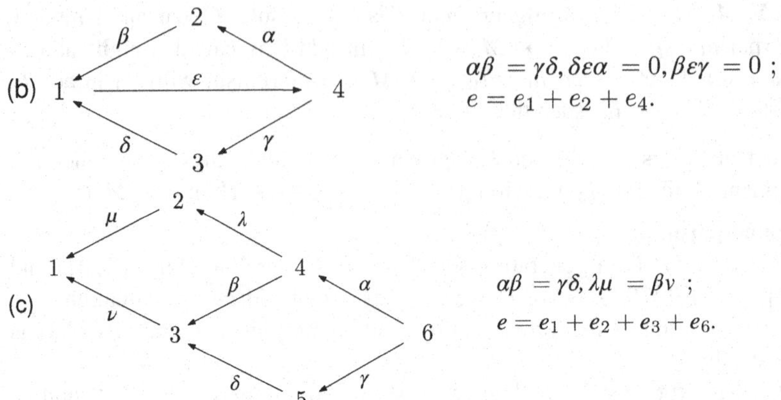

V.2 Tilting theory

V.2.1 Tilting modules

Tilting theory is at present one of the most active areas of research in the representation theory of algebras, with applications in several other parts of mathematics. Our purpose here is not to give an overview, but rather a short introduction. The main idea of tilting theory is to projectivise a module that is "close enough" to a progenerator of the module category, and then to compare the module category of the original algebra with that of the endomorphism algebra of the projectivised module.

Definition V.2.1. Let A be an algebra, an A-module T is called a **partial tilting module** if it satisfies the following conditions:

(a) $\operatorname{pd} T \leq 1$; and
(b) $\operatorname{Ext}^1_A(T, T) = 0$.

In addition, it is a **tilting module** if it also satisfies the third condition:

(c) There is a short exact sequence of the form

$$0 \longrightarrow A_A \longrightarrow T_0 \longrightarrow T_1 \longrightarrow 0$$

with T_0, T_1 in $\operatorname{add} T$.

Thus, every projective module is a partial tilting module and every progenerator is a tilting module. All three conditions above express in some way that tilting modules are close to progenerators. In addition:

(a) Because of Proposition III.1.11, the first condition is equivalent to saying that $\operatorname{Hom}_A(DA, \tau T) = 0$, that is, no injective maps nontrivially to τT. Because of

Theorem III.2.4, the second condition in the presence of the first is equivalent to saying that $\mathrm{Hom}_A(T, \tau T) = 0$, see Exercise III.2.3(a).

(b) To verify the third condition, it suffices to construct, for each indecomposable projective A-module P_x, a short exact sequence $0 \longrightarrow P_x \longrightarrow T_x^0 \longrightarrow T_x^1 \longrightarrow 0$, with T_x^0, T_x^1 in $\mathrm{add}\, T$. Indeed, the direct sum of such sequences yields the required sequence for A_A.

(c) Every tilting module T is faithful, that is, its annihilator $\mathrm{Ann}\, T = \{a \in A : Ta = 0\}$ vanishes. Indeed, because of condition (c) of the definition, there exists a monomorphism $j : A \longrightarrow T_0$ with T_0 in $\mathrm{add}\, T$. Let $a \in \mathrm{Ann}\, T$, then $j(a) = j(1)a \in T_0 a = 0$. Because j is injective, we infer that $a = 0$.

We give an example of a tilting module.

Example V.2.2. Let A be given by the quiver

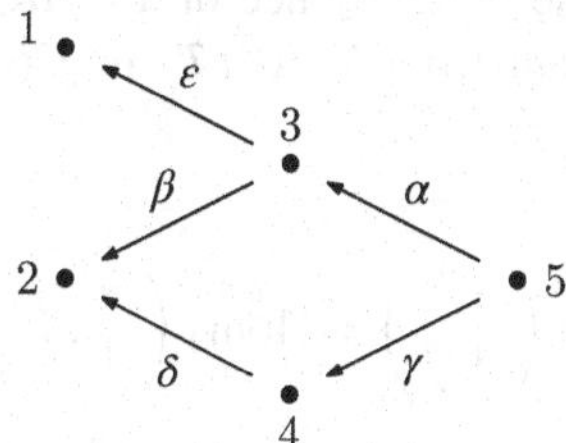

bound by $\alpha\beta = \gamma\delta$ and $\alpha\varepsilon = 0$. Then, $\Gamma(\mathrm{mod}\, A)$ is

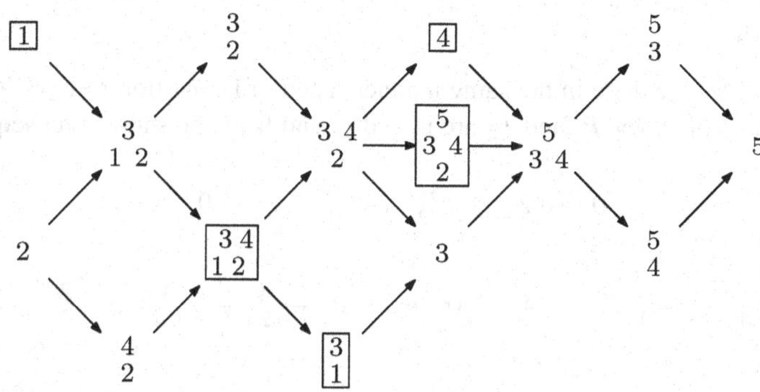

We claim that

$$T = 1 \oplus \begin{smallmatrix} 3\,4 \\ 1\,2 \end{smallmatrix} \oplus \begin{smallmatrix} 3 \\ 1 \end{smallmatrix} \oplus 4 \oplus \begin{smallmatrix} & 5 & \\ 3 & & 4 \\ & 2 & \end{smallmatrix}$$

is a tilting module. The short exact sequences

$$0 \longrightarrow 2 \longrightarrow \begin{smallmatrix}3\\1\,2\end{smallmatrix} \oplus \begin{smallmatrix}4\\2\end{smallmatrix} \longrightarrow \begin{smallmatrix}3\,4\\1\,2\end{smallmatrix} \longrightarrow 0,$$

$$0 \longrightarrow 2 \longrightarrow \begin{smallmatrix}3\\1\,2\end{smallmatrix} \longrightarrow \begin{smallmatrix}3\\1\end{smallmatrix} \longrightarrow 0 \text{ and } 0 \longrightarrow 2 \longrightarrow \begin{smallmatrix}4\\2\end{smallmatrix} \longrightarrow 4 \longrightarrow 0$$

and the projectivity of

$$P_1 = 1 \quad \text{and} \quad P_5 = \begin{smallmatrix}5\\3\,4\\2\end{smallmatrix}$$

yield that $\operatorname{pd} T = 1$. To prove that $\operatorname{Ext}^1_A(T, T) = D\operatorname{Hom}_A(T, \tau T) = 0$, we need to show that, for any indecomposable summands T_i, T_j of T, we have $\operatorname{Hom}_A(T_i, \tau T_j) = 0$. Because the existence of a nonzero morphism $T_i \longrightarrow \tau T_j$ implies the existence of a path from T_i to τT_j in $\Gamma(\operatorname{mod} A)$, we see easily, for instance, that

$$\operatorname{Hom}_A\left(\begin{smallmatrix}3\\1\end{smallmatrix}, \tau\left(\begin{smallmatrix}3\\1\end{smallmatrix}\right)\right) = \operatorname{Hom}_A\left(\begin{smallmatrix}3\\1\end{smallmatrix}, \begin{smallmatrix}4\\2\end{smallmatrix}\right) = 0 \text{ and}$$

$$\operatorname{Hom}_A\left(\begin{smallmatrix}3\\1\end{smallmatrix}, \tau(4)\right) = \operatorname{Hom}_A\left(\begin{smallmatrix}3\\1\end{smallmatrix}, \begin{smallmatrix}3\\2\end{smallmatrix}\right) = 0.$$

The other cases are done in the same manner. The third condition follows from the fact that the projectives P_1 and P_5 are in $\operatorname{add} T$, and from the short exact sequences

$$0 \longrightarrow 2 \longrightarrow \begin{smallmatrix}3\,4\\1\,2\end{smallmatrix} \longrightarrow \begin{smallmatrix}3\\1\end{smallmatrix} \oplus 4 \longrightarrow 0,$$

$$0 \longrightarrow \begin{smallmatrix}3\\1\,2\end{smallmatrix} \longrightarrow \begin{smallmatrix}3\,4\\1\,2\end{smallmatrix} \longrightarrow 4 \longrightarrow 0 \quad \text{and} \quad 0 \longrightarrow \begin{smallmatrix}4\\2\end{smallmatrix} \longrightarrow \begin{smallmatrix}3\,4\\1\,2\end{smallmatrix} \longrightarrow \begin{smallmatrix}3\\1\end{smallmatrix} \longrightarrow 0.$$

We justify now the name of partial tilting module by proving that every partial tilting module can be completed to (that is, is a direct summand of) a tilting module. We need a homological lemma.

Lemma V.2.3. *Let T, M be A-modules. Then, there exists a short exact sequence*

$$0 \longrightarrow M \longrightarrow E \longrightarrow T_0 \longrightarrow 0$$

with T_0 in $\operatorname{add} T$, such that the connecting morphism $\delta\colon \operatorname{Hom}_A(T, T_0) \longrightarrow \operatorname{Ext}^1_A(T, M)$ is surjective.

Proof. If $\mathrm{Ext}_A^1(T, M) = 0$, there is nothing to prove. Otherwise, let $\{\xi_1, \ldots, \xi_d\}$ be a basis of the $\mathbf{k}$-vector space $\mathrm{Ext}_A^1(T, M)$, where each ξ_i is represented by a short exact sequence

$$0 \longrightarrow M \xrightarrow{f_i} E_i \xrightarrow{g_i} T \longrightarrow 0.$$

Let $c = (1, 1, \ldots, 1)\colon M^d \longrightarrow M$ be the codiagonal morphism. There exists a commutative diagram with exact rows:

$$\begin{array}{ccccccccc}
0 & \longrightarrow & M & \xrightarrow{f_i} & E_i & \xrightarrow{g_i} & T & \longrightarrow & 0 \\
 & & \downarrow{\scriptstyle u_i} & & \downarrow{\scriptstyle v_i} & & \downarrow{\scriptstyle w_i} & & \\
0 & \longrightarrow & M^d & \xrightarrow{\oplus f_i} & \oplus E_i & \xrightarrow{\oplus g_i} & T^d & \longrightarrow & 0 \\
 & & \downarrow{\scriptstyle c} & & \downarrow{\scriptstyle h} & & \| & & \\
0 & \longrightarrow & M & \xrightarrow{f} & E & \xrightarrow{g} & T^d & \longrightarrow & 0
\end{array}$$

where $\oplus f_i$, $\oplus g_i$ are the morphisms induced by passing to the direct sums, u_i, v_i, w_i are the respective inclusion morphisms into the ith coordinate space and E is the amalgamated sum of the morphisms c and $\oplus f_i$. Because $cu_i = 1_M$, for every i, we deduce another commutative diagram with exact rows

$$\begin{array}{ccccccccc}
0 & \longrightarrow & M & \xrightarrow{f_i} & E_i & \xrightarrow{g_i} & T & \longrightarrow & 0 \\
 & & \| & & \downarrow{\scriptstyle hv_i} & & \downarrow{\scriptstyle w_i} & & \\
0 & \longrightarrow & M & \xrightarrow{f} & E & \xrightarrow{g} & T^d & \longrightarrow & 0
\end{array}$$

Let ξ be the element of $\mathrm{Ext}_A^1(T^d, M)$ represented by the lower sequence. The previous diagram says that $\xi_i = \mathrm{Ext}_A^1(w_i, M)(\xi)$ for every i, that is, ξ_i lies in the image of the connecting morphism δ. Because the ξ_i generate $\mathrm{Ext}_A^1(T, M)$, we infer that δ is surjective. □

We state and prove the announced result, known as Bongartz' lemma.

Proposition V.2.4. *Let T be a partial tilting A-module. Then, there exists a module E such that $T \oplus E$ is a tilting module.*

Proof. Because of Lemma V.2.3 above, there exists a short exact sequence

$$(*) \qquad 0 \longrightarrow A \longrightarrow E \longrightarrow T_0 \longrightarrow 0$$

with T_0 in $\mathrm{add}\, T$ such that the connecting morphism $\mathrm{Hom}_A(T, T_0) \longrightarrow \mathrm{Ext}_A^1(T, A)$ is surjective. Applying $\mathrm{Hom}_A(T, -)$ to $(*)$ yields an exact cohomology sequence

$$\ldots \longrightarrow \mathrm{Hom}_A(T, T_0) \longrightarrow \mathrm{Ext}^1_A(T, A) \longrightarrow \mathrm{Ext}^1_A(T, E) \longrightarrow \mathrm{Ext}^1_A(T, T_0) = 0$$

where the last equality follows from the definition of a partial tilting module. Surjectivity of the connecting morphism yields $\mathrm{Ext}^1_A(T, E) = 0$.

Applying successively the functors $\mathrm{Hom}_A(-, T)$ and $\mathrm{Hom}_A(-, E)$ to the sequence $(*)$, we get the exact sequences

$$0 = \mathrm{Ext}^1_A(T_0, T) \longrightarrow \mathrm{Ext}^1_A(E, T) \longrightarrow \mathrm{Ext}^1_A(A, T) = 0 \text{ and}$$

$$0 = \mathrm{Ext}^1_A(T_0, E) \longrightarrow \mathrm{Ext}^1_A(E, E) \longrightarrow \mathrm{Ext}^1_A(A, E) = 0.$$

Therefore, $\mathrm{Ext}^1_A(E, T) = 0 = \mathrm{Ext}^1_A(E, E)$, so that $\mathrm{Ext}^1_A(T \oplus E, T \oplus E) = 0$. Because $\mathrm{pd}\, T \leq 1$, the sequence $(*)$ yields $\mathrm{pd}\, E \leq 1$ and so $\mathrm{pd}(T \oplus E) \leq 1$. The third condition of the definition of a tilting module is satisfied because of the exact sequence $(*)$. □

V.2.2 *A torsion pair in* mod *A*

Because we want to projectivise tilting modules, it is useful, as in Subsection V.1.1 to consider the full subcategory $\mathrm{pres}\, T$ of $\mathrm{mod}\, A$ consisting of all A-modules M such that there exists an exact sequence

$$T_1 \longrightarrow T_0 \longrightarrow M \longrightarrow 0$$

with T_0, T_1 in $\mathrm{add}\, T$. Clearly, in this case, M is generated by T, that is, there exist $m > 0$ and an epimorphism $T^m \longrightarrow M$. The surprising fact is that the converse also holds true. This will be proven in the proposition following the next two lemmata, the first of which should be compared with Exercise V.1.4.

Lemma V.2.5. *Let T, M be A-modules, $\{f_1, \ldots, f_d\}$ a* **k***-basis of* $\mathrm{Hom}_A(T, M)$ *and $f = (f_1, \ldots, f_d)\colon T^d \longrightarrow M$. Then:*

(a) *For every morphism $g\colon T_0 \longrightarrow M$, with T_0 in* $\mathrm{add}\, T$*, there exists $h\colon T_0 \longrightarrow T^d$ such that $g = fh$.*
(b) *The morphism* $\mathrm{Hom}_A(T, f)\colon \mathrm{Hom}_A(T, T^d) \longrightarrow \mathrm{Hom}_A(T, M)$ *is surjective.*
(c) *The morphism f is surjective if and only if M is generated by T.*

Proof.

(a) It suffices to prove the statement when $T_0 = T$. In this case, g is a linear combination of the f_i, that is, there exist $\lambda_1, \ldots, \lambda_d \in \mathbf{k}$ such that $g = \sum_{i=1}^{d} \lambda_i f_i$. But then $g = fh$ with $h = \begin{pmatrix} \lambda_1 \\ \vdots \\ \lambda_d \end{pmatrix}$:

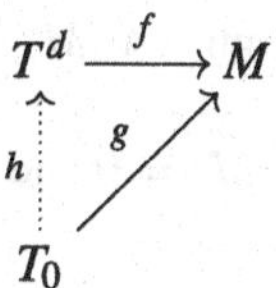

(b) This follows from (a).
(c) Clearly, if f is surjective, then M is generated by T. Conversely, assume M to be generated by T. There exist T_0 in $\operatorname{add} T$ and an epimorphism $g: T_0 \longrightarrow M$. Because of (a), g factors through f. But g is surjective. Hence, so is f.

□

We define the notion of **trace** of a subcategory $\mathscr{C}$ of mod A on a module M. This is the submodule of M given by

$$t_{\mathscr{C}}M = \sum\{\operatorname{Im} f \mid f: X \longrightarrow M \text{ for some object } X \text{ in } \mathscr{C}\}.$$

Thus, $t_{\mathscr{C}}M$ is generated by the objects of $\mathscr{C}$ and, because of its definition, it is the largest submodule of M to be generated by the objects of $\mathscr{C}$. If, in particular, $\mathscr{C} = \operatorname{add} T$ for some module T, then we write $t_{\mathscr{C}}M = t_T M$. Thus, if T is a module, then $t_T M$ is the largest submodule of M generated by T.

Lemma V.2.6. *Let T be a tilting A-module. Then, an A-module M is generated by T if and only if* $\operatorname{Ext}^1_A(T, M) = 0$.

Proof. Assume that M is generated by T. There exist T_0 in $\operatorname{add} T$ and an epimorphism $T_0 \longrightarrow M$. Because $\operatorname{pd} T \leq 1$, this epimorphism induces an epimorphism $\operatorname{Ext}^1_A(T, T_0) \longrightarrow \operatorname{Ext}^1_A(T, M)$. Because $\operatorname{Ext}^1_A(T, T_0) = 0$, we get $\operatorname{Ext}^1_A(T, M) = 0$. Conversely, let M be such that $\operatorname{Ext}^1_A(T, M) = 0$. Because we have already proven that modules generated by T have no extension with T, we have $\operatorname{Ext}^1_A(T, t_T M) = 0$. Also, because of the definition of the trace, we have $\operatorname{Hom}_A(T, t_T M) = \operatorname{Hom}_A(T, M)$. Therefore, applying $\operatorname{Hom}_A(T, -)$ to the short exact sequence

$$0 \longrightarrow t_T M \longrightarrow M \longrightarrow \frac{M}{t_T M} \longrightarrow 0$$

yields $\operatorname{Hom}_A(T, M/t_T M) = 0$. On the other hand, $\operatorname{pd} T \leq 1$ implies the existence of an epimorphism $\operatorname{Ext}^1_A(T, M) \longrightarrow \operatorname{Ext}^1_A(T, M/t_T M)$. Because $\operatorname{Ext}^1_A(T, M) = 0$ by hypothesis, we get $\operatorname{Ext}^1_A(T, M/t_T M) = 0$. Finally, applying $\operatorname{Hom}_A(-, M/t_T M)$ to the short exact sequence

$$0 \longrightarrow A_A \longrightarrow T_0 \longrightarrow T_1 \longrightarrow 0$$

with T_0, T_1 in $\operatorname{add} T$, whose existence is asserted in the definition of tilting module, we get an exact sequence

$$0 = \mathrm{Hom}_A(T_0, M/t_T M) \longrightarrow \mathrm{Hom}_A(A, M/t_T M) \longrightarrow \mathrm{Ext}^1_A(T_1, M/t_T M) = 0.$$

Thus, $M/t_T M \cong \mathrm{Hom}_A(A, M/t_T M) = 0$ and so, $M = t_T M$ is generated by T. □

We next prove that the equivalent properties of Lemma V.2.6 characterise the subcategory $\mathrm{pres}\, T$.

Proposition V.2.7. *Let T be a tilting A-module. Then,* $\mathrm{pres}\, T$ *coincides with the full subcategory of* $\mathrm{mod}\, A$ *consisting of all A-modules generated by T.*

Proof. Because every T-presented A-module is obviously generated by T, it suffices to prove the reverse implication. Let M be generated by T and $f\colon T^d \longrightarrow M$ be as in Lemma V.2.5. Then, f is surjective and we have a short exact sequence

$$0 \longrightarrow L \xrightarrow{j} T^d \xrightarrow{f} M \longrightarrow 0$$

where $L = \mathrm{Ker}\, f$. Applying $\mathrm{Hom}_A(T, -)$, we get an exact sequence

$$0 \longrightarrow \mathrm{Hom}_A(T, L) \longrightarrow \mathrm{Hom}_A(T, T^d) \xrightarrow{\mathrm{Hom}_A(T,f)} \mathrm{Hom}_A(T, M)$$
$$\longrightarrow \mathrm{Ext}^1_A(T, L) \longrightarrow 0$$

because $\mathrm{Ext}^1_A(T, T) = 0$. Because of Lemma V.2.5, $\mathrm{Hom}_A(T, f)$ is surjective. Therefore, $\mathrm{Ext}^1_A(T, L) = 0$. Because of Lemma V.2.6, L is generated by T and thus, there exist $m > 0$ and an epimorphism $p\colon T^m \longrightarrow L$. We deduce the required presentation

$$T^m \xrightarrow{jp} T^d \longrightarrow M \longrightarrow 0.$$

□

As seen in the proof, the fact that M lies in $\mathrm{pres}\, T$ can be expressed equivalently by stating that there exists a short exact sequence

$$0 \longrightarrow L \longrightarrow T_0 \longrightarrow M \longrightarrow 0$$

with T_0 in $\mathrm{add}\, T$ and L generated by T. We shall use this fact repeatedly in the sequel.

The main consequence of the proposition is the existence of a torsion pair in $\mathrm{mod}\, A$. Torsion pairs in $\mathrm{mod}\, A$ say roughly how the morphisms go in this category.

Definition V.2.8. A **torsion pair** $(\mathcal{T}, \mathcal{F})$ in $\mathrm{mod}\, A$ is a pair of $\mathbf{k}$-linear full subcategories such that:

(a) $\mathrm{Hom}_A(M, N) = 0$ for all M in $\mathcal{T}$ and N in $\mathcal{F}$.

(b) $\mathscr{T}$ and $\mathscr{F}$ are maximal for (a), that is:

(i) $\mathrm{Hom}_A(M, -)|_{\mathscr{F}} = 0$ implies that M lies in $\mathscr{T}$.
(ii) $\mathrm{Hom}_A(-, N)|_{\mathscr{T}} = 0$ implies that N lies in $\mathscr{F}$.

The class $\mathscr{T}$ is the **torsion class**, and the modules in it are **torsion modules**, whereas the class $\mathscr{F}$ is the **torsion-free class**, and the modules in that class are **torsion-free modules**.

Clearly, if $(\mathscr{T}, \mathscr{F})$ is a torsion pair, then $\mathscr{T} \cap \mathscr{F} = 0$. We list a few properties of torsion pairs.

Lemma V.2.9.

(a) *Let $\mathscr{T}$ be a* **k***-linear full subcategory of* mod A. *There exists a subcategory $\mathscr{F}$ such that $(\mathscr{T}, \mathscr{F})$ is a torsion pair if and only if $\mathscr{T}$ is closed under quotients and extensions.*
(b) *Let $\mathscr{F}$ be a* **k***-linear full subcategory of* mod A. *There exists a subcategory $\mathscr{T}$ such that $(\mathscr{T}, \mathscr{F})$ is a torsion pair if and only if $\mathscr{F}$ is closed under submodules and extensions.*
(c) *Let $(\mathscr{T}, \mathscr{F})$ be a torsion pair in* mod A. *For each A-module M, there exists a unique short exact sequence $0 \longrightarrow L \longrightarrow M \longrightarrow N \longrightarrow 0$ such that L lies in $\mathscr{T}$ and N lies in $\mathscr{F}$.*

Proof.

(a) Let $0 \longrightarrow L \longrightarrow M \longrightarrow N \longrightarrow 0$ be a short exact sequence in mod A. Then there exists a left exact sequence of functors

$$0 \longrightarrow \mathrm{Hom}_A(N, -)|_{\mathscr{F}} \longrightarrow \mathrm{Hom}_A(M, -)|_{\mathscr{F}} \longrightarrow \mathrm{Hom}_A(L, -)|_{\mathscr{F}}.$$

Then L, N in $\mathscr{T}$ imply M in $\mathscr{T}$, and M in $\mathscr{T}$ implies N in $\mathscr{T}$. That is, $\mathscr{T}$ is closed under quotients and extensions.

Conversely, assume that $\mathscr{T}$ satisfies this condition and consider the trace $t_{\mathscr{T}}M$.

We claim that $t_{\mathscr{T}}(M/t_{\mathscr{T}}M) = 0$. Indeed, there exists a submodule L of M containing $t_{\mathscr{T}}M$ such that $t_{\mathscr{T}}(M/t_{\mathscr{T}}M) = L/t_{\mathscr{T}}M$. Because both $L/t_{\mathscr{T}}M$ and $t_{\mathscr{T}}M$ lie in $\mathscr{T}$, the short exact sequence

$$0 \longrightarrow t_{\mathscr{T}}M \longrightarrow L \longrightarrow \frac{L}{t_{\mathscr{T}}M} \longrightarrow 0$$

gives L in $\mathscr{T}$, because $\mathscr{T}$ is closed under extensions. Hence, $L \subseteq t_{\mathscr{T}}M$ and so $t_{\mathscr{T}}(M/t_{\mathscr{T}}M) = 0$, as required.

Let now $\mathscr{F}$ be the **k**-linear full subcategory of mod A defined by:

$$\mathscr{F} = \{M : t_{\mathscr{T}}M = 0\}.$$

In particular, for each A-module M, we have $M/t_{\mathscr{T}}M \in \mathscr{F}$. We claim that $(\mathscr{T}, \mathscr{F})$ is a torsion pair. Let $f: M \longrightarrow N$ be a nonzero morphism with M in $\mathscr{T}$ and N in $\mathscr{F}$. The induced epimorphism $t_{\mathscr{T}}M = M \longrightarrow \operatorname{Im} f$ is also nonzero. Hence, $t_{\mathscr{T}}N \neq 0$, a contradiction. Therefore, $\operatorname{Hom}_A(M, N) = 0$. Assume $\operatorname{Hom}_A(M, -)|_{\mathscr{F}} = 0$. In particular, $\operatorname{Hom}_A(M, M/t_{\mathscr{T}}M) = 0$, which implies $M/t_{\mathscr{T}}M = 0$ and thus $M = t_{\mathscr{T}}M$ lies in $\mathscr{T}$. Similarly, $\operatorname{Hom}_A(-, N)|_{\mathscr{T}} = 0$ implies that N lies in $\mathscr{F}$.

(b) is dual to (a).

(c) We claim that the required sequence is

$$0 \longrightarrow t_{\mathscr{T}}M \longrightarrow M \longrightarrow \frac{M}{t_{\mathscr{T}}M} \longrightarrow 0$$

as constructed in (a). Because $t_{\mathscr{T}}M$ lies in $\mathscr{T}$ and $M/t_{\mathscr{T}}M$ lies in $\mathscr{F}$, we just have to prove uniqueness. Let $0 \longrightarrow L \longrightarrow M \longrightarrow N \longrightarrow 0$ be a short exact sequence with L in $\mathscr{T}$ and N in $\mathscr{F}$. Because $t_{\mathscr{T}}M$ is the largest submodule of M to lie in $\mathscr{T}$, we have $L \subseteq t_{\mathscr{T}}M$. We get a commutative diagram with exact rows

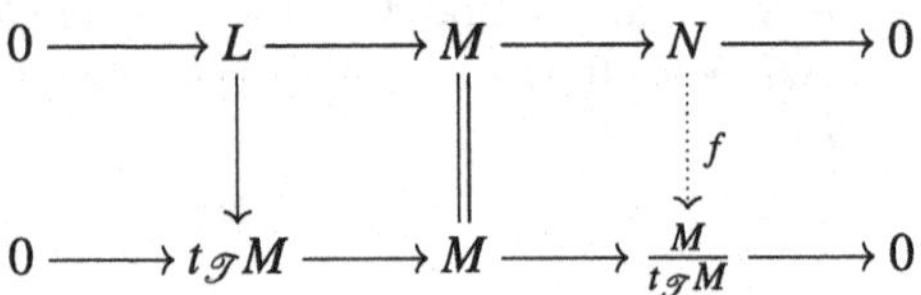

where f is deduced by passing to cokernels. The snake lemma gives that f is surjective with kernel $t_{\mathscr{T}}M/L$. Because the kernel is a submodule of N, which lies in $\mathscr{F}$, then $t_{\mathscr{T}}M/L$ also lies in $\mathscr{F}$. On the other hand, $t_{\mathscr{T}}M$ lies in $\mathscr{T}$; therefore, $t_{\mathscr{T}}M/L$ lies in $\mathscr{T}$. Because $\mathscr{T} \cap \mathscr{F} = 0$, we get $t_{\mathscr{T}}M/L = 0$, that is, $L = t_{\mathscr{T}}M$. Consequently, $N \cong M/t_{\mathscr{T}}M$.

□

The short exact sequence of (c) is called the **canonical sequence** for the module M. An easy consequence of the existence of the canonical sequence is that every simple A-module lies either in $\mathscr{T}$ or in $\mathscr{F}$. We are now able to prove the wanted corollary of Proposition V.2.7.

Corollary V.2.10. *Let T be a tilting A-module. Then, $\mathscr{T}(T) = \operatorname{pres} T$ is a torsion class and the corresponding torsion-free class is*

$$\mathscr{F}(T) = \{N\colon \operatorname{Hom}_A(T, N) = 0\}.$$

Proof. Because $\operatorname{pres} T$ coincides with the class of modules generated by T, it is closed under quotients. We now prove that it is closed under extensions. Applying $\operatorname{Hom}_A(T, -)$ to a short exact sequence $0 \longrightarrow L \longrightarrow M \longrightarrow N \longrightarrow 0$ yields an exact sequence

$$\mathrm{Ext}_A^1(T, L) \longrightarrow \mathrm{Ext}_A^1(T, M) \longrightarrow \mathrm{Ext}_A^1(T, N).$$

If L, N lie in $\mathrm{pres}\, T$, then $\mathrm{Ext}_A^1(T, L) = 0$ and $\mathrm{Ext}_A^1(T, N) = 0$ and so $\mathrm{Ext}_A^1(T, M) = 0$. Therefore, $\mathscr{T}(T) = \mathrm{pres}\, T$ is a torsion class.

Let M be torsion-free. Because T is torsion, we have $\mathrm{Hom}_A(T, M) = 0$. Therefore, M lies in $\mathscr{F}(T)$. Conversely, let M lie in $\mathscr{F}(T)$. For every L in $\mathscr{F}(T)$, there exist T_0 in $\mathrm{add}\, T$ and an epimorphism $T_0 \longrightarrow L$. Therefore, $\mathrm{Hom}_A(L, M) = 0$ and so M is torsion-free. □

It follows from Lemma V.2.6 and Corollary V.2.10 that all injective A-modules lie in $\mathscr{T}(T)$ because they annihilate the functor $\mathrm{Ext}_A^1(T, -)$. Also, if P is a projective module lying in $\mathscr{T}(T)$, then it must belong to $\mathrm{add}\, T$. Indeed, every epimorphism from $\mathrm{add}\, T$ to P must split. In particular, every indecomposable projective–injective A-module is a direct summand of every tilting A-module.

Corollary V.2.11. *With the above notation,* $\mathrm{Hom}_A(T, -)|_{\mathscr{T}(T)}$ *is an exact functor.*

Proof. Let $0 \longrightarrow L \longrightarrow M \longrightarrow N \longrightarrow 0$ be a short exact sequence in $\mathscr{T}(T)$. Because $\mathrm{Ext}_A^1(T, L) = 0$, we deduce an exact sequence

$$0 \longrightarrow \mathrm{Hom}_A(T, L) \longrightarrow \mathrm{Hom}_A(T, M) \longrightarrow \mathrm{Hom}_A(T, N) \longrightarrow 0. \qquad \square$$

We have proved in Lemma V.1.6 that the morphism $\varepsilon_M \colon \mathrm{Hom}_A(T, M) \otimes_B T \longrightarrow M$ given by $f \otimes t \longmapsto f(t)$ is an isomorphism whenever M lies in $\mathrm{add}\, T$. We now prove that, if T is tilting, then ε_M is an isomorphism for every M in $\mathrm{pres}\, T$.

Corollary V.2.12. *An A-module M lies in $\mathscr{T}(T)$ if and only if the morphism ε_M: $\mathrm{Hom}_A(T, M) \otimes_B T \longrightarrow M$ given by $f \otimes t \mapsto f(t)$ is an isomorphism.*

Proof. Assume that ε_M is an isomorphism, then $M \cong \mathrm{Hom}_A(T, M) \otimes_B T$ lies in $\mathscr{T}(T)$ because of Lemma V.1.5. Conversely, let M belong to $\mathscr{T}(T)$. The proof of this implication is similar to that of Proposition V.1.7. There exists an exact sequence

$$T_1 \longrightarrow T_0 \longrightarrow M \longrightarrow 0$$

with T_0, T_1 in $\mathrm{add}\, T$. Applying Corollary V.2.11, we have an exact sequence

$$\mathrm{Hom}_A(T, T_1) \longrightarrow \mathrm{Hom}_A(T, T_0) \longrightarrow \mathrm{Hom}_A(T, M) \longrightarrow 0.$$

Applying $- \otimes_B T$ yields a commutative diagram with exact rows

$$\begin{array}{ccccccc}
\mathrm{Hom}_A(T, T_1) \otimes_B T & \longrightarrow & \mathrm{Hom}_A(T, T_0) \otimes_B T & \longrightarrow & \mathrm{Hom}_A(T, M) \otimes_B T & \longrightarrow & 0 \\
\downarrow{\scriptstyle \varepsilon_{T_1}} & & \downarrow{\scriptstyle \varepsilon_{T_0}} & & \downarrow{\scriptstyle \varepsilon_M} & & \\
T_1 & \longrightarrow & T_0 & \longrightarrow & M & \longrightarrow & 0.
\end{array}$$

Because of Lemma V.1.6, ε_{T_1} and ε_{T_0} are isomorphisms. Hence, so is ε_M. □

Example V.2.13. It is easy to compute in the Auslander–Reiten quiver of an algebra the subcategories $\mathscr{T}(T)$ and $\mathscr{F}(T)$. For instance, in Example V.2.2,

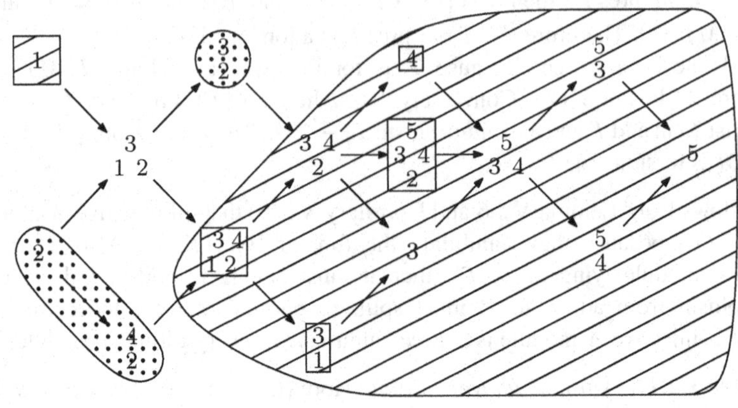

$\mathscr{T}(T)$ is illustrated by the hatched area and $\mathscr{F}(T)$ by the dotted area.

One particular class of tilting modules is the class of so-called APR-tilting modules where the letters A, P and R stand for Auslander, Platzeck and Reiten.

Lemma V.2.14. *Let S_x be a simple projective noninjective module. Then:*

(a) $T_x = \tau^{-1} S_x \oplus (\bigoplus_{y \neq x} P_y)$ *is a tilting module.*
(b) $\mathscr{F}(T_x) = \operatorname{add} S_x$ *while* $\mathscr{T}(T_x) = \operatorname{add}(\operatorname{ind} A \setminus \{S_x\})$.

Proof.

(a) Because of Lemma III.2.10, there exists an almost split sequence $0 \longrightarrow S_x \longrightarrow P \longrightarrow \tau^{-1} S_x \longrightarrow 0$ with P projective. This proves the first and third conditions of the definition of a tilting module. In addition,

$$\operatorname{Ext}^1_A(T_x, T_x) \cong \operatorname{D} \operatorname{Hom}_A(T_x, \tau T_x) \cong \operatorname{D} \operatorname{Hom}_A(T_x, S_x) = 0$$

because the simple projective module S_x is not a summand of T_x.
(b) Let M be indecomposable. Then, M lies in $\mathscr{T}(T_x)$ if and only if $\operatorname{Ext}^1_A(T_x, M) = 0$, that is, $\operatorname{Hom}_A(M, S_x) = 0$ or equivalently, $M \ncong S_x$. Similarly, $\operatorname{Hom}_A(T_x, S_x) = 0$ implies that S_x lies in $\mathscr{F}(T_x)$. Therefore, $\mathscr{F}(T_x) = \operatorname{add} S_x$. □

Because of (b), every indecomposable A-module belongs either to $\mathscr{T}(T_x)$ or to $\mathscr{F}(T_x)$. If a torsion pair $(\mathscr{T}, \mathscr{F})$ in mod A is such that every indecomposable module lies either in $\mathscr{T}$ or in $\mathscr{F}$, then the pair $(\mathscr{T} \mathscr{F})$ is said to be **split**. Thus, an APR-tilting module T_A induces a split torsion pair $(\mathscr{T}(T_x), \mathscr{F}(T_x))$ in mod A.

V.2.3 The main theorems

We now prove the two main results of tilting theory, called the reciprocity theorem and the tilting theorem. We keep the notations above, that is, let A be an algebra, T a tilting A-module and $B = \operatorname{End} T_A$. We first show that the functor $\operatorname{Hom}_A(T, -) : \operatorname{mod} A \longrightarrow \operatorname{mod} B$ preserves both the morphism spaces and the first extension spaces between torsion modules.

Lemma V.2.15. *Let M, N be A-modules lying in $\mathscr{T}(T)$, then*

(a) $\operatorname{Hom}_A(M, N) \cong \operatorname{Hom}_B(\operatorname{Hom}_A(T, M), \operatorname{Hom}_A(T, N))$.
(b) $\operatorname{Ext}^1_A(M, N) \cong \operatorname{Ext}^1_B(\operatorname{Hom}_A(T, M), \operatorname{Hom}_A(T, N))$.

Proof.

(a) Because M lies in $\mathscr{T}(T)$, there is an exact sequence

$$T_1 \longrightarrow T_0 \longrightarrow M \longrightarrow 0$$

with T_1, T_0 in $\operatorname{add} T$. Because of Corollary V.2.11, it induces an exact sequence

$$\operatorname{Hom}_A(T, T_1) \longrightarrow \operatorname{Hom}_A(T, T_0) \longrightarrow \operatorname{Hom}_A(T, M) \longrightarrow 0.$$

Applying $\operatorname{Hom}_B(-, \operatorname{Hom}_A(T, N))$ yields a commutative diagram with exact rows

$$\begin{array}{ccccccc} 0 \to & \operatorname{Hom}_B(\operatorname{Hom}_A(T,M),\operatorname{Hom}_A(T,N)) & \to & \operatorname{Hom}_B(\operatorname{Hom}_A(T,T_0),\operatorname{Hom}_A(T,N)) & \to & \operatorname{Hom}_B(\operatorname{Hom}_A(T,T_1),\operatorname{Hom}_A(T,N)) \\ & & & \downarrow \cong & & \downarrow \cong \\ 0 \to & \operatorname{Hom}_A(M,N) & \longrightarrow & \operatorname{Hom}_A(T_0,N) & \longrightarrow & \operatorname{Hom}_A(T_1,N) \end{array}$$

where the lower exact sequence comes from the application of $\operatorname{Hom}_A(-, N)$ to the given T-presentation of M, and the vertical isomorphisms come from Corollary V.1.4(a). The statement follows.

(b) Let $T_1 \xrightarrow{d_1} T_0 \xrightarrow{d_0} M \longrightarrow 0$ be exact with T_0, T_1 in $\operatorname{add} T$. Because $L = \operatorname{Im} d_1$ is generated by T, an obvious induction yields a resolution

$$\ldots \longrightarrow T_2 \xrightarrow{d_2} T_1 \xrightarrow{d_1} T_0 \xrightarrow{d_0} M \longrightarrow 0$$

with all T_i in $\operatorname{add} T$. Because of Corollary V.2.11, we deduce an exact sequence

$$\ldots \longrightarrow \operatorname{Hom}_A(T, T_2) \longrightarrow \operatorname{Hom}_A(T, T_1) \longrightarrow \operatorname{Hom}_A(T, T_0)$$
$$\longrightarrow \operatorname{Hom}_A(T, M) \longrightarrow 0$$

which is a projective resolution of $\operatorname{Hom}_A(T, M)$ in $\operatorname{mod} B$.

Let $d_1 = jp$ be the canonical factorisation of d_1 through its image L. The short exact sequence

$$0 \longrightarrow L \xrightarrow{j} T_0 \xrightarrow{d_0} M \longrightarrow 0$$

induces an exact sequence

$$0 \longrightarrow \mathrm{Hom}_A(M,N) \longrightarrow \mathrm{Hom}_A(T_0,N) \xrightarrow{\mathrm{Hom}_A(j,N)} \mathrm{Hom}_A(L,N) \\ \longrightarrow \mathrm{Ext}^1_A(M,N) \longrightarrow \mathrm{Ext}^1_A(T_0,N) = 0$$

where $\mathrm{Ext}^1_A(T_0, N) = 0$ follows from the fact that N lies in $\mathscr{T}(T)$. Therefore, $\mathrm{Ext}^1_A(M,N) \cong \mathrm{Coker}\,\mathrm{Hom}_A(j,N)$.

On the other hand, the exact sequence

$$T_2 \xrightarrow{d_2} T_1 \xrightarrow{p} L \longrightarrow 0$$

induces an exact sequence

$$0 \longrightarrow \mathrm{Hom}_A(L,N) \longrightarrow \mathrm{Hom}_A(T_1,N) \xrightarrow{\mathrm{Hom}_A(d_2,N)} \mathrm{Hom}_A(T_2,N)$$

so that $\mathrm{Hom}_A(L,N) \cong \mathrm{Ker}\,\mathrm{Hom}_A(d_2,N)$.

By definition, $\mathrm{Ext}^1_B(\mathrm{Hom}_A(T,M), \mathrm{Hom}_A(T,N))$ is the first cohomology group of the complex on the upper row of the following commutative diagram

$$\begin{array}{ccccccc}
0 \to & \mathrm{Hom}_B(\mathrm{Hom}_A(T,T_0),\mathrm{Hom}_A(T,N)) & \longrightarrow & \mathrm{Hom}_B(\mathrm{Hom}_A(T,T_1),\mathrm{Hom}_A(T,N)) & \longrightarrow & \mathrm{Hom}_B(\mathrm{Hom}_A(T,T_2),\mathrm{Hom}_A(T,N)) \\
& \downarrow \cong & & \downarrow \cong & & \downarrow \cong \\
0 \longrightarrow & \mathrm{Hom}_A(T_0,N) & \xrightarrow{\mathrm{Hom}_A(d_1,N)} & \mathrm{Hom}_A(T_1,N) & \xrightarrow{\mathrm{Hom}_A(d_2,N)} & \mathrm{Hom}_A(T_2,N)
\end{array}$$

where the vertical isomorphisms follow from (a). Thus,

$$\mathrm{Ext}^1_B(\mathrm{Hom}_A(T,M)), \mathrm{Hom}_A(T,N)) \cong \frac{\mathrm{Ker}\,\mathrm{Hom}_A(d_2,N)}{\mathrm{Im}\,\mathrm{Hom}_A(d_1,N)} \cong \frac{\mathrm{Hom}_A(L,N)}{\mathrm{Im}\,\mathrm{Hom}_A(d_1,N)}.$$

Now, $\mathrm{Hom}_A(d_1,N) = \mathrm{Hom}_A(p,N)\,\mathrm{Hom}_A(j,N)$. The injectivity of $\mathrm{Hom}_A(p,N)$ implies $\mathrm{Im}\,\mathrm{Hom}_A(d_1,N) \cong \mathrm{Im}\,\mathrm{Hom}_A(j,N)$. Then we get

$$\begin{aligned}
\mathrm{Ext}^1_B(\mathrm{Hom}_A(T,M), \mathrm{Hom}_A(T,N)) &\cong \frac{\mathrm{Hom}_A(L,N)}{\mathrm{Im}\,\mathrm{Hom}_A(j,N)} \\
&\cong \mathrm{Coker}\,\mathrm{Hom}_A(j,N) \\
&\cong \mathrm{Ext}^1_A(M,N).
\end{aligned}$$

□

Before continuing, we need a pair of homological lemmata. The first one asserts that duality interchanges homology and cohomology.

Lemma V.2.16. *Let* $C^\bullet : \ldots \longrightarrow C^{n-1} \xrightarrow{d^n} C^n \xrightarrow{d^{n+1}} C^{n+1} \longrightarrow \ldots$ *be a complex. Then, for every* $n \geq 0$, *we have a functorial isomorphism* $\mathrm{H}_n(\mathrm{D}C^\bullet) \cong \mathrm{D}\,\mathrm{H}^n(C^\bullet)$.

Proof. We have, for every $n \geq 0$, a short exact sequence

$$0 \longrightarrow \operatorname{Im} d^n \longrightarrow \operatorname{Ker} d^{n+1} \longrightarrow \mathrm{H}^n(C^\bullet) \longrightarrow 0$$

hence, an exact sequence

$$0 \longrightarrow \mathrm{D}\,\mathrm{H}^n(C^\bullet) \longrightarrow \mathrm{D}(\operatorname{Ker} d^{n+1}) \longrightarrow \mathrm{D}(\operatorname{Im} d^n) \longrightarrow 0.$$

Now,

$$\mathrm{D}(\operatorname{Ker} d^{n+1}) \cong \mathrm{Coker}\mathrm{D}d^{n+1} \cong \frac{\mathrm{D}C^n}{\mathrm{Im}\mathrm{D}d^{n+1}} \text{ and}$$

$$\mathrm{D}(\operatorname{Im} d^n) \cong \operatorname{Im} \mathrm{D}d^n \cong \frac{\mathrm{D}C^n}{\operatorname{Ker} \mathrm{D}d^n}$$

thus, comparing the previous sequence with the sequence

$$0 \longrightarrow \frac{\operatorname{Ker} \mathrm{D}d^n}{\operatorname{Im} \mathrm{D}d^{n+1}} \longrightarrow \frac{\mathrm{D}C^n}{\operatorname{Im} \mathrm{D}d^{n+1}} \longrightarrow \frac{\mathrm{D}C^n}{\operatorname{Ker} \mathrm{D}d^n} \longrightarrow 0$$

yields

$$\mathrm{D}\,\mathrm{H}^n(C^\bullet) \cong \frac{\operatorname{Ker} \mathrm{D}d^n}{\operatorname{Im} \mathrm{D}d^{n+1}} \cong \mathrm{H}_n(\mathrm{D}C^\bullet).$$

□

We deduce functorial isomorphisms whose existence was asserted in Subsection IV.1.5.

Proposition V.2.17. *Let* L, M *be* A*-modules. We have functorial isomorphisms*

(a) $L \otimes_A \mathrm{D}M \cong \mathrm{D}\,\mathrm{Hom}_A(L, M)$, *and*
(b) $\mathrm{Tor}_n^A(L, \mathrm{D}M) \cong \mathrm{D}\,\mathrm{Ext}_A^n(L, M)$, *for every* $n \geq 0$.

Proof.

(a) Follows from the adjunction formula. Indeed,

$$\begin{aligned}\mathrm{D}(L \otimes_A \mathrm{D}M) &= \mathrm{Hom}_{\mathbf{k}}(L \otimes_A \mathrm{D}M, \mathbf{k}) \\ &\cong \mathrm{Hom}_A(L, \mathrm{Hom}_{\mathbf{k}}(\mathrm{D}M, \mathbf{k}))\end{aligned}$$

$$\cong \mathrm{Hom}_A(L, \mathrm{D}^2 M)$$
$$\cong \mathrm{Hom}_A(L, M).$$

(b) Let $P_\bullet$ be a projective resolution of L in mod A. Applying (a) above to each P_i yields a functorial isomorphism $P_i \otimes_A \mathrm{D}M \cong \mathrm{D}\,\mathrm{Hom}_A(P_i, M)$. Thus, we have an isomorphism of complexes $P_\bullet \otimes_A \mathrm{D}M \cong \mathrm{D}\,\mathrm{Hom}_A(P_\bullet, M)$. Hence, for every $n \geq 0$,

$$\begin{aligned}\mathrm{Tor}_n^A(L, \mathrm{D}M) &\cong \mathrm{H}_n(P_\bullet \otimes_A \mathrm{D}M)\\ &\cong \mathrm{H}_n(\mathrm{D}\,\mathrm{Hom}_A(P_\bullet, M))\\ &\cong \mathrm{D}\,\mathrm{H}^n\,\mathrm{Hom}_A(P_\bullet, M)\\ &\cong \mathrm{D}\,\mathrm{Ext}_A^n(L, M)\end{aligned}$$

where the third isomorphism follows from Lemma V.2.16 above.

□

We next prove the reciprocity theorem.

Theorem V.2.18 (Reciprocity theorem). *Let A be an algebra, T a tilting A-module and $B = \mathrm{End}\,T_A$. Then, ${}_BT$ is a tilting left B-module and we have an isomorphism $A \cong (\mathrm{End}_B\,T)^{op}$ given by $a \longmapsto (t \mapsto ta)$.*

Proof. Applying $\mathrm{Hom}_A(-, {}_BT_A)$ to a short exact sequence

$$0 \longrightarrow A_A \longrightarrow T_0 \longrightarrow T_1 \longrightarrow 0$$

with T_0, T_1 in add T, and using that $\mathrm{Hom}_A(A, {}_BT_A) \cong {}_BT$ whereas $\mathrm{Ext}_A^1(T, T) = 0$, we get an exact sequence

$$0 \longrightarrow \mathrm{Hom}_A(T_1, {}_BT_A) \longrightarrow \mathrm{Hom}_A(T_0, {}_BT_A) \longrightarrow {}_BT \longrightarrow 0$$

so that $\mathrm{pd}_B\,T \leq 1$.

Next, $\mathrm{D}({}_BT) \cong \mathrm{D}({}_BT \otimes_A A) \cong \mathrm{Hom}_A({}_BT_A, \mathrm{D}A)$ where we have applied Proposition V.2.17(a). Then, Lemma V.2.15(b) yields

$$\mathrm{Ext}_B^1(\mathrm{D}T, \mathrm{D}T) \cong \mathrm{Ext}_B^1(\mathrm{Hom}_A(T, \mathrm{D}A), \mathrm{Hom}_A(T, \mathrm{D}A)) \cong \mathrm{Ext}_A^1(\mathrm{D}A, \mathrm{D}A) = 0$$

because $\mathrm{D}A$ lies in $\mathscr{T}(T)$. Therefore $\mathrm{Ext}_{B^{op}}^1(T, T) = 0$.

Finally, applying the functor $\mathrm{Hom}_A(-, {}_BT_A)$ to a projective resolution $0 \longrightarrow P_1 \longrightarrow P_0 \longrightarrow T \longrightarrow 0$ in mod A, we get an exact sequence

$$0 \longrightarrow {}_BB \longrightarrow \mathrm{Hom}_A(P_0, T) \longrightarrow \mathrm{Hom}_A(P_1, T) \longrightarrow 0$$

because $\mathrm{Ext}_A^1(T, T) = 0$. Now, $\mathrm{Hom}_A(P_0, T)$, $\mathrm{Hom}_A(P_1, T)$ both belong to add T. This completes the proof that ${}_BT$ is a tilting module.

For each $a \in A$, the map $\rho_a : t \mapsto ta$ is an endomorphism of ${}_BT$. Also, $\eta : a \mapsto \rho_a$ is an algebra morphism from A to $(\mathrm{End}_B\, T)^{op}$. If $a \in \mathrm{Ker}\,\eta$, then $Ta = 0$ and hence $a = 0$, because T is faithful. Then, η is injective. On the other hand, the isomorphism $\mathrm{D}T \cong \mathrm{Hom}_A(T, \mathrm{D}A)$ and Lemma V.2.15(a) yield isomorphisms of vector spaces

$$A \cong \mathrm{End}\,\mathrm{D}A \cong \mathrm{End}\,\mathrm{Hom}_A(T, \mathrm{D}A) \cong \mathrm{End}\,\mathrm{D}T \cong \mathrm{End}\,T.$$

Therefore, η is an isomorphism of algebras. □

Corollary V.2.19. *Let $\mathscr{X}(T_A)$, $\mathscr{Y}(T_A)$ be the* **k**-*linear full subcategories defined by $\mathscr{X}(T_A) = \{X_B : X \otimes_B T = 0\}$ and $\mathscr{Y}(T_A) = \{Y_B : \mathrm{Tor}_1^B(Y, T) = 0\}$ respectively. Then, $(\mathscr{X}(T_A), \mathscr{Y}(T_A))$ is a torsion pair in* mod B.

Proof. Because ${}_BT$ is tilting, it induces in mod B^{op} a torsion pair $(\mathscr{T}({}_BT), \mathscr{F}({}_BT))$ with $\mathscr{T}({}_BT) = \{{}_BU : \mathrm{Ext}^1_{B^{op}}(T, U) = 0\}$ and $\mathscr{F}({}_BT) = \{{}_BV : \mathrm{Hom}_{B^{op}}(T, V) = 0\}$. Therefore, $(\mathrm{D}\mathscr{F}({}_BT), \mathrm{D}\mathscr{T}({}_BT))$ is a torsion pair in mod B. Now, X lies in $\mathrm{D}\mathscr{F}({}_BT)$ if and only if $\mathrm{D}X$ lies in $\mathscr{F}({}_BT)$, that is, $\mathrm{Hom}_{B^{op}}(T, \mathrm{D}X) = 0$, and this is equivalent to $X \otimes_B T = 0$, or in other words X lies in $\mathscr{X}(T_A)$, because of Proposition V.2.17(a).

Similarly, Y lies in $\mathrm{D}\mathscr{T}({}_BT)$ if and only if $\mathrm{D}Y$ lies in $\mathscr{T}({}_BT)$, that is, if and only if $\mathrm{Tor}_1^B(Y, T) \cong \mathrm{D}\,\mathrm{Ext}_B^1(Y, \mathrm{D}T) \cong \mathrm{D}\,\mathrm{Ext}^1_{B^{op}}(T, \mathrm{D}Y) = 0$, because of Proposition V.2.17(b). □

Corollary V.2.20. *If Y lies in $\mathscr{Y}(T)$, then δ_Y is an isomorphism.*

Proof. Let $P_1 \xrightarrow{p_1} P_0 \xrightarrow{p_0} Y \longrightarrow 0$ be a projective presentation of Y. Setting $Z_0 = \mathrm{Ker}\, p_0$, $Z_1 = \mathrm{Ker}\, p_1$, we get short exact sequences

$$0 \longrightarrow Z_0 \longrightarrow P_0 \longrightarrow Y \longrightarrow 0 \quad \text{and} \quad 0 \longrightarrow Z_1 \longrightarrow P_1 \longrightarrow Z_0 \longrightarrow 0.$$

Because $\mathrm{Tor}_1^B(Y, T) = 0$, we deduce exact sequences

$$0 \longrightarrow Z_0 \otimes_B T \longrightarrow P_0 \otimes_B T \longrightarrow Y \otimes_B T \longrightarrow 0$$

and

$$Z_1 \otimes_B T \longrightarrow P_1 \otimes_B T \longrightarrow Z_0 \otimes_B T \longrightarrow 0$$

from which we get an exact sequence

$$P_1 \otimes_B T \longrightarrow P_0 \otimes_B T \longrightarrow Y \otimes_B T \longrightarrow 0.$$

Because of Lemma V.1.5, this sequence lies in $\mathscr{T}(T)$ and therefore, because of Corollary V.2.11, the lower row in the commutative diagram below is exact

$$\begin{array}{ccccccc}
P_1 & \longrightarrow & P_0 & \longrightarrow & Y & \longrightarrow & 0 \\
\downarrow \delta_{P_1} & & \downarrow \delta_{P_0} & & \downarrow \delta_Y & & \\
\mathrm{Hom}_A(T, P_1 \otimes_B T) & \longrightarrow & \mathrm{Hom}_A(T, P_0 \otimes_B T) & \longrightarrow & \mathrm{Hom}_A(T, Y \otimes_B T) & \longrightarrow & 0
\end{array}$$

Because of Lemma V.1.6, $\delta_{P_0}, \delta_{P_1}$ are isomorphisms. Therefore, so is δ_Y. □

We now prove our second main result, known as the tilting theorem or the **Brenner–Butler theorem**.

Theorem V.2.21 (Tilting theorem). *Let A be an algebra, T a tilting A-module and $B = \mathrm{End}\, T_A$. Then,*

(a) *The functors $\mathrm{Hom}_A(T, -)$ and $- \otimes_B T$ induce quasi-inverse equivalences between $\mathscr{T}(T)$ and $\mathscr{Y}(T)$.*
(b) *The functors $\mathrm{Ext}^1_A(T, -)$ and $\mathrm{Tor}^B_1(-, T)$ induce quasi-inverse equivalences between $\mathscr{F}(T)$ and $\mathscr{X}(T)$.*

Proof.

(a) If M lies in $\mathscr{T}(T)$, there is an exact sequence $0 \longrightarrow L \longrightarrow T_0 \longrightarrow M \longrightarrow 0$ with T_0 in $\mathrm{add}\, T$ and L in $\mathscr{T}(T)$. Because of Corollary V.2.11, we have a short exact sequence

$$0 \longrightarrow \mathrm{Hom}_A(T, L) \longrightarrow \mathrm{Hom}_A(T, T_0) \longrightarrow \mathrm{Hom}_A(T, M) \longrightarrow 0$$

Applying $- \otimes_B T$ and using Corollary V.2.12, we get a commutative diagram with exact rows

$$\begin{array}{ccccccccc}
0 \to \mathrm{Tor}^B_1(\mathrm{Hom}_A(T, M), T) & \to & \mathrm{Hom}_A(T, L) \otimes_B T & \to & \mathrm{Hom}_A(T, T_0) \otimes_B T & \to & \mathrm{Hom}_A(T, M) \otimes_B T & \to & 0 \\
& & \cong \downarrow \varepsilon_L & & \cong \downarrow \varepsilon_{T_0} & & \cong \downarrow \varepsilon_M & & \\
0 & \longrightarrow & L & \longrightarrow & T_0 & \longrightarrow & M & \longrightarrow & 0
\end{array}$$

where the vertical arrows are isomorphisms and we also use the projectivity of $\mathrm{Hom}_A(T, T_0)$ in $\mathrm{mod}\, B$. Therefore, $\mathrm{Tor}^B_1(\mathrm{Hom}_A(T, M), T) = 0$ and so $\mathrm{Hom}_A(T, M)$ lies in $\mathscr{Y}(T)$. Thus, the functor $\mathrm{Hom}_A(T, -)\colon \mathscr{T}(T) \longrightarrow \mathscr{Y}(T)$ is well-defined. Because of Lemma V.1.5, the functor $- \otimes_B T\colon \mathscr{Y}(T) \longrightarrow \mathscr{T}(T)$ is also well-defined. Finally, Corollary V.2.12 and Corollary V.2.20 above state that, if M lies in $\mathscr{T}(T)$, then $\mathrm{Hom}_A(T, M) \otimes_B T \cong M$ and, if Y lies in $\mathscr{Y}(T)$, then $Y \cong \mathrm{Hom}_A(T, Y \otimes_B T)$.

(b) Let N belong to $\mathscr{F}(T)$. There exists an exact sequence

$$0 \longrightarrow N \longrightarrow I \longrightarrow N' \longrightarrow 0$$

with I injective. Then, I lies in $\mathscr{T}(T)$. Because $\mathscr{T}(T)$ is closed under quotients, also N' lies in $\mathscr{T}(T)$. Applying $\mathrm{Hom}_A(T, -)$ yields an exact sequence

$$0 \longrightarrow \mathrm{Hom}_A(T, I) \longrightarrow \mathrm{Hom}_A(T, N') \longrightarrow \mathrm{Ext}^1_A(T, N) \longrightarrow 0.$$

Because $\mathrm{Hom}_A(T, N')$ lies in $\mathscr{Y}(T)$, we have $\mathrm{Tor}_1^B(\mathrm{Hom}_A(T, N'), T) = 0$. We deduce a commutative diagram with exact rows

$$\begin{array}{ccccccccc}
0 \to & \mathrm{Tor}_1^B(\mathrm{Ext}^1_A(T,N),T) & \to & \mathrm{Hom}_A(T,I)\otimes_B T & \to & \mathrm{Hom}_A(T,N')\otimes_B T & \to & \mathrm{Ext}^1_A(T,N)\otimes_B T & \to 0 \\
 & & & \cong \downarrow \varepsilon_I & & \cong \downarrow \varepsilon_{N'} & & & \\
0 \to & N & \to & I & \to & N' & \to & 0 &
\end{array}$$

where the vertical isomorphisms are those of Corollary V.2.12.

Therefore, $\mathrm{Ext}^1_A(T, N) \otimes_B T = 0$, that is, $\mathrm{Ext}^1_A(T, N)$ lies in $\mathscr{X}(T)$. Also $\mathrm{Tor}_1^B(\mathrm{Ext}^1_A(T, N), T) \cong N$.

Similarly, let X belong to $\mathscr{X}(T)$. There exists an exact sequence

$$0 \longrightarrow X' \longrightarrow P \longrightarrow X \longrightarrow 0$$

with P projective. Then, P lies in $\mathscr{Y}(T)$. Because $\mathscr{Y}(T)$ is closed under taking submodules, X' is in $\mathscr{Y}(T)$ as well. Applying $- \otimes_B T$ yields an exact sequence

$$0 \longrightarrow \mathrm{Tor}_1^B(X, T) \longrightarrow X' \otimes_B T \longrightarrow P \otimes_B T \longrightarrow 0.$$

Because $X' \otimes_B T$ lies in $\mathscr{T}(T)$, we have $\mathrm{Ext}^1_A(T, X' \otimes_B T) = 0$. We get a commutative diagram with exact rows

$$\begin{array}{ccccccccc}
 & 0 & \to & X' & \to & P & \to & X & \to 0 \\
 & & & \delta_{X'} \downarrow \cong & & \delta_P \downarrow \cong & & & \\
0 \to & \mathrm{Hom}_A(T,\mathrm{Tor}_1^B(X,T)) & \to & \mathrm{Hom}_A(T,X'\otimes_B T) & \to & \mathrm{Hom}_A(T,P\otimes_B T) & \to & \mathrm{Ext}^1_A(T,\mathrm{Tor}_1^B(X,T)) & \to 0
\end{array}$$

where the vertical isomorphisms are those of Corollary V.2.20. We deduce that $\mathrm{Hom}_A(T, \mathrm{Tor}_1^B(X, T)) = 0$ so that $\mathrm{Tor}_1^B(X, T)$ belongs to $\mathscr{F}(T_A)$. Also $X \cong \mathrm{Ext}^1_A(T, \mathrm{Tor}_1^B(X, T))$.

□

Example V.2.22. In Example V.2.2, a quick calculation shows that B is given by the quiver

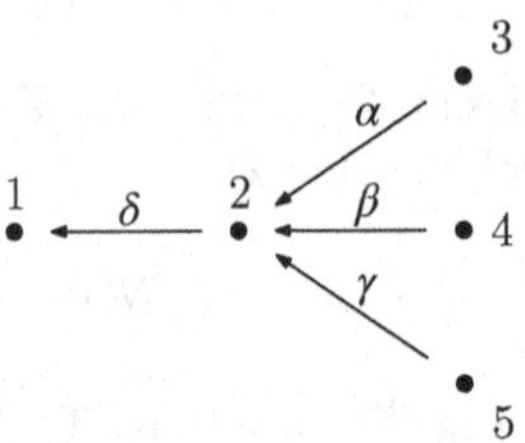

bound by $\alpha\delta = 0$, $\gamma\delta = 0$. While drawing this quiver, we should remember that endomorphisms of a module compose in the reverse way to arrows in a quiver, and thus the arrows are drawn in the opposite direction to morphisms. Then, $\Gamma(\text{mod}\, B)$ is given by

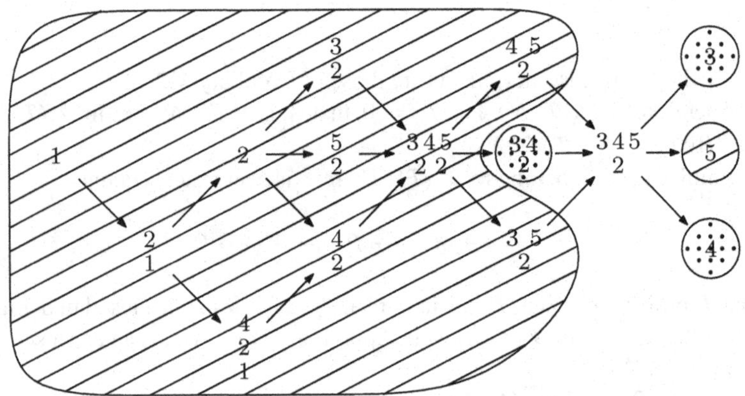

where $\mathscr{Y}(T)$ is illustrated by the hatched area, whereas $\mathscr{X}(T)$ is illustrated by the dotted area. It is easy to compute the image of every indecomposable A-module under the tilting functors $\text{Hom}_A(T, -)$ and $\text{Ext}^1_A(T, -)$. For instance, we have $\text{Hom}_A(T, 3) = {}^{5}_{2}$ while $\text{Ext}^1_A(T, 2) \cong \text{D}\,\text{Hom}_A(2, \tau T) \cong {}^{3\,4}_{\ 2}$.

V.2.4 Consequences of the main results

As usual, we denote by A an algebra, T a tilting A-module and $B = \text{End}\, T_A$. We start by comparing the global dimensions of A and B.

Lemma V.2.23. *Assume M lies in $\mathscr{T}(T)$, then* $\text{pd}\, \text{Hom}_A(T, M) \leq \text{pd}\, M$.

Proof. By induction on $d = \operatorname{pd} M$. If $d = 0$, then M_A is projective and so, because it lies in $\mathscr{T}(T)$, it must belong to $\operatorname{add} T$ so that $\operatorname{Hom}_A(T, M)$ is a projective B-module.

Assume $d \geq 1$. Because M lies in $\mathscr{T}(T)$, there exists an exact sequence $0 \longrightarrow L \longrightarrow T_0 \longrightarrow M \longrightarrow 0$, with T_0 in $\operatorname{add} T$ and L in $\mathscr{T}(T)$. Because of Corollary V.2.11, we have a short exact sequence

$$(*) \qquad 0 \longrightarrow \operatorname{Hom}_A(T, L) \longrightarrow \operatorname{Hom}_A(T, T_0) \longrightarrow \operatorname{Hom}_A(T, M) \longrightarrow 0.$$

Also, applying $\operatorname{Hom}_A(-, N)$ where N is an arbitrary module yields an epimorphism $\operatorname{Ext}^d_A(T_0, N) \longrightarrow \operatorname{Ext}^d_A(L, N)$, because $d = \operatorname{pd} M$.

Assume first $d = 1$. Because L is in $\mathscr{T}(T)$, there is an exact sequence $0 \longrightarrow N \longrightarrow T_1 \longrightarrow L \longrightarrow 0$ with T_1 in $\operatorname{add} T$ and N lying in $\mathscr{T}(T)$. In particular, $\operatorname{Ext}^1_A(L, N) = 0$; consequently, this sequence splits and L is a summand of T, hence lies in $\operatorname{add} T$. The sequence $(*)$ then gives $\operatorname{pd} \operatorname{Hom}_A(T, M) \leq 1$.

If $d > 1$ and N is arbitrary, then $\operatorname{pd} T_0 \leq 1$ gives $\operatorname{Ext}^d_A(L, N) = 0$ so that $\operatorname{pd} L \leq d - 1$. Because of the induction hypothesis, $\operatorname{pd} \operatorname{Hom}_A(T, L) \leq d - 1$ and so $\operatorname{pd} \operatorname{Hom}_A(T, M) \leq 1 + (d - 1) = d$. □

Theorem V.2.24. *With the above notations,* $|\operatorname{gl.dim.} A - \operatorname{gl.dim.} B| \leq 1$.

Proof. Let Z be an arbitrary B-module. There exists a short exact sequence $0 \longrightarrow Y \longrightarrow P \longrightarrow Z \longrightarrow 0$, with P projective. Because P lies in $\mathscr{Y}(T)$, so does Y. Because of Lemma V.2.23, we have $\operatorname{pd} Y \leq \operatorname{gl.dim.} A$ and so $\operatorname{pd} Z \leq 1 + \operatorname{pd} Y \leq 1 + \operatorname{gl.dim.} A$. Therefore, $\operatorname{gl.dim.} B \leq 1 + \operatorname{gl.dim.} A$. The reciprocity theorem V.2.18 implies that $\operatorname{gl.dim.} A \leq 1 + \operatorname{gl.dim.} B$. □

The next theorem says that the numbers of isoclasses of indecomposable projective A-modules and of indecomposable projective B-modules are equal. Equivalently, the quivers of A and B have the same number of points.

Theorem V.2.25. *The map* $f\colon \mathrm{K}_0(A) \longrightarrow \mathrm{K}_0(B)$ *given by*

$$M \mapsto \underline{\dim} \operatorname{Hom}_A(T, M) - \underline{\dim} \operatorname{Ext}^1_A(T, M)$$

is an isomorphism of abelian groups.

Proof. Because $\operatorname{pd} T \leq 1$, a short exact sequence $0 \longrightarrow L \longrightarrow M \longrightarrow N \longrightarrow 0$ in $\operatorname{mod} A$ induces a long exact cohomology sequence in $\operatorname{mod} B$

$$\begin{aligned} 0 \longrightarrow \operatorname{Hom}_A(T, L) \longrightarrow \operatorname{Hom}_A(T, M) \longrightarrow \operatorname{Hom}_A(T, N) \\ \longrightarrow \operatorname{Ext}^1_A(T, L) \longrightarrow \operatorname{Ext}^1_A(T, M) \longrightarrow \operatorname{Ext}^1_A(T, N) \longrightarrow 0. \end{aligned}$$

Taking dimensions shows that $f\colon \mathrm{K}_0(A) \longrightarrow \mathrm{K}_0(B)$ defined as in the statement is a morphism of groups.

Let S be a simple B-module. Either S lies in $\mathscr{Y}(T)$, or in $\mathscr{X}(T)$ (see the remark after Lemma V.2.9). In the former case, $S \cong \operatorname{Hom}_A(T, S \otimes_B T)$ whereas $\operatorname{Ext}^1_A(T, S \otimes_B T) = 0$, and in the latter, $S \cong \operatorname{Ext}^1_A(T, \operatorname{Tor}^B_1(S, T))$ whereas $\operatorname{Hom}_A(T, \operatorname{Tor}^B_1(S, T)) = 0$. In either case, $\underline{\dim}\, S$ lies in the image of f. Thus, the canonical basis of $K_0(B) \cong \mathbb{Z}^n$ lies in $\operatorname{Im} f$. Hence, f is surjective. Therefore, $\operatorname{rk} K_0(A) \geq \operatorname{rk} K_0(B)$. The reciprocity theorem V.2.18 implies that $\operatorname{rk} K_0(B) \geq \operatorname{rk} K_0(A)$. We get $\operatorname{rk} K_0(A) = \operatorname{rk} K_0(B)$; thus, f is an isomorphism. □

The following corollary, due to Bongartz, simplifies considerably the task of verifying whether a given module T is tilting or not.

Corollary V.2.26. *Let $T = \oplus_{i=1}^m T_i^{m_i}$ where the T_i are indecomposable and $T_i \ncong T_j$ for $i \neq j$. Then, T is a tilting module if and only if it is a partial tilting module and $m = \operatorname{rk} K_0(A)$.*

Proof. Necessity follows directly from Theorem V.2.25; thus, we prove sufficiency. Because T is a partial tilting module, there exists E such that $T \oplus E$ is tilting, see Proposition V.2.4. But then the theorem implies that the number of isoclasses of indecomposable summands of $T \oplus E$ equals $\operatorname{rk} K_0(A)$. Because of the hypothesis, $\operatorname{rk} K_0(A)$ equals the number of isoclasses of indecomposable summands of T. Hence, E belongs to $\operatorname{add} T$ and T is tilting. □

Thus, a partial tilting A-module is tilting if and only if the number of isoclasses of indecomposable summands of T equals the number of isoclasses of simple A-modules, that is, the number of points in the quiver of A.

In the language used at the end of Subsection V.2.3, the next proposition states that, if A is hereditary and T is a tilting A-module, then the torsion pair $(\mathscr{X}(T), \mathscr{Y}(T))$ in $\operatorname{mod} B$ is split.

Proposition V.2.27. *If A is hereditary, then every indecomposable B-module either lies in $\mathscr{X}(T)$ or in $\mathscr{Y}(T)$.*

Proof. We claim first that $\operatorname{Ext}_B^1(Y, X) = 0$ for all Y in $\mathscr{Y}(T)$ and all X in $\mathscr{X}(T)$. Indeed, because of the tilting theorem, there exist M in $\mathscr{T}(T)$ and N in $\mathscr{F}(T)$ such that $Y \cong \operatorname{Hom}_A(T, M)$ and $X \cong \operatorname{Ext}_A^1(T, N)$. Applying $\operatorname{Hom}_A(T, -)$ to an injective coresolution $0 \longrightarrow N \longrightarrow I_0 \longrightarrow I_1 \longrightarrow 0$ yields a short exact sequence

$$0 \longrightarrow \operatorname{Hom}_A(T, I_0) \longrightarrow \operatorname{Hom}_A(T, I_1) \longrightarrow \operatorname{Ext}_A^1(T, N) \longrightarrow 0.$$

Because M lies in $\mathscr{T}(T)$, we have $\operatorname{pd} \operatorname{Hom}_A(T, M) \leq 1$, see Lemma V.2.23; therefore, there is an epimorphism

$$\operatorname{Ext}_B^1(\operatorname{Hom}_A(T, M), \operatorname{Hom}_A(T, I_1)) \longrightarrow$$
$$\operatorname{Ext}_B^1(\operatorname{Hom}_A(T, M), \operatorname{Ext}_A^1(T, N)) = \operatorname{Ext}_B^1(Y, X).$$

Lemma V.2.15 gives

$$\operatorname{Ext}_B^1(\operatorname{Hom}_A(T, M), \operatorname{Hom}_A(T, I_1)) \cong \operatorname{Ext}_A^1(M, I_1) = 0.$$

Therefore, $\operatorname{Ext}_B^1(Y, X) = 0$, as required.

Let now Z be an arbitrary indecomposable B-module. We have just proved that its canonical sequence in the torsion pair $(\mathscr{X}(T), \mathscr{Y}(T))$ splits. The indecomposability of Z implies that Z lies either in $\mathscr{X}(T)$ or in $\mathscr{Y}(T)$. □

Exercises for Section V.2

Exercise V.2.1. Prove that the following conditions are equivalent for a module M_A:

(a) M is faithful,
(b) A_A is cogenerated by M,
(c) $\mathrm{D}A_A$ is generated by M,
(d) Every left add M-approximation $A_A \longrightarrow T^d$ is injective, see Exercise V.1.5.

Exercise V.2.2. Prove that a partial tilting module T is tilting if and only if, for every indecomposable projective A-module P_x, there exists a short exact sequence $0 \longrightarrow P_x \longrightarrow T_x^0 \longrightarrow T_x^1 \longrightarrow 0$ with T_x^0, T_x^1 in add T.

Exercise V.2.3. Let T be an A-module such that $\mathrm{pd}\, T \leq 1$. Prove that T is a partial tilting module if and only if $\mathrm{Ext}^1_A(T, M) = 0$ for every module M generated by T.

Exercise V.2.4. Let $\mathscr{C}$ be a full subcategory of mod A, **k**-linear and closed under extensions. An object M in $\mathscr{C}$ is called **Ext-projective** in $\mathscr{C}$ if $\mathrm{Ext}^1_A(M, -)|_{\mathscr{C}} = 0$ and **Ext-injective** in $\mathscr{C}$ if $\mathrm{Ext}^1_A(-, M)|_{\mathscr{C}} = 0$.

(a) Let $(\mathscr{T}, \mathscr{F})$ be a torsion pair. Prove that M in $\mathscr{T}$ is Ext-projective in $\mathscr{T}$ if and only if τM lies in $\mathscr{F}$, and that M in $\mathscr{F}$ is Ext-injective in $\mathscr{F}$ if and only if $\tau^{-1} M$ lies in $\mathscr{T}$.
(b) Let T be a tilting module. Prove that M is Ext-projective in $\mathscr{T}(T)$ if and only if M lies in add T, and that M is Ext-injective in $\mathscr{T}(T)$ if and only if M is an injective A-module.

Exercise V.2.5. Prove that a partial tilting module T is tilting if and only if, for every E such that $T \oplus E$ is partial tilting, we have E in add T.

Exercise V.2.6. Let T be an A-module.

(a) If T is a faithful module and such that $\mathrm{Hom}_A(T, \tau T) = 0$, then T is a partial tilting module.
(b) Prove that T is tilting if and only if $\mathrm{Hom}_A(T, \tau T) = 0$ and there is an exact sequence $0 \longrightarrow A_A \longrightarrow T_0 \longrightarrow T_1 \longrightarrow 0$ with T_0, T_1 in add T.

Exercise V.2.7. Let T be a tilting module and $0 \longrightarrow L \longrightarrow M \longrightarrow N \longrightarrow 0$ an exact sequence with L in $\mathscr{T}(T)$. Prove that $\mathrm{Ext}^1_A(T, M) \cong \mathrm{Ext}^1_A(T, N)$.

Exercise V.2.8. For each of the following bound quiver algebras, verify that the given module T is tilting, compute the bound quivers of $B = \mathrm{End}\, T$ and the torsion pairs $(\mathscr{T}(T), \mathscr{F}(T))$ and $(\mathscr{X}(T), \mathscr{Y}(T))$.

(a) $4 \xrightarrow{\gamma} 2$, $1 \xleftarrow{\beta} 2 \xleftarrow{\alpha} 3$

$\alpha\beta = 0$

$T = P_1 \oplus P_4 \oplus \tau^{-2} P_1 \oplus S_4$

(b) $1 \xleftarrow{\gamma} 2 \xleftarrow{\beta} 3 \xleftarrow{\alpha} 4$

$\alpha\beta\gamma = 0$

$T = P_1 \oplus \tau^{-2} P_1 \oplus P_3 \oplus P_4$

(c) $4 \xrightarrow{\lambda} 1$, $4 \xrightarrow{\beta} 2$, $6 \xrightarrow{\alpha} 4$, $5 \xrightarrow{\delta} 2$, $6 \xrightarrow{\gamma} 5$, $5 \xrightarrow{\mu} 3$

$\alpha\beta = \gamma\delta,\ \alpha\lambda = 0,\ \gamma\mu = 0$

$T = P_1 \oplus P_3 \oplus P_6 \oplus \tau^{-1} P_2 \oplus \tau^{-2} P_3 \oplus \tau^{-2} P_1$

(d) $3 \xrightarrow{\gamma} 2$, $5 \xrightarrow{\beta} 3$, $6 \xrightarrow{\alpha} 5$, $4 \xrightarrow{\varepsilon} 2$, $5 \xrightarrow{\delta} 4$, $4 \xrightarrow{\mu} 1$

$\alpha\delta = 0,\ \delta\mu = 0,\ \beta\gamma = \delta\varepsilon$

$T = \tau^{-1} P_3 \oplus S_4 \oplus P_5 \oplus P_6 \oplus S_3 \oplus (P_5/S_2)$

(e) $2 \longrightarrow 1$, $5 \longrightarrow 3 \longrightarrow 1 \longleftarrow 4 \longleftarrow 6$

$T = P_2 \oplus \tau^{-1} P_5 \oplus \tau^{-3} P_5 \oplus \tau^{-1} P_6 \oplus \tau^{-3} P_6 \oplus \tau^{-2} P_2$

(f) $2 \xrightarrow{\lambda} 1$, $5 \xrightarrow{\alpha} 2$, $3 \xrightarrow{\mu} 1$, $5 \xrightarrow{\beta} 3$, $4 \xrightarrow{\nu} 1$, $5 \xrightarrow{\gamma} 4$

$\alpha\lambda = \beta\mu = \gamma\nu$

$T = P_4 \oplus \tau^{-1} P_2 \oplus \tau^{-1} P_3 \oplus \tau^{-2} P_4 \oplus P_5$

(g) $2 \xrightarrow{\beta} 1$, $3 \xrightarrow{\alpha} 2$, $3 \xrightarrow{\gamma} 1$

$\alpha\beta = 0$

$T = \tau^{-1} P_1 \oplus P_2 \oplus P_3$

(h) $2 \xrightarrow{\beta} 1$, $5 \xrightarrow{\alpha} 2$, $5 \xrightarrow{\gamma} 4$, $4 \xrightarrow{\delta} 3$, $3 \xrightarrow{\varepsilon} 1$

$\alpha\beta = 0, \gamma\delta = 0, \delta\varepsilon = 0$

$T = \tau^{-1} P_1 \oplus P_2 \oplus P_3 \oplus P_4 \oplus P_5$

Exercise V.2.9. Let A be as in Example V.2.2. Find a tilting module T such that $B = \operatorname{End} T$ is hereditary and compute the torsion pairs $(\mathscr{T}(T), \mathscr{F}(T))$ and $(\mathscr{X}(T), \mathscr{Y}(T))$.

Exercise V.2.10. Let A be given by the quiver

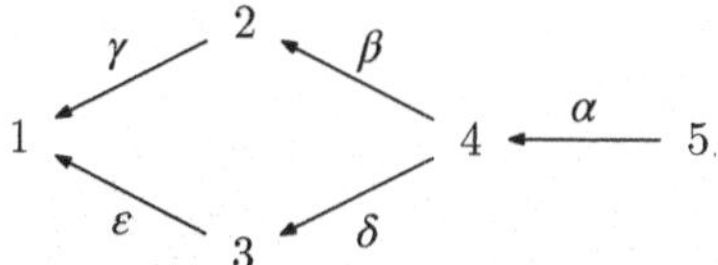

bound by $\alpha\delta = 0$ and $\beta\gamma = \delta\varepsilon$. Prove that $T = {}^{\;4}_{2\,3} \oplus 3$ is a partial tilting module. Find an A-module E such that $T' = T \oplus E$ is tilting and compute the torsion pair $(\mathscr{T}(T'), \mathscr{F}(T'))$.

Exercise V.2.11. Let T be a tilting module and, for a module M, denote by $t_T M$ the trace of T in M. Prove that $\mathrm{Hom}_A(T, M) \cong \mathrm{Hom}_A(T, t_T M)$ and $\mathrm{Ext}^1_A(T, M) \cong \mathrm{Ext}^1_A(T, M/t_T M)$. Deduce that, if $B = \mathrm{End}\, T_A$, then $t_T M \cong \mathrm{Hom}_A(T, M) \otimes_B T$ and $M/t_T M \cong \mathrm{Tor}^B_1(\mathrm{Ext}^1_A(T, M), T)$.

Exercise V.2.12. Let T be a tilting module and N in $\mathscr{F}(T)$ such that $\mathrm{id}\, N = 1$. Prove that $\mathrm{Ext}^1_B(Y, \mathrm{Ext}^1_A(T, N)) = 0$ for all Y in $\mathscr{Y}(T)$.

Exercise V.2.13. Prove that the following conditions are equivalent for a torsion pair $(\mathscr{T}, \mathscr{F})$ in $\mathrm{mod}\, A$:

(a) $(\mathscr{T}, \mathscr{F})$ is split,
(b) For every A-module, the canonical sequence splits,
(c) For every M in $\mathscr{T}$, we have $\tau^{-1} M$ in $\mathscr{T}$,
(d) For every N in $\mathscr{F}$, we have τN in $\mathscr{F}$.

Exercise V.2.14. If T_A is a tilting module, show that the functor $\mathrm{Ext}^1_A(T, -)|_{\mathscr{F}(T)}$ is exact.

Exercise V.2.15. Let T_A be a partial tilting module and $\mathscr{T}(T), \mathscr{F}(T)$ the **k**-linear full subcategories of $\mathrm{mod}\, A$ defined by $\mathscr{T}(T) = \{M \mid M \text{ is generated by } T\}$ and $\mathscr{F}(T) = \{M \mid \mathrm{Hom}_A(T, M) = 0\}$.

(a) Prove that $(\mathscr{T}(T), \mathscr{F}(T))$ is a torsion pair.
(b) Let $e \in A$ be an idempotent and $T = eA$. Prove that $\mathscr{F}(T)$ is equivalent to $\mathrm{mod}\,(A/AeA)$.

Exercise V.2.16. Let T_A be a tilting module and N such that $\mathrm{Hom}_A(T, N) = 0$. Prove that $\mathrm{pd}\, \mathrm{Ext}^1_A(T, N) \leq 1 + \max\{1, \mathrm{pd}\, N\}$.

Chapter VI
Representation-finite algebras

For a long time, researchers in the representation theory of algebras concentrated on finding criteria allowing us to verify whether a given algebra is representation-finite or not, and, if this was the case, of computing all its (isoclasses of) indecomposable modules. Indeed, it was believed that this class of algebras would be relatively easy to classify and that their indecomposable modules have a relatively simple structure. This approach was largely successful. Actually, one of the first important results of modern-day representation theory was Gabriel's theorem, which says that a hereditary algebra over an algebraically closed field is representation-finite if and only if it is the path algebra of a quiver whose underlying graph is one of the well-known Dynkin diagrams $\mathbb{A}$, $\mathbb{D}$ or $\mathbb{E}$. Nowadays, there exists a reasonable global theory of representation-finite algebras. At present, we do not have a similar theory for studying representation-infinite algebras, but the ideas and techniques developed for representation-finite algebras still show their usefulness when applied to the understanding of new classes. The aim of this chapter is to prove some of the most important known results on representation-finite algebras highlighting the methods that led to their proofs.

We start by showing how the representation-finiteness of an algebra is reflected by the finiteness properties of the radical of its module category, then study the Auslander algebra of a representation-finite algebra, and end the chapter proving the so-called Four Terms in the Middle theorem, which gives a bound on the number of indecomposable middle terms of almost split sequences over a representation-finite algebra.

I. Assem, F. U. Coelho, *Basic Representation Theory of Algebras*, Graduate Texts in Mathematics 283, https://doi.org/10.1007/978-3-030-35118-2_6

VI.1 The Auslander–Reiten quiver and the radical

VI.1.1 The Harada–Sai lemma

A first attempt to characterise representation-finiteness can be made using the Auslander–Reiten quiver. Indeed, an algebra is representation-finite if and only if its Auslander-Reiten quiver is finite. As we have seen in Corollary II.4.8, the finiteness of nonzero paths in Auslander–Reiten quivers (which is certainly implied by the finiteness of the quiver itself) is closely related to the vanishing of some power of the radical of the module category, and thus to the vanishing of the infinite radical. This led to the question whether an algebra would be representation-finite provided some (large enough) power of the radical would equal zero. Our objective in this subsection is to prove this statement. We start with a useful result, which will be applied to radical computations inside the module category.

Lemma VI.1.1 (Harada–Sai lemma). *Let $m \geq 0$ and*

$$M_1 \xrightarrow{f_1} M_2 \xrightarrow{f_2} \dots \xrightarrow{f_{2^m-1}} M_{2^m}$$

be a radical path where each M_i has composition length at most equal to m. Then, $f_{2^m-1} \dots f_2 f_1 = 0$.

Proof. We prove by induction on n that, if

$$M_1 \xrightarrow{f_1} M_2 \xrightarrow{f_2} \dots \xrightarrow{f_{2^n-1}} M_{2^n}$$

is a radical path, with all M_i satisfying $l(M_i) \leq m$, then $l(\mathrm{Im}(f_{2^n-1} \dots f_2 f_1)) \leq m - n$. This immediately implies the statement upon setting $m = n$.

Let $n = 1$. Because $f_1 \in \mathrm{rad}_A(M_1, M_2)$, it is not an isomorphism. Therefore $l(\mathrm{Im}\, f_1) \leq m - 1$ which proves the statement in this case.

Assume that the statement holds true for a given n and consider a radical path

$$M_1 \xrightarrow{f_1} M_2 \longrightarrow \dots \xrightarrow{f_{2^n-1}} M_{2^n} \xrightarrow{f_{2^n}} M_{2^n+1} \longrightarrow \dots \xrightarrow{f_{2^{n+1}-1}} M_{2^{n+1}}$$

with $l(M_i) \leq m$ for all i. To simplify notation, we set $f = f_{2^n-1} \dots f_2 f_1$, $g = f_{2^n}$ and $h = f_{2^{n+1}-1} \dots f_{2^n+2} f_{2^n+1}$. The induction hypothesis implies that $l(\mathrm{Im}\, f) \leq m - n$ and $l(\mathrm{Im}\, h) \leq m - n$. If at least one of the inequalities is strict, then $l(\mathrm{Im}\, hgf) \leq m - n - 1$ and we have finished. We may thus restrict to the case where $l(\mathrm{Im}\, h) = l(\mathrm{Im}\, f) = m - n$.

We assume that $l(\mathrm{Im}\, hgf) > m - n - 1$ and reach a contradiction. Because $l(\mathrm{Im}\, hgf) \leq l(\mathrm{Im}\, h) \leq m - n$, the only possibility is that $l(\mathrm{Im}\, hgf) = m - n$. However, it is well-known, see Exercise VI.1.1 that

$$\mathrm{Im}\, hgf = \frac{\mathrm{Im}\, f}{\mathrm{Im}\, f \cap \mathrm{Ker}\, hg}.$$

Therefore, $l(\mathrm{Im}\, hgf) = l(\mathrm{Im}\, f)$ implies that $\mathrm{Im}\, f \cap \mathrm{Ker}\, hg = 0$. On the other hand, $\mathrm{Im}\, hgf \subseteq \mathrm{Im}\, hg \subseteq \mathrm{Im}\, h$ and $l(\mathrm{Im}\, hgf) = l(\mathrm{Im}\, h) = m - n$ imply $l(\mathrm{Im}\, hg) = m - n = l(\mathrm{Im}\, f)$ as well. Therefore, $l(\mathrm{Ker}\, hg) = l(M_{2^n}) - l(\mathrm{Im}\, hg) = l(M_{2^n}) - l(\mathrm{Im}\, f)$ and so $M_{2^n} = \mathrm{Ker}\, hg \oplus \mathrm{Im}\, f$. However, M_{2^n} is indecomposable and f is nonzero (because the length of its image is $m - n$). Therefore, $\mathrm{Ker}\, hg = 0$. This shows that hg is a monomorphism. Hence, so is g.

Similarly, one can show that $M_{2^n+1} = \mathrm{Ker}\, h \oplus \mathrm{Im}\, gf$. Because M_{2^n+1} is indecomposable and $gf \neq 0$, we get $M_{2^n+1} = \mathrm{Im}(gf)$. Therefore, gf is an epimorphism and hence so is g.

We have shown that g is an isomorphism, contrary to the hypothesis that it belongs to the radical. This is the required contradiction. □

Exercise VI.1.2 below gives an example in which the bound in the Harada–Sai lemma is sharp.

VI.1.2 The infinite radical and representation-finiteness

We now prove that an algebra is representation-finite if and only if the infinite radical of its module category (or, equivalently, a power large enough of the radical) vanishes. One implication is easy.

Lemma VI.1.2. *Let A be a representation-finite algebra and m a bound on the length of indecomposable A-modules. Then, $\mathrm{rad}_A^{2^m-1} = 0$. In particular, $\mathrm{rad}_A^{\infty} = 0$.*

Proof. This follows immediately from the Harada–Sai lemma VI.1.1. □

The Harada–Said lemma may be rephrased as follows: given a representation-finite algebra A, there exists (at least) an $n > 0$ such that $\mathrm{rad}_A^n = 0$, which clearly implies that $\mathrm{rad}_A^{\infty} = 0$. We shall see in the sequel that if, conversely, $\mathrm{rad}_A^{\infty} = 0$, then A is representation-finite, but before that, we look at an easy consequence.

Corollary VI.1.3. *Let A be a representation-finite algebra, M, N indecomposable A-modules and $f \in \mathrm{rad}_A(M, N)$ a nonzero morphism. Then:*

(a) *f is a sum of compositions of irreducible morphisms.*
(b) *There exists a path $M \rightsquigarrow N$ of irreducible morphisms.*

Proof. This follows from Lemma VI.1.2, Corollary II.4.6 and Corollary II.4.8. □

In particular, if A is representation-finite, and M, N are indecomposable modules such that there is no path $M \rightsquigarrow N$ of irreducible morphisms, then $\mathrm{Hom}_A(M, N) = 0$.

We now set out to prove the converse of Lemma VI.1.2. We need one further result. For an A-module M and every $m \geq 1$, we denote by $\mathrm{rad}_A^m(M, -)$ and $\mathrm{rad}_A^m(-, M)$ the obvious subfunctors of $\mathrm{Hom}_A(-, M)$ and $\mathrm{Hom}_A(M, -)$ respectively, defined as we did for $\mathrm{rad}_A(-, M)$ and $\mathrm{rad}_A(M, -)$ in Subsection II.1.3.

Lemma VI.1.4. *Let A be an algebra such that* $\operatorname{rad}_A^\infty = 0$, *and M an indecomposable A-module. Then:*

(a) *There exists $m_M > 0$ such that* $\operatorname{rad}_A^{m_M}(M, -) = 0$.
(b) *There exists $n_M > 0$ such that* $\operatorname{rad}_A^{n_M}(-, M) = 0$.

Proof. We only prove (a), because the proof of (b) is dual.

Let $\mathrm{D}A$ be the minimal injective cogenerator of $\operatorname{mod} A$. Because of Lemma II.4.1, there exists a least $m > 0$ such that $\operatorname{rad}_A^m(M, \mathrm{D}A) = \operatorname{rad}_A^\infty(M, \mathrm{D}A) = 0$. We claim that, for every indecomposable A-module N, we have $\operatorname{rad}_A^m(M, N) = 0$. Because $\mathrm{D}A$ is the minimal injective cogenerator, there exist $t > 0$ and a monomorphism $j \colon N \longrightarrow (\mathrm{D}A)^t$. If $f \in \operatorname{rad}_A^m(M, N)$, then $jf \in \operatorname{rad}_A^m(M, (\mathrm{D}A)^t)$. However, $\operatorname{rad}_A^m(M, \mathrm{D}A) = 0$. Hence, $jf = 0$. But j is a monomorphism. Thus, $f = 0$. □

Theorem VI.1.5. *A finite dimensional algebra A is representation-finite if and only if* $\operatorname{rad}_A^\infty = 0$.

Proof. Because of Lemma VI.1.2, if A is representation-finite, then $\operatorname{rad}_A^\infty = 0$. Conversely, if $\operatorname{rad}_A^\infty = 0$, then, because of Lemma VI.1.4, for each indecomposable projective A-module P, there exists a least $m_P > 0$ such that $\operatorname{rad}_A^{m_P}(P, -) = 0$. Let m be the maximum of all m_P, as P runs through the isoclasses of indecomposable projective A-modules. Let M be an indecomposable A-module. There exists an indecomposable projective A-module P_0 such that $\operatorname{Hom}_A(P_0, M) \neq 0$. However, $\operatorname{rad}_A^m(P_0, M) = 0$. Therefore, because of Corollary II.4.8, there exists a path $P_0 \rightsquigarrow M$ of irreducible morphisms of length at most $m - 1$ with nonzero composition. We observe that given a module L and a positive integer l, there are only finitely many indecomposable modules that have a path of irreducible morphisms from L, with length at most l: this is because the Auslander–Reiten quiver is locally finite, see Definition IV.1.17. Thus, because there are only finitely many isoclasses of indecomposable projective A-modules, we infer that A is representation-finite. □

VI.1.3 Auslander's theorem

We have already stated and used in several examples Auslander's theorem, which asserts that if the Auslander–Reiten quiver $\Gamma(\operatorname{mod} A)$ of an algebra A admits a finite connected component Γ, then $\Gamma(\operatorname{mod} A) = \Gamma$ and hence A is representation-finite. To prove it, we start by interpreting in terms of the Auslander–Reiten quiver some results of Chapter II.

Because irreducible morphisms correspond to arrows in the Auslander–Reiten quiver, there is a path of irreducible morphisms between indecomposables from M to N if and only if there is a path in $\Gamma(\operatorname{mod} A)$ from the point M to the point N. In particular, if this is the case, then M and N belong to the same connected component of $\Gamma(\operatorname{mod} A)$. We may then restate Corollary II.4.8 and Proposition II.4.9 as follows: let M, N be indecomposable A-modules and assume that $\operatorname{rad}_A(M, N) \neq 0$. Then, we have two cases to consider.

If $\operatorname{rad}_A^\infty(M, N) = 0$, then there exists a path $M \rightsquigarrow N$ in $\Gamma(\operatorname{mod} A)$. In addition, there exists $t \geq 1$ such that $\operatorname{rad}_A^t(M, N) \setminus \operatorname{rad}_A^{t+1}(M, N) \neq 0$. In this case, there exists a path $M \rightsquigarrow N$ in $\Gamma(\operatorname{mod} A)$ of length t.

If $\operatorname{rad}_A^\infty(M, N) \neq 0$, then, for every $i > 0$, one can find a path of length i in $\Gamma(\operatorname{mod} A)$ from M to some M' with nonzero composition, and a morphism in $\operatorname{rad}_A^\infty(M', N)$ whose composition with the previous path is still nonzero. Dually, there exists a path of length i in $\Gamma(\operatorname{mod} A)$ from some N' to N with nonzero composition and a morphism in $\operatorname{rad}_A^\infty(M, N')$ whose composition with the previous path is still nonzero.

We need one additional lemma.

Lemma VI.1.6. *Let Γ be a connected component of $\Gamma(\operatorname{mod} A)$, all of whose modules have length bounded by m, and M, N indecomposable A-modules such that $\operatorname{Hom}_A(M, N) \neq 0$. Then, M lies in Γ if and only if N does too, and, if this is the case, then there exists a path of irreducible morphisms $M \rightsquigarrow N$.*

Proof. We may clearly assume $M \not\cong N$ and thus $\operatorname{rad}_A(M, N) \neq 0$. Assume that M lies in Γ. To prove that N also lies in Γ, it suffices to prove that there exists a path of irreducible morphisms

$$M = M_0 \longrightarrow M_1 \longrightarrow \dots \longrightarrow M_l = N$$

of length $l < 2^m - 1$. Assume that this is not the case. Because of Corollary II.4.8 and Proposition II.4.9, there exists a path of irreducible morphisms

$$M = M_0 \longrightarrow M_1 \longrightarrow \dots \longrightarrow M_{2^m-1}$$

with nonzero composition, and this contradicts the Harada–Sai lemma VI.1.1. Therefore, N belongs to Γ.

Dually, if N lies in Γ, then so does M. □

We now prove Auslander's theorem.

Theorem VI.1.7. *Let A be a finite dimensional algebra such that $\Gamma(\operatorname{mod} A)$ has a connected component Γ whose modules have bounded composition length. Then, $\Gamma(\operatorname{mod} A) = \Gamma$ and A is representation-finite.*

Proof. Let M be an indecomposable module lying in Γ and P an indecomposable projective A-module P such that $\operatorname{Hom}_A(P, M) \neq 0$. Because of Lemma VI.1.6, P lies in Γ. Now, let P' be an arbitrary indecomposable projective A-module. Because the algebra A is connected, there exists a sequence of indecomposable projective A-modules $P = P_0, P_1, \dots, P_t = P'$ such that, for each i, we have

$\mathrm{Hom}_A(P_i, P_{i+1}) \neq 0$ or $\mathrm{Hom}_A(P_{i+1}, P_i) \neq 0$. Because of Lemma VI.1.6 and an easy induction, P' lies in Γ. Thus, all indecomposable projective A-modules lie in Γ. Now, let N be an arbitrary indecomposable A-module. Then, there exists an indecomposable projective A-module P' such that $\mathrm{Hom}_A(P', N) \neq 0$. But P' in Γ implies that N also lies in Γ. This proves that $\Gamma(\mathrm{mod}\, A) = \Gamma$.

It remains to show that Γ is finite. For every indecomposable module N, there exists an indecomposable projective module P' such that $\mathrm{Hom}_A(P', N) \neq 0$. Therefore, as observed in the proof of Lemma VI.1.6, there exists a path $P' \rightsquigarrow N$ of irreducible morphisms of length smaller than $2^m - 1$, where m is a bound on the length of modules in Γ. Because there are only finitely many isoclasses of indecomposable projectives, and $\Gamma(\mathrm{mod}\, A)$ is locally finite, we deduce that Γ is finite. Therefore, A is representation-finite. □

As a first and obvious corollary, we get the following statement, which was the statement used effectively in the examples of Chapter IV.

Corollary VI.1.8. *If $\Gamma(\mathrm{mod}\, A)$ admits a finite connected component Γ, then $\Gamma(\mathrm{mod}\, A) = \Gamma$. In particular, A is representation-finite.* □

The next corollary is of great historical importance, as it answered positively a conjecture that motivated several of the developments of modern-day representation theory. This conjecture is known as the first **Brauer–Thrall conjecture** (now a theorem).

Corollary VI.1.9. *An algebra A is either representation-finite or there exist indecomposable A-modules that have arbitrarily large length.*

Proof. Indeed, if the indecomposable A-modules have bounded length, then Auslander's theorem implies that A is representation-finite. □

Exercises for Section VI.1

Exercise VI.1.1. Let $f: L \longrightarrow M, g: M \longrightarrow N$ be morphisms of modules. Prove that

$$\mathrm{Im}\, gf = \frac{\mathrm{Im}\, f}{\mathrm{Im}\, f \cap \mathrm{Ker}\, g}.$$

Exercise VI.1.2. Let A be given by the quiver

$$\alpha \circlearrowleft \bullet \circlearrowright \beta$$

bound by $\alpha^2 = 0$, $\beta^2 = 0$, $\alpha\beta = 0$, $\beta\alpha = 0$, and S denote the unique simple A-module. Construct morphisms

$$A_A \xrightarrow{f_1} \frac{A_A}{S} \xrightarrow{f_2} (DA)_A \xrightarrow{f_3} S \xrightarrow{f_4} A_A \xrightarrow{f_5} \frac{A_A}{S} \xrightarrow{f_6} (DA)_A$$

such that $f_6 f_5 f_4 f_3 f_2 f_1 \neq 0$ (thus, the Harada–Sai bound is sharp in this example).

Exercise VI.1.3. Prove that the following conditions are equivalent for an algebra A:

(a) A is representation-finite.
(b) for every indecomposable module M, there exists $m_M > 0$ such that $\mathrm{rad}_A^{m_M}(M, -) = 0$.
(c) for every indecomposable module N, there exists $n_N > 0$ such that $\mathrm{rad}_A^{n_N}(-, N) = 0$.
(d) for every indecomposable projective module P, there exists $m_P > 0$ such that $\mathrm{rad}_A^{m_P}(P, -) = 0$.
(e) for every indecomposable injective module I, there exists $n_I > 0$ such that $\mathrm{rad}_A^{n_I}(-, I) = 0$.

Exercise VI.1.4. Let $F\colon \mathrm{mod}\, A \longrightarrow \mathrm{mod}\, \mathbf{k}$ be a functor. Its **support** $\mathrm{Supp}\, F$ is the full subcategory of $\mathrm{ind}\, A$ consisting of the objects M such that $F(M) \neq 0$. Prove that the following conditions are equivalent for an algebra A.

(a) A is representation-finite.
(b) For every M in $\mathrm{ind}\, A$, $\mathrm{Supp}\, \mathrm{Hom}_A(M, -)$ is finite.
(c) For every N in $\mathrm{ind}\, A$, $\mathrm{Supp}\, \mathrm{Hom}_A(-, N)$ is finite.
(d) For every indecomposable projective P, $\mathrm{Supp}\, \mathrm{Hom}_A(P, -)$ is finite.
(e) For every indecomposable injective I, $\mathrm{Supp}\, \mathrm{Hom}_A(-, I)$ is finite.

Exercise VI.1.5.

(a) Let A be an algebra such that all indecomposable projectives belong to one component Γ of $\Gamma(\mathrm{mod}\, A)$. Prove that A is representation-finite and $\Gamma(\mathrm{mod}\, A) = \Gamma$ if and only if the modules in Γ have bounded length.
(b) Give an example of a representation-infinite algebra, all of whose indecomposable projective modules lie in the same component Γ of $\Gamma(\mathrm{mod}\, A)$ and prove that, in this example, the modules in Γ have unbounded length.

Exercise VI.1.6. Let A be a finite dimensional algebra. Prove that the following conditions are equivalent:

(a) A is representation-finite,
(b) $\Gamma(\mathrm{mod}\, A)$ admits one connected component, all of whose indecomposable modules are of bounded length,
(c) $\Gamma(\mathrm{mod}\, A)$ admits one finite connected component,
(d) $\mathrm{rad}_A^{\infty} = 0$,
(e) There exists $m > 0$ such that $\mathrm{rad}_A^m = 0$.

VI.2 Representation-finiteness using depths

VI.2.1 A characterisation using depths

We give another characterisation of representation-finiteness for an algebra A in terms of the radical of its module category.

In this section, for each simple module S, we fix a projective cover $p_S : P_S \longrightarrow S$ and an injective envelope $i_S : S \longrightarrow I_S$.

Theorem VI.2.1. *The following statements are equivalent for an algebra A:*

(a) *A is representation-finite.*
(b) *The depth of every nonzero morphism is finite.*
(c) *The depth of p_S is finite for every simple module S.*
(d) *The depth of ι_S is finite for every simple module S.*

Proof. Because (d) is dual to (c), it suffices to prove the equivalence of the first three conditions.

(a) implies (b). Assume that A is representation-finite. We have seen in Lemma VI.1.2 that there exists $m > 0$ such that $\mathrm{rad}_A^m = 0$. Then, the definition of depth implies that every morphism has a depth of at most $m - 1$.

(b) implies (c). This is trivial.

(c) implies (a). Suppose that the depth of p_S is finite for every simple module S. Denote by d the maximal depth of the p_S when S runs through all isoclasses of simple A-modules.

Let M be an indecomposable A-module, and Γ the component of the Auslander–Reiten quiver of A containing M. Let S be any simple summand of the top of M. Then, there exists a surjective morphism $q : M \longrightarrow S$. Because q is surjective and P_S is indecomposable projective, the lifting property of projectives yields $f : P_S \longrightarrow M$ such that $qf = p_S$. Then, $\mathrm{dp}(f) \leq \mathrm{dp}(p_S) \leq d$. Because of Corollary II.4.8, there exists a path of irreducible morphisms $P_S \rightsquigarrow M$ of length at most d. Now, Γ is locally finite and contains at most finitely many isoclasses of indecomposable projective modules. Therefore, Γ is finite. Because of Auslander's theorem VI.1.7, A is representation-finite and the result is proven. □

VI.2.2 The nilpotency index

In this subsection, we let A be a representation-finite algebra. Because of Lemma VI.1.2, there exists $m > 0$ such that $\mathrm{rad}_A^m = 0$. The least such m is called the **nilpotency index of the radical of the module category**. We show how one can compute this index in terms of the depths of the morphisms of the form $f_S = i_S p_S : P_S \longrightarrow I_S$, from each indecomposable projective P_S to the corresponding injective I_S and having as an image the simple module S. We need a couple of lemmata.

Lemma VI.2.2. *Let S be a simple A-module. If a morphism $f\colon P_S \longrightarrow I_S$ is nonzero, then there exist $g \in \operatorname{End}(P_S)$ and $h \in \operatorname{End}(I_S)$ such that $hfg = f_S$.*

Proof. Because of the Harada–Sai lemma VI.1.1, there exists a maximal integer $r \geq 0$ for which one can find a radical path

$$I_S = I_{S_0} \xrightarrow{h_1} I_{S_1} \longrightarrow \dots \longrightarrow I_{S_{r-1}} \xrightarrow{h_r} I_{S_r}$$

where the S_i are simple modules and $h_r \dots h_1 f \neq 0$. Writing $h = h_r \dots h_1$ if $r > 0$ and $h = 1$ if $r = 0$, it yields $hf \neq 0$.

Now consider the short exact sequence

$$0 \longrightarrow S_r \xrightarrow{i_{S_r}} I_{S_r} \xrightarrow{q} \frac{I_{S_r}}{S_r} \longrightarrow 0$$

where q is the cokernel of i_{S_r}.

We claim that qhf is zero. Indeed, if this is not the case, then there exists a simple module S' and a morphism $p\colon I_{S_r}/S_r \longrightarrow I_{S'}$ such that $p(qhf) \neq 0$. However, $pq \in \operatorname{rad}(I_{S_r}, I_{S'})$ because q is radical. This contradicts the maximality of r and establishes our claim.

As a consequence, hf factors through i_{S_r}, which is the kernel of q. That is, there exists $u\colon P_S \longrightarrow S_r$ such that $hf = i_{S_r} u$. Because $hf \neq 0$, we have $u \neq 0$; therefore, u is an epimorphism and $S = S_r$. Because $\operatorname{Hom}_A(P_S, S)$ is generated by p_S as right $\operatorname{End} P_S$-module, there exists $g \in \operatorname{End}(P_S)$ such that $hfg = i_S ug = i_S p_S = f_S$, as required. □

Lemma VI.2.3. *Let S be a simple A-module. If $f\colon M \longrightarrow I_S$ is a nonzero morphism, then there exists a morphism $g\colon P_S \longrightarrow M$ such that $fg \neq 0$.*

Proof. Let $j\colon K \longrightarrow M$ be the kernel of f. We have an induced sequence

$$0 \longrightarrow K \xrightarrow{j} M \xrightarrow{q} \frac{M}{K} \longrightarrow 0.$$

Because $fj = 0$, there exists $q'\colon M/K \longrightarrow I_S$ such that $f = q'q$. Then, $f \neq 0$ implies $\operatorname{Hom}_A(M/K, I_S) \neq 0$. In particular, M/K has the simple S as a composition factor and thus there exist submodules L, N of M with $K \subseteq N \subseteq L \subseteq M$ and $L/N \cong S$. Recalling that P_S is projective, we get a commutative diagram with an exact row

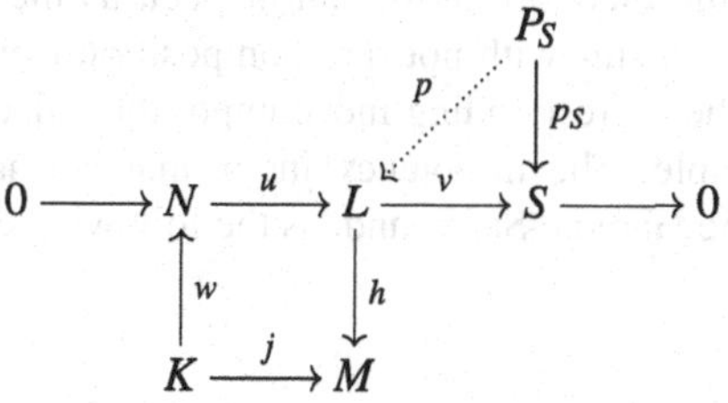

$$\begin{array}{ccccccccc}
 & & & & & & P_S & & \\
 & & & & & \overset{p}{\swarrow} & \downarrow{\scriptstyle p_S} & & \\
0 & \longrightarrow & N & \xrightarrow{u} & L & \xrightarrow{v} & S & \longrightarrow & 0 \\
 & & \uparrow{\scriptstyle w} & & \downarrow{\scriptstyle h} & & & & \\
 & & K & \xrightarrow{j} & M & & & &
\end{array}$$

where w, u, h are inclusion maps and v is the cokernel of u. Suppose $fhp = 0$. Because j is the kernel of f, there exists $k: P_S \longrightarrow K$ such that $hp = jk = huwk$. Because h is a monomorphism, we get $p = uwk$. Therefore, $p_S = vp = vuwk = 0$, a contradiction. Thus, $f(hp) \neq 0$, and the result is proven. □

Theorem VI.2.4. *Let A be a representation-finite algebra and m the maximal depth of the f_S with S ranging over all isoclasses of simple modules. Then, the nilpotency index of* rad_A *is $m+1$.*

Proof. Because of the definition of m, we have $\operatorname{rad}_A^m \neq 0$. Then let

$$M = M_0 \xrightarrow{f_1} M_1 \longrightarrow \dots \xrightarrow{f_{m+1}} M_{m+1} = N$$

be a radical path of length $m+1$ in $\operatorname{ind} A$. We must prove that $f = f_{m+1} \dots f_1$ is zero. Assume that this is not the case; then, $L = \operatorname{Im}(f)$ contains a simple module S as a composition factor.

We first claim that there exist morphisms $g: P_S \longrightarrow M$ and $h: N \longrightarrow I_S$ such that $hfg = f_S$. Let $f = qp$ be the canonical factorisation of f through its image L. Because S is a composition factor of L, there exists a nonzero morphism $u: L \longrightarrow I_S$. Because I_S is injective and q a monomorphism, there exists $v: N \longrightarrow I_S$ such that $u = vq$. Because $u \neq 0$ and p is an epimorphism, we get

$$vf = vqp = up \neq 0.$$

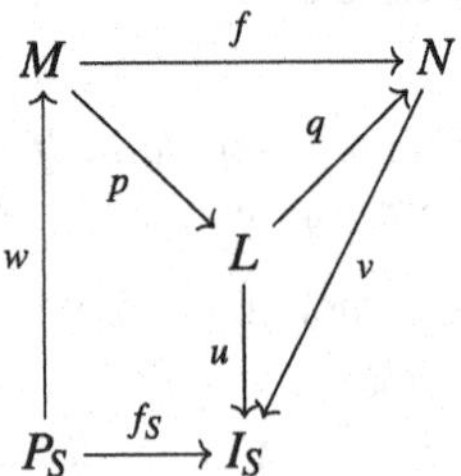

Applying Lemma VI.2.3 above yields a morphism $w: P_S \longrightarrow M$ such that $vfw \neq 0$. Because of Lemma VI.2.2, there exist morphisms g', h' such that $h'(vfw)g' = f_S$. Setting $h = h'v$ and $g = wg'$, we get $hfg = f_S$. This establishes our claim.

But then $hfg = f_S$ implies that $\operatorname{dp}(f_S) \geq \operatorname{dp}(f) \geq m+1$, and this contradicts the definition of m. The proof is now complete. □

The number m in the theorem above can be seen as the length of a maximal path of irreducible morphisms with nonzero composition from an indecomposable projective module to the corresponding indecomposable injective, passing through the corresponding simple. The nilpotency index that we have just computed is usually smaller than the Harada–Sai bound, as the following example shows.

Example VI.2.5. Let A be given by the quiver

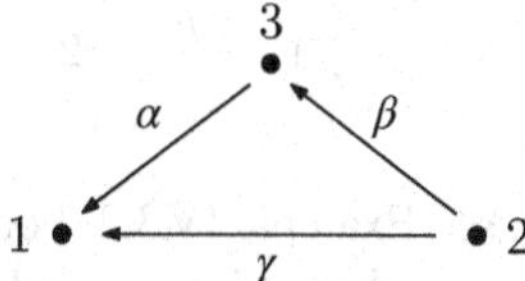

bound by $\beta\alpha = 0$. As already seen, its Auslander–Reiten quiver is

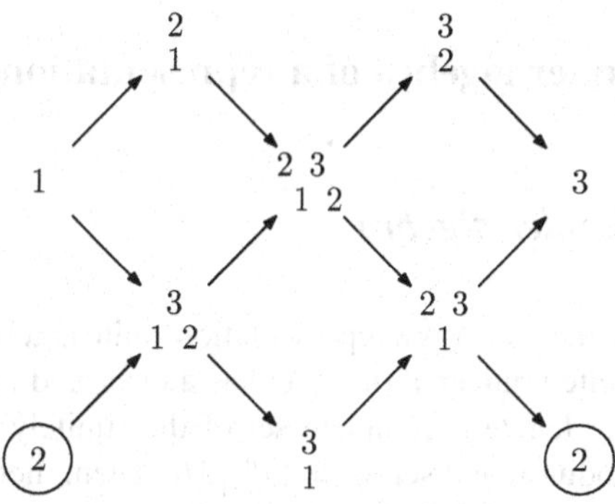

where one has to identify the two copies of the simple S_2. The maximal length of an indecomposable module is equal to 4; thus, the bound given by the Harada–Sai lemma is $2^4 - 1 = 15$. In particular, $\mathrm{rad}_A^{15} = 0$. However, the lengths of the maximal paths from P_x to I_x passing through S_x with nonzero composition are 3, 6 and 3 for $x = 1, 2$ and 3 respectively. Therefore, 7 is the nilpotency index of A (and so, $\mathrm{rad}_A^7 = 0$).

Exercises for Section VI.2

Exercise VI.2.1. Let A be given by the quiver

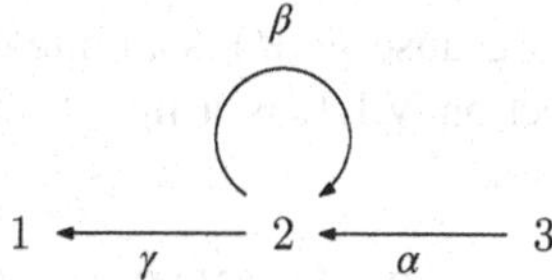

bound by $\alpha\beta = 0$, $\beta^2 = 0$, $\beta\gamma = 0$, $\alpha\gamma = 0$ (that is, $\mathrm{rad}^2 A = 0$).

(a) Compute the Auslander–Reiten quiver of A.
(b) For each isoclass of simple module S, compute the depths of i_S, p_S, f_S.
(c) Deduce the nilpotency index of rad_A.

Exercise VI.2.2. Let A be given by the quiver

$$\alpha \circlearrowleft 1 \xrightarrow{\beta} 2$$

bound by $\alpha\beta = 0$ and $\alpha^2 = 0$, see Example IV.3.2. For each of S_1 and S_2, compute the depth of the corresponding morphisms i_{S_i}, p_{S_i} and f_{S_i} for $i = 1, 2$, and deduce the nilpotency index of the radical of mod A.

VI.3 The Auslander algebra of a representation-finite algebra

VI.3.1 The Auslander algebra

In this section, we assume that A is a representation-finite algebra. Then, as we shall see, the Auslander–Reiten quiver $\Gamma(\text{mod}\, A)$ has a clear and interesting interpretation. Let $M_1, \ldots, M_m$ denote a complete set of the (finitely many) isoclasses of indecomposable A-modules, and set $M = \oplus_{i=1}^m M_i$. Then, mod A = add M, which is expressed by saying that M is an **additive generator** of the category mod A. The algebra $\mathscr{A} = \text{End}_A M$ is called the **Auslander algebra** of A. The main result of this subsection states that the ordinary quiver $Q_{\mathscr{A}}$ of $\mathscr{A}$ is precisely the Auslander–Reiten quiver $\Gamma(\text{mod}\, A)$ of A.

We start by looking at the evaluation functor $\mathscr{E}\colon \text{Fun}\, A \longrightarrow \text{mod}\, \mathscr{A}$, defined by $F \mapsto F(M)$, considered in Subsection V.1.1, when we assume additionally that A is representation-finite.

Lemma VI.3.1. *If A is representation-finite, then* $\text{Fun}\, A \cong \text{mod}\, \mathscr{A}$.

Proof. In view of the projectivisation theorem V.1.2, it suffices to prove that pres $\mathscr{P}(M)$ coincides with Fun A. We recall that pres $\mathscr{P}(M)$ consists of all F in Fun A such that there exist M_0, M_1 in add M and an exact sequence

$$\text{Hom}_A(-, M_1) \longrightarrow \text{Hom}_A(-, M_0) \longrightarrow F \longrightarrow 0.$$

Let F be an object in Fun A. Because $F(M)$ is a finite dimensional $\mathbf{k}$-vector space, and hence, as seen in Subsection V.1.1, is a finitely generated $\mathscr{A}$-module, there exists an exact sequence in mod $\mathscr{A}$

$$P_1 \longrightarrow P_0 \longrightarrow F(M) \longrightarrow 0$$

with P_0, P_1 projective $\mathscr{A}$-modules. Therefore, because of Corollary V.1.4, there exist M_0, M_1 in add M such that the previous sequence becomes

$$\text{Hom}_A(M, M_1) \longrightarrow \text{Hom}_A(M, M_0) \longrightarrow F(M) \longrightarrow 0.$$

Because mod A = add M and the functors considered are **k**-functors, this yields an exact sequence in Fun A

$$\mathrm{Hom}_A(-, M_1) \longrightarrow \mathrm{Hom}_A(-, M_0) \longrightarrow F \longrightarrow 0,$$

that is, F lies in pres $\mathscr{P}(M)$, as required. □

We prove the main result of this subsection.

Theorem VI.3.2. *If A is a representation-finite algebra, then the Auslander–Reiten quiver of A is isomorphic to the ordinary quiver of the Auslander algebra $\mathscr{A}$ of A.*

Proof. Because of Corollary V.1.4(b), the $\mathscr{A}$-modules $P_i = \mathrm{Hom}_A(M, M_i)$ with M_i an indecomposable A-module and $1 \le i \le m$ form a complete set of representatives of the isoclasses of indecomposable projective $\mathscr{A}$-modules. Thus, the map $P_i \longrightarrow M_i$ induces a bijection between the sets of points $(Q_{\mathscr{A}})_0$ and $\Gamma(\mathrm{mod}\, A)_0$ of the two quivers. To prove that there is a bijection between the arrows, let M_i, M_j be indecomposable A-modules, and $P_i = \mathrm{Hom}_A(M, M_i)$, $P_j = \mathrm{Hom}_A(M, M_j)$ the corresponding indecomposable projective $\mathscr{A}$-modules. Applying Lemma V.1.1, we get

$$\begin{aligned}\mathrm{rad}_A(M_i, M_j) &\cong \mathrm{rad}(\mathrm{Hom}_A(-, M_i), \mathrm{Hom}_A(-, M_j))\\ &\cong \mathrm{Hom}(\mathrm{Hom}_A(-, M_i), \mathrm{rad}_A(-, M_j))\\ &\cong \mathrm{Hom}_{\mathscr{A}}(P_i, \mathrm{rad}\, P_j)\end{aligned}$$

and, similarly

$$\begin{aligned}\mathrm{rad}^2_A(M_i, M_j) &\cong \mathrm{Hom}(\mathrm{Hom}_A(-, M_i), \mathrm{rad}^2_A(-, M_j))\\ &\cong \mathrm{Hom}_{\mathscr{A}}(P_i, \mathrm{rad}^2\, P_j).\end{aligned}$$

Let e_i, e_j be the idempotents of $\mathscr{A}$ such that $P_i = e_i\mathscr{A}$, $P_j = e_j\mathscr{A}$. Then, we have

$$\frac{\mathrm{rad}_A(M_i, M_j)}{\mathrm{rad}^2_A(M_i, M_j)} \cong \frac{\mathrm{Hom}_A(P_i, \mathrm{rad}\, P_j)}{\mathrm{Hom}_A(P_i, \mathrm{rad}^2\, P_j)} \cong e_i \left(\frac{\mathrm{rad}\,\mathscr{A}}{\mathrm{rad}^2\,\mathscr{A}}\right) e_j.$$

This proves the required statement. □

Although this theorem characterises the ordinary quiver of an Auslander algebra, it does not give a complete system of relations on this quiver. There are, however, obvious relations in the quiver of an Auslander algebra. Indeed, let N be an indecomposable nonprojective A-module. There exists an almost split sequence

$$0 \longrightarrow L \xrightarrow{f} \oplus_{i=1}^{t} M_i \xrightarrow{g} N \longrightarrow 0$$

with the M_i indecomposable and pairwise nonisomorphic (because A is representation-finite and so Proposition IV.1.7 holds). Exactness of the sequence yields $gf = 0$, which means that in the corresponding mesh

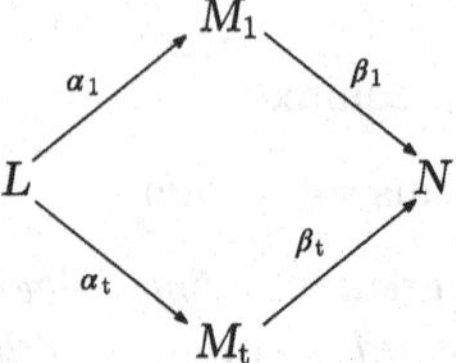

we have $\sum_{i=1}^{t} \alpha_i \beta_i = 0$. This is clearly a relation in $Q_{\mathscr{A}}$. In several important cases, such as those in which the Auslander–Reiten quiver of $\mathscr{A}$ is acyclic, relations of this form constitute a complete set of relations on $Q_{\mathscr{A}}$. But in general, this is not true. The proof lies beyond the scope of this text, but we give an example.

Example VI.3.3. Let A be given by the quiver

$$\underset{1}{\bullet} \xleftarrow{\ \mu\ } \underset{2}{\bullet} \xleftarrow{\ \lambda\ } \underset{3}{\bullet}$$

with $\lambda\mu = 0$. Then $\Gamma(\operatorname{mod} A)$ is

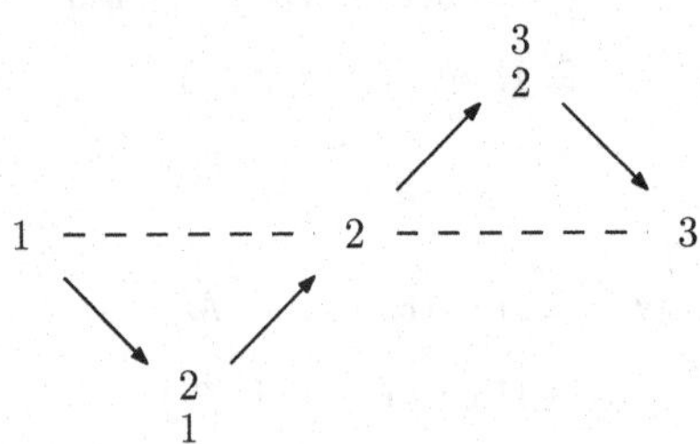

Thus, the ordinary quiver of the Auslander algebra $\mathscr{A}$ is

$$\underset{1}{\bullet} \xleftarrow{\ \delta\ } \underset{2}{\bullet} \xleftarrow{\ \gamma\ } \underset{3}{\bullet} \xleftarrow{\ \beta\ } \underset{4}{\bullet} \xleftarrow{\ \alpha\ } \underset{5}{\bullet}$$

As mentioned above, we have the relations $\alpha\beta = 0, \gamma\delta = 0$ on $Q_{\mathscr{A}}$ induced from the almost split sequences in mod A. We prove that these are all the possible relations on $Q_{\mathscr{A}}$. Indeed, $Q_{\mathscr{A}}$ is a tree; hence, the only possible relations are zero-relations. Now, the only path of length at least two in $Q_{\mathscr{A}}$, which is possibly nonzero, is $\beta\gamma$. The path $\beta\gamma$ corresponds to the composed morphism

$$\begin{smallmatrix}2\\1\end{smallmatrix} \longrightarrow 2 \longrightarrow \begin{smallmatrix}3\\2\end{smallmatrix}$$

in mod A. The latter being nonzero, we have $\beta\gamma \neq 0$ and so the given relations $\alpha\beta = 0, \gamma\delta = 0$ are the only ones on $Q_{\mathscr{A}}$.

VI.3.2 Characterisation of the Auslander algebra

One may ask: is there a criterion allowing us to verify whether or not a given algebra is the Auslander algebra of some representation-finite algebra? And, if this is the case, can one explicitly compute this representation-finite algebra?

Our objective in this subsection is to answer these two questions. Surprisingly, the criterion sought is of a purely homological nature. We start with two lemmata.

Lemma VI.3.4. *Let A be an algebra, and I an indecomposable injective A-module, then* $\mathrm{Hom}_A(-, I)$ *is an indecomposable projective–injective object in* Fun A.

Proof. Clearly, $\mathrm{Hom}_A(-, I)$ is indecomposable and projective. We just have to prove its injectivity.

Let $e \in A$ be a primitive idempotent such that $I = \mathrm{D}(Ae)$. We have a canonical isomorphism

$$\mathrm{Hom}_A(-, \mathrm{D}(Ae)) \cong \mathrm{D}\,\mathrm{Hom}_A(eA, -),$$

see Lemma I.1.19. Let F be an object in Fun A. We have isomorphisms of functors

$$\begin{aligned}\mathrm{Hom}(F, \mathrm{Hom}_A(-, \mathrm{D}(Ae)) &\cong \mathrm{Hom}(F, \mathrm{D}\,\mathrm{Hom}_A(eA, -))\\ &\cong \mathrm{Hom}(\mathrm{Hom}_A(eA, -), \mathrm{D}F)\\ &\cong \mathrm{D}F(eA)\end{aligned}$$

where we applied the covariant version of Yoneda's lemma. Applying the contravariant version, we get

$$F(eA) \cong \mathrm{Hom}(\mathrm{Hom}_A(-, eA), F).$$

Therefore,

$$\mathrm{Hom}(F, \mathrm{Hom}_A(-, \mathrm{D}(Ae)) \cong \mathrm{D}\,\mathrm{Hom}(\mathrm{Hom}_A(-, eA), F).$$

Because $\mathrm{Hom}(-, \mathrm{Hom}_A(-, \mathrm{D}(Ae))$ is of the form $\mathrm{D}\,\mathrm{Hom}(G, -)$, where $G = \mathrm{Hom}_A(-, eA)$ is a projective functor, then $\mathrm{Hom}_A(-, \mathrm{D}(Ae))$ is injective. □

Let now $\mathscr{I}$ denote a full subcategory of an abelian category $\mathscr{C}$ consisting of injective objects. We define a new category $m(\mathscr{I})$ as follows. The objects of $m(\mathscr{I})$ are morphisms between objects of $\mathscr{I}$. A morphism from an object $f : J_0 \longrightarrow J_1$ to an object $f'\colon J_0' \longrightarrow J_1'$ in $\mathscr{I}$ is a pair (u_0, u_1) of morphisms in $\mathscr{C}$, with $u_0\colon J_0 \longrightarrow J_0'$, $u_1\colon J_1 \longrightarrow J_1'$ satisfying $u_1 f = f' u_0$, that is, the square below commutes

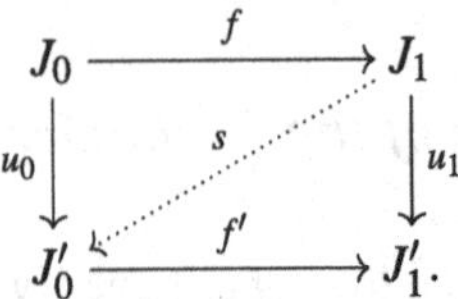

Finally, the composition of morphisms in $m(\mathscr{I})$ is induced from the composition in $\mathscr{C}$.

A morphism $(u_0, u_1)\colon f \longrightarrow f'$ as above is called **cancellable** if there exists $s\colon J_1 \longrightarrow J_0'$ such that $u_0 = sf$. Clearly, this implies $f'u_0 = u_1 f = f'sf$. Thus, if we set $\mathscr{C} = \operatorname{mod} A$, cancellable morphisms are negligible in the sense of Subsection III.1.2. The converse is clearly not true. We denote by $\mathscr{N}_0(f, f')$ the set of cancellable morphisms from f to f'. One can prove directly that the sets $\mathscr{N}_0(f, f')$ constitute an ideal $\mathscr{N}_0$ in $m(\mathscr{I})$, see Exercise VI.3.1 below. However, this follows from the next lemma, which the reader should compare with Theorems III.1.4 and III.1.5.

Lemma VI.3.5. *With the above notation, the functor $\mathscr{K}\colon m(\mathscr{I}) \longrightarrow \mathscr{C}$, which maps each object f of $m(\mathscr{I})$ to its kernel, induces a full and faithful functor*

$$\overline{\mathscr{K}}\colon \frac{m(\mathscr{I})}{\mathscr{N}_0} \longrightarrow \mathscr{C}.$$

Proof. We first prove that $\mathscr{K}$ induces a faithful functor $\overline{\mathscr{K}}\colon \frac{m(\mathscr{I})}{\mathscr{N}_0} \longrightarrow \mathscr{C}$ by proving that $\mathscr{N}_0$ is the kernel of $\mathscr{K}$ (and so, an ideal of $m(\mathscr{I})$).

Let (u_0, u_1) be in $\mathscr{N}_0(f, f')$. We have a commutative diagram with exact rows

$$\begin{array}{ccccccc}
0 \longrightarrow & \mathscr{K}(f) & \xrightarrow{\ j\ } & J_0 & \xrightarrow{\ f\ } & J_1 \\
 & \downarrow{\scriptstyle u} & & \downarrow{\scriptstyle u_0} & & \downarrow{\scriptstyle u_1} \\
0 \longrightarrow & \mathscr{K}(f') & \xrightarrow{\ j'\ } & J_0' & \xrightarrow{\ f'\ } & J_1'
\end{array}$$

We claim that $u = 0$. Indeed, because (u_0, u_1) is cancellable, there exists $s\colon J_1 \longrightarrow J_0'$ such that $u_0 = sf$. Then, $j'u = u_0 j = sfj = 0$ because $fj = 0$. Now, j' is a monomorphism. Hence, $u = 0$, as required.

Conversely, let $(u_0, u_1)\colon f \longrightarrow f'$ be a morphism such that, in the diagram above, we have $u = 0$. Then, $u_0 j = 0$ and so u_0 factors through $\operatorname{Im} f$. Letting $f = ip\colon J_0 \longrightarrow \operatorname{Im} f \longrightarrow J_1$ be the canonical factorisation of f, there exists $s'\colon \operatorname{Im} f \longrightarrow J_0'$ such that $u_0 = s'p$. Because J_0' is injective and i is a monomorphism, there exists $s\colon J_1 \longrightarrow J_0'$ such that $si = s'$. Hence, $sf = sip = s'p = u_0$ and so (u_0, u_1) is cancellable. This establishes our claim.

Fullness follows from the injectivity of J_0' and J_1'. Indeed, if $u\colon \mathscr{K}(f) \longrightarrow \mathscr{K}(f')$ is given, then, because J_0' is injective, there exists $u_0\colon J_0 \longrightarrow J_0'$ such that

$u_0 j = j'u$. Letting i, p be as before and $f' = i'p'$ the canonical factorisation of f', the morphisms u and u_0 imply the existence of $u'\colon \operatorname{Im} f \longrightarrow \operatorname{Im} f'$ such that $u'p = p'u_0$. Then, injectivity of J_1' yields a morphism $u_1\colon J_1 \longrightarrow J_1'$ such that $i'u' = u_1 i$. Therefore,

$$u_1 f = u_1 i p = i'u'p = i'p'u_0 = f'u_0$$

and we have finished. □

We now state and prove our homological characterisation of Auslander algebras.

Theorem VI.3.6. *An algebra $\mathscr{A}$ is the Auslander algebra of some representation-finite algebra if and only if the following conditions are satisfied:*

(a) gl. dim. $\mathscr{A} \leq 2$ *;*
(b) *For every indecomposable projective $\mathscr{A}$-module P, there exists an exact sequence*

$$0 \longrightarrow P \longrightarrow J_0 \longrightarrow J_1$$

where J_0 and J_1 are projective–injective.

Proof. Necessity. Assume first that $\mathscr{A}$ is the Auslander algebra of some representation-finite algebra A. Because of Lemma VI.3.1, we have a category equivalence $\operatorname{mod}\mathscr{A} \cong \operatorname{Fun} A$ that restricts to an equivalence $\operatorname{proj}\mathscr{A} \cong \operatorname{add} M = \operatorname{mod} A$ given by $\operatorname{Hom}_A(-, M_0) \mapsto M_0$. Now, to prove that gl. dim. $\mathscr{A} \leq 2$, it suffices to prove that the kernel of every morphism between projectives is projective. Let $g\colon P_1 \longrightarrow P_0$ be a morphism between projective $\mathscr{A}$-modules. There exist A-modules M_0, M_1 and a morphism $f\colon M_1 \longrightarrow M_0$ such that we have a commutative square

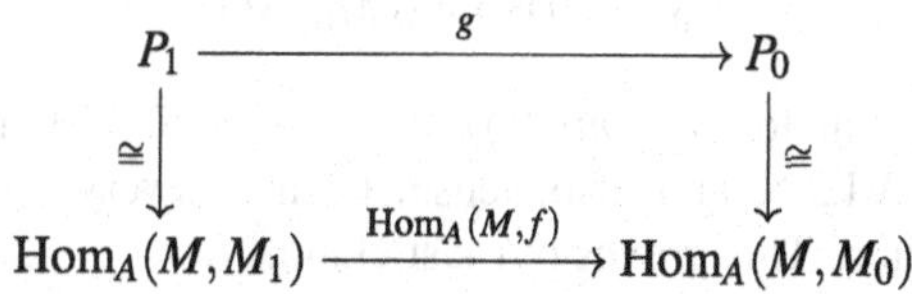

so $\operatorname{Ker} g \cong \operatorname{Ker}\operatorname{Hom}_A(M, f) \cong \operatorname{Hom}_A(M, \operatorname{Ker} f)$ because the Hom-functors are left exact. Hence, $\operatorname{Ker} g$ is projective. This proves (a).

Let P be any projective $\mathscr{A}$-module. Then there exists an A-module U such that $P \cong \operatorname{Hom}_A(M, U)$. We have an injective copresentation

$$0 \longrightarrow U \longrightarrow I_0 \longrightarrow I_1$$

in $\operatorname{mod} A$. Because the Hom-functors are left exact, it induces an exact sequence

$$0 \longrightarrow \operatorname{Hom}_A(-, U) \longrightarrow \operatorname{Hom}_A(-, I_0) \longrightarrow \operatorname{Hom}_A(-, I_1)$$

in $\operatorname{Fun} A$ and then also in $\operatorname{proj}\mathscr{A}$. Because of Lemma VI.3.4, $\operatorname{Hom}_A(-, I_0)$ and $\operatorname{Hom}_A(-, I_1)$ are projective–injective in $\operatorname{Fun} A$. This establishes (b).

Sufficiency. Conversely, assume that the algebra $\mathscr{A}$ satisfies both conditions (a) and (b) and let $\mathscr{I}$ denote the full subcategory of mod $\mathscr{A}$ consisting of all projective–injective $\mathscr{A}$-modules. We claim that the kernel functor $\mathscr{K} : m(\mathscr{I}) \longrightarrow \operatorname{mod}\mathscr{A}$ induces an equivalence

$$\overline{\mathscr{K}} : \frac{m(\mathscr{I})}{\mathscr{N}_0} \xrightarrow{\cong} \operatorname{proj}\mathscr{A}.$$

Indeed, because of condition (a), the kernel of every morphism between objects in $\mathscr{I}$ lies in proj $\mathscr{A}$ so that $\mathscr{K} : m(\mathscr{I}) \longrightarrow \operatorname{proj}\mathscr{A}$ is a well-defined functor. Lemma VI.3.5 states that there is an induced full and faithful functor

$$\overline{\mathscr{K}} : \frac{m(\mathscr{I})}{\mathscr{N}_0} \longrightarrow \operatorname{proj}\mathscr{A}.$$

It remains to show that this functor is dense. Let P be any projective $\mathscr{A}$-module. Because of (b), there exists an exact sequence

$$0 \longrightarrow P \longrightarrow J_0 \xrightarrow{f} J_1$$

where J_0, J_1 are projective–injective. Therefore, f lies in $m(\mathscr{I})$ and $P = \overline{\mathscr{K}}(f)$. This establishes our claim.

Let ind $\mathscr{I}$ denote a complete set of representatives of the isoclasses of the indecomposable objects in $\mathscr{I}$. Because $\mathscr{A}$ is a finite dimensional algebra, ind $\mathscr{I}$ is a finite set. Let A be the endomorphism algebra of the direct sum of all X in ind $\mathscr{I}$. We claim that $m(\mathscr{I})/\mathscr{N}_0 \cong \operatorname{mod} A$.

Because of Corollary V.1.4(c), we have an equivalence between $\mathscr{I} = \operatorname{add}(\operatorname{ind}\mathscr{I})$ and inj A given by

$$M_0 \mapsto \operatorname{D}\operatorname{Hom}_A(M_0, M).$$

Consider the kernel functor $\overline{\mathscr{K}} : m(\mathscr{I})/\mathscr{N}_0 \longrightarrow \operatorname{mod} A$. It is full and faithful because of Lemma VI.3.5, and also dense, because every A-module admits an injective copresentation. This proves our last claim.

Composing the equivalences in our two claims, we get an equivalence proj $\mathscr{A} \cong \operatorname{mod} A$. This implies that A is representation-finite, because there are only finitely many isoclasses of indecomposable projective $\mathscr{A}$-modules. In addition, $\mathscr{A}$, which is the endomorphism algebra of a complete set of representatives of the isoclasses of the indecomposable projective $\mathscr{A}$-modules is isomorphic to the endomorphism algebra of a complete set of representatives of the isoclasses of all indecomposable A-modules. That is, $\mathscr{A}$ is the Auslander algebra of A. □

The proof just given is constructive: namely, starting from an Auslander algebra $\mathscr{A}$, the proof shows how to recover A as the endomorphism algebra of a complete set of representatives of the isoclasses of indecomposable projective–injective modules.

Condition (b) in the theorem is sometimes expressed by saying that $\mathscr{A}$ has **dominant dimension at least two.** Because of condition (a) and the equivalence $\operatorname{Fun} A = \operatorname{mod} \mathscr{A}$, for every indecomposable A-module N, the simple functor S_N admits a projective resolution of length at most two

$$0 \longrightarrow \operatorname{Hom}_A(-, L) \longrightarrow \operatorname{Hom}_A(-, M) \longrightarrow \operatorname{Hom}_A(-, N) \longrightarrow S_N \longrightarrow 0.$$

We thus recover, for representation-finite algebras, the result of Theorem II.3.10. Recall from Proposition II.3.11 that, if $L \neq 0$, then N is nonprojective and the sequence

$$0 \longrightarrow L \longrightarrow M \longrightarrow N \longrightarrow 0$$

is almost split in $\operatorname{mod} A$.

Example VI.3.7. Consider the algebra $\mathscr{A}$ given by the quiver

$$\underset{1}{\bullet} \xleftarrow{\ \delta\ } \underset{2}{\bullet} \xleftarrow{\ \gamma\ } \underset{3}{\bullet} \xleftarrow{\ \beta\ } \underset{4}{\bullet} \xleftarrow{\ \alpha\ } \underset{5}{\bullet}$$

bound by $\alpha\beta = 0, \gamma\delta = 0$. Its global dimension equals two: indeed, we have projective resolutions of the simple modules

$$0 \longrightarrow \begin{smallmatrix}3\\2\end{smallmatrix} \longrightarrow \begin{smallmatrix}4\\3\\2\end{smallmatrix} \longrightarrow \begin{smallmatrix}5\\4\end{smallmatrix} \longrightarrow 5 \longrightarrow 0$$

$$0 \longrightarrow \begin{smallmatrix}3\\2\end{smallmatrix} \longrightarrow \begin{smallmatrix}4\\3\\2\end{smallmatrix} \longrightarrow 4 \longrightarrow 0$$

$$0 \longrightarrow 1 \longrightarrow \begin{smallmatrix}2\\1\end{smallmatrix} \longrightarrow \begin{smallmatrix}3\\2\end{smallmatrix} \longrightarrow 3 \longrightarrow 0$$

$$0 \longrightarrow 1 \longrightarrow \begin{smallmatrix}2\\1\end{smallmatrix} \longrightarrow 2 \longrightarrow 0$$

whereas 1 is simple projective. Also, three indecomposable projectives, namely

$$P_2 = \begin{smallmatrix}2\\1\end{smallmatrix}, \qquad P_5 = \begin{smallmatrix}5\\4\end{smallmatrix} \quad \text{and} \quad P_4 = \begin{smallmatrix}4\\3\\2\end{smallmatrix}$$

are projective–injective, whereas there exist exact sequences for the others

$$0 \longrightarrow P_1 = 1 \longrightarrow \begin{smallmatrix}2\\1\end{smallmatrix} \longrightarrow \begin{smallmatrix}4\\3\\2\end{smallmatrix}$$

$$0 \longrightarrow P_3 = \begin{smallmatrix}3\\2\end{smallmatrix} \longrightarrow \begin{smallmatrix}4\\3\\2\end{smallmatrix} \longrightarrow \begin{smallmatrix}5\\4\end{smallmatrix}$$

with both the second and the third terms projective–injective. Therefore, $\mathscr{A}$ is the Auslander algebra of some representation-finite algebra A. To find the latter, we take a complete set of representatives of isoclasses of indecomposable projective–injective modules, that is,

$$\{P_2 = I_1 = \begin{smallmatrix} 2 \\ 1 \end{smallmatrix}, P_4 = I_2 = \begin{smallmatrix} 4 \\ 3 \\ 2 \end{smallmatrix}, P_5 = I_4 = \begin{smallmatrix} 5 \\ 4 \end{smallmatrix}\}$$

and let $A = \mathrm{End}_{\mathscr{A}}(P_2 \oplus P_4 \oplus P_5)$. We have morphisms

$$\begin{smallmatrix} 2 \\ 1 \end{smallmatrix} \xrightarrow{f} \begin{smallmatrix} 4 \\ 3 \\ 2 \end{smallmatrix} \xrightarrow{g} \begin{smallmatrix} 5 \\ 4 \end{smallmatrix}$$

where each of f, g is nonzero, but $gf = 0$. Thus, $\mathscr{A}$ is the Auslander algebra of the algebra A given by the quiver

$$\overset{1}{\bullet} \xleftarrow{\mu} \overset{2}{\bullet} \xleftarrow{\lambda} \overset{3}{\bullet}$$

bound by $\lambda\mu = 0$.

VI.3.3 The representation dimension

As an application, we give a criterion of representation-finiteness using a homological invariant called representation dimension. This invariant was introduced by Auslander who expected it to be a measure of how far an algebra would be from being representation-finite. Let A be an algebra. An A-module T is called a **generator–cogenerator** of $\mathrm{mod}\, A$ if it is at the same time a generator and a cogenerator of the module category, that is, if for every A-module M, there exist T_0, T_1 in $\mathrm{add}\, T$, an epimorphism $T_0 \longrightarrow M$ (so that T generates M) and a monomorphism $M \longrightarrow T_1$ (so that T cogenerates M). Because every indecomposable projective module is a direct summand of any generator of $\mathrm{mod}\, A$, and dually, every indecomposable injective module is a direct summand of any cogenerator, we infer that both A_A and $(DA)_A$ are direct summands of any generator–cogenerator.

Definition VI.3.8. Let A be a nonsemisimple algebra. The **representation dimension** of A, denoted as $\mathrm{rep.dim.}\, A$, is the minimum of the global dimensions of $\mathrm{End}\, T_A$, as T ranges over all generator–cogenerators of $\mathrm{mod}\, A$.

The representation dimension is meant to measure the complexity of the morphisms in $\mathrm{mod}\, A$. Indeed, a nonzero morphism from M to N, say, preserves at least one common simple composition factor S of both M and N. But then, there exists a nonzero morphism from the projective cover of S to M, and another one from N

to the injective envelope of S. These three morphisms have a nonzero composition, which is a morphism from A_A to $(DA)_A$. In this sense, all nonzero morphisms in $\operatorname{mod} A$ are counted inside the endomorphism algebra of any generator–cogenerator.

Our objective is to prove that, for any nonsemisimple algebra A, we have rep. dim. $A \geq 2$ and that the least value 2 occurs if and only if A is representation-finite. Our first lemma can be thought of as a partial converse of Lemma VI.3.4.

Lemma VI.3.9. *Let T be a generator–cogenerator of* $\operatorname{mod} A$ *and* $B = \operatorname{End} T_A$. *If* $\operatorname{Hom}_A(T, M)$ *is injective in* $\operatorname{mod} B$, *then M is injective in* $\operatorname{mod} A$.

Proof. Let L_A be a submodule of A_A. Because T is a generator, there exists an exact sequence $T_1 \xrightarrow{f_1} T_0 \xrightarrow{p} L \longrightarrow 0$ with T_0, T_1 in $\operatorname{add} T$. Composing p with the inclusion $j : L \longrightarrow A$ yields an exact sequence

$$T_1 \xrightarrow{f_1} T_0 \xrightarrow{f_0} A$$

with $f_0 = jp$. Because this exact sequence lies in $\operatorname{add} T$, applying $\operatorname{Hom}_A(T, -)$ gives an exact sequence in $\operatorname{mod} B$

$$\operatorname{Hom}_A(T, T_0) \xrightarrow{\operatorname{Hom}_A(T,f_1)} \operatorname{Hom}_A(T, T_0) \xrightarrow{\operatorname{Hom}_A(T,f_0)} \operatorname{Hom}_A(T, A)$$

using Corollary V.1.4. Because $\operatorname{Hom}_A(T, M)$ is injective, we get a commutative diagram with exact rows

$$\begin{array}{ccccc}
\operatorname{Hom}_B(\operatorname{Hom}_A(T,A),\operatorname{Hom}_A(T,M)) & \longrightarrow & \operatorname{Hom}_B(\operatorname{Hom}_A(T,T_0),\operatorname{Hom}_A(T,M)) & \longrightarrow & \operatorname{Hom}_B(\operatorname{Hom}_A(T,T_1),\operatorname{Hom}_A(T,M)) \\
\downarrow \cong & & \downarrow \cong & & \downarrow \cong \\
\operatorname{Hom}_A(A,M) & \xrightarrow{\operatorname{Hom}_A(f_0,M)} & \operatorname{Hom}_A(T_0,M) & \xrightarrow{\operatorname{Hom}_A(f_1,M)} & \operatorname{Hom}_A(T_1,M)
\end{array}$$

where we again used Corollary V.1.4. On the other hand, applying $\operatorname{Hom}_A(-, M)$ to the exact sequence $T_1 \xrightarrow{f_1} T_0 \longrightarrow L \longrightarrow 0$ yields $\operatorname{Hom}_A(L, M) = \operatorname{Ker} \operatorname{Hom}_A(f_1, M)$.

Now, the morphism $\operatorname{Hom}_A(j, M) : \operatorname{Hom}_A(A, M) \longrightarrow \operatorname{Hom}_A(L, M)$ is surjective: let $u \in \operatorname{Hom}_A(L, M)$. Then, $pf_1 = 0$ implies that

$$up \in \operatorname{Hom}_A(p, M)(u) \in \operatorname{Ker} \operatorname{Hom}_A(f_1, M) = \operatorname{Im} \operatorname{Hom}_A(f_0, M)$$

thus, there exists $v : A \longrightarrow M$ such that $up = vf_0 = vjp$. Because p is an epimorphism, $u = vj = \operatorname{Hom}_A(j, M)(v)$; that is, $\operatorname{Hom}_A(j, M)$ is indeed surjective. But this surjectivity means exactly that the module M_A is injective. □

The second lemma says that, in the terminology used in Subsection VI.3.2, endomorphism rings of generator–cogenerators have dominant dimension at least two.

Lemma VI.3.10. *Let T be a generator–cogenerator of* $\operatorname{mod} A$ *and* $B = \operatorname{End} T_A$. *Then, there exists a minimal injective copresentation*

$$0 \longrightarrow B_B \longrightarrow J_0 \longrightarrow J_1$$

in $\operatorname{mod} B$ *such that* J_0, J_1 *are projective–injective and* $\operatorname{End} J_0$ *is Morita equivalent to* A.

Proof. Take a minimal injective copresentation of T in $\operatorname{mod} A$

$$0 \longrightarrow T \longrightarrow I_0 \longrightarrow I_1.$$

Because this sequence lies in $\operatorname{add} T$, applying $\operatorname{Hom}_A(T, -)$ to it gives an exact sequence of B-modules

$$0 \longrightarrow \operatorname{Hom}_A(T, T) \longrightarrow \operatorname{Hom}_A(T, I_0) \longrightarrow \operatorname{Hom}_A(T, I_1)$$

using Corollary V.1.4. Now, because of Lemma I.1.19, we have $\operatorname{Hom}_A(T, \mathrm{D}A) \cong \mathrm{D}\operatorname{Hom}_A(A, T)$. The module A being a direct summand of T, the left B-module $\operatorname{Hom}_A(A, T)$ is projective and therefore the right B-module $\mathrm{D}\operatorname{Hom}_A(A, T)$ is injective. This proves that the last exact sequence is an injective copresentation of B_B. On the other hand, the fact that $\mathrm{D}A$ lies in $\operatorname{add} T$ implies that $J_0 = \operatorname{Hom}_A(T, I_0)$, $J_1 = \operatorname{Hom}_A(T, I_1)$ are also projective and therefore projective–injective. Finally, the resulting injective copresentation $0 \longrightarrow B_B \longrightarrow J_0 \longrightarrow J_1$ is minimal because the original injective copresentation of T is also minimal.

The embedding $B_B \longrightarrow J_0$ implies that every indecomposable projective–injective B-module occurs as a direct summand of J_0 and, because J_0 itself is projective–injective, then $\operatorname{add} J_0$ is the full subcategory of $\operatorname{mod} B$ consisting of all the projective–injective B-modules. Because of Lemma VI.3.5, such B-modules are of the form $\operatorname{Hom}_A(T, I)$ with I an injective A-module. Therefore, $\operatorname{End} J_0$ is Morita equivalent to $\operatorname{End}_B \operatorname{Hom}_A(T, \mathrm{D}A) \cong \operatorname{End}_A \mathrm{D}A$, and thus to A. □

Corollary VI.3.11. *For every nonsemisimple algebra A, we have* $\operatorname{rep.dim.} A \geq 2$.

Proof. Let T be a generator–cogenerator of $\operatorname{mod} A$ and $B = \operatorname{End} T_A$. We assume that $\operatorname{gl.dim.} B \leq 1$ and we reach a contradiction. Indeed, in this case, a minimal injective copresentation of B is a short exact sequence

$$0 \longrightarrow B_B \longrightarrow J_0 \longrightarrow J_1 \longrightarrow 0.$$

Because of Lemma VI.3.10, J_1 is projective; thus, the sequence splits. Therefore, B_B is a direct summand of J_0; hence, is injective, that is, B is a selfinjective algebra. But the global dimension of a nonsemisimple selfinjective algebra is infinite. Therefore, B is semisimple and we have $B_B = J_0$ (because the above copresentation is minimal). Because of Lemma VI.3.10, A is Morita equivalent to $\operatorname{End} J_0 = \operatorname{End} B_B \cong B$. Therefore, A itself is semisimple, a contradiction. □

We now state and prove the aforementioned criterion of representation-finiteness.

Theorem VI.3.12. *Let A be a nonsemisimple algebra. Then, A is representation-finite if and only if* $\operatorname{rep.dim.} A = 2$.

Proof. Let A be representation-finite, T the direct sum of a complete set of representatives of the isoclasses of indecomposable A-modules and $B = \operatorname{End} T_A$. Because of Theorem VI.3.6, we have $\operatorname{gl.dim.} B \leq 2$. Therefore, $\operatorname{rep.dim.} A \leq 2$. Corollary VI.3.11 gives $\operatorname{rep.dim.} A = 2$.

Conversely, if $\operatorname{rep.dim.} A = 2$, there exists a generator–cogenerator T of $\operatorname{mod} A$ such that $B = \operatorname{End} T_A$ has global dimension 2. We claim that B satisfies the conditions of Theorem VI.3.6. As condition (a) is granted by the hypothesis, we prove (b).

Let P^* be a projective B-module. Applying Theorem V.1.2, there exists T^* in $\operatorname{add} T$ such that $P^* = \operatorname{Hom}_A(T, T^*) = \mathscr{E} \operatorname{Hom}_A(-, T^*)$, where $\mathscr{E}$ is the evaluation functor of Subsection V.1.1. But now T^* has a minimal injective copresentation

$$0 \longrightarrow T^* \longrightarrow I_0 \longrightarrow I_1$$

in $\operatorname{mod} A$. Because T is a generator–cogenerator, I_0, I_1 are in $\operatorname{add} T$; thus, the previous sequence lies completely in $\operatorname{add} T$. Applying Theorem V.1.2 again, an exact sequence in $\operatorname{Fun} A$

$$0 \longrightarrow \operatorname{Hom}_A(-, T^*) \longrightarrow \operatorname{Hom}_A(-, I_0) \longrightarrow \operatorname{Hom}_A(-, I_1).$$

corresponds to the previous sequence. Because of Lemma VI.3.4, $\operatorname{Hom}_A(-, I_0)$ and $\operatorname{Hom}_A(-, I_1)$ are projective–injective. Setting $J_0 = \mathscr{E} \operatorname{Hom}_A(-, I_0)$, $J_1 = \mathscr{E} \operatorname{Hom}_A(-, I_1)$, we get an injective copresentation

$$0 \longrightarrow P^* \longrightarrow J_0 \longrightarrow J_1$$

in $\operatorname{mod} B$ with J_0, J_1 projective–injective. This proves condition (b).

Applying Theorem VI.3.6, there exists a representation-finite algebra A' such that B is the Auslander algebra of A'. In addition, A' is the endomorphism algebra of the direct sum of a complete set of representatives of isoclasses of indecomposable projective–injective B-modules. To complete the proof, we just need to show that A' is Morita equivalent to A.

Because of Lemma VI.3.5, every projective–injective B-module is of the form $\operatorname{Hom}_A(T, I)$ with I an injective A-module. Therefore, A' is Morita equivalent to $\operatorname{End}_B \operatorname{Hom}_A(T, DA) \cong \operatorname{End}_A(DA)$ and thus to A. □

Computing the representation dimension of a given algebra is usually very difficult. The main tool is the following result.

Theorem VI.3.13. *Let A be a nonsemisimple algebra. Then* $\operatorname{rep.dim.} A \leq d+2$ *if and only if there exists a generator–cogenerator T of* $\operatorname{mod} A$ *that has the property that for any A-module M, there exists an exact sequence*

$$0 \longrightarrow T_d \longrightarrow \dots \longrightarrow T_0 \longrightarrow M \longrightarrow 0$$

with the T_i in $\operatorname{add} T$*, such that the induced sequence*

$$0 \longrightarrow \operatorname{Hom}_A(T, T_d) \longrightarrow \dots \longrightarrow \operatorname{Hom}_A(T, T_0) \longrightarrow \operatorname{Hom}_A(T, M) \longrightarrow 0$$

is also exact.

Proof. *Sufficiency.* Let T be a generator–cogenerator satisfying the stated property and $B = \operatorname{End} T$. We claim that gl. dim. $B \leq d + 2$.

Let X be a B-module. Because of Theorem V.1.2, there exist F in Fun A and T', T'' in add T such that $X = \mathscr{E}F = F(T)$ and we have an exact sequence

$$\operatorname{Hom}_A(-, T'') \longrightarrow \operatorname{Hom}_A(-, T') \longrightarrow F \longrightarrow 0$$

in Fun A. Because of Yoneda's lemma, there exists a morphism $f : T'' \longrightarrow T'$ such that $F = \operatorname{Coker} \operatorname{Hom}_A(-, f)$. Let $M = \operatorname{Ker} f$. The hypothesis implies the existence of an exact sequence

$$0 \longrightarrow T_d \longrightarrow \ldots \longrightarrow T_0 \longrightarrow M \longrightarrow 0$$

with the T_i in add T such that the induced sequence

$$0 \longrightarrow \operatorname{Hom}_A(T, T_d) \longrightarrow \ldots \longrightarrow \operatorname{Hom}_A(T, T_0) \longrightarrow \operatorname{Hom}_A(T, M) \longrightarrow 0$$

is exact in mod B. On the other hand, applying the evaluation functor $\mathscr{E}$ to the above exact sequence in Fun A yields another exact sequence

$$0 \longrightarrow \operatorname{Hom}_A(T, M) \longrightarrow \operatorname{Hom}_A(T, T'') \longrightarrow \operatorname{Hom}_A(T, T') \longrightarrow F(T) \cong X \longrightarrow 0.$$

Splicing both sequences, we get an exact sequence

$$\begin{aligned} 0 \longrightarrow \operatorname{Hom}_A(T, T_d) \longrightarrow \ldots \longrightarrow \operatorname{Hom}_A(T, T_0) \longrightarrow \operatorname{Hom}_A(T, T'') \\ \longrightarrow \operatorname{Hom}_A(T, T') \longrightarrow X \longrightarrow 0. \end{aligned}$$

which is a projective resolution of X in mod B. Therefore, pd $X \leq d + 2$. This establishes our claim that gl. dim. $B \leq d + 2$ which, in turn, implies that rep. dim. $A \leq d + 2$.

Necessity. Assume that rep. dim. $A = s \leq d + 2$. Then, there exists a generator–cogenerator T of mod A such that $B = \operatorname{End} T_A$ satisfies gl. dim. $B = s$. Let M be an A-module. We wish to prove the existence of an exact sequence as in the statement. Without loss of generality, we may assume that M does not lie in add T. Then a minimal injective coresolution

$$0 \longrightarrow M \longrightarrow I_0 \xrightarrow{f} I_1$$

induces an exact sequence of functors

$$0 \longrightarrow \operatorname{Hom}_A(-, M) \longrightarrow \operatorname{Hom}_A(-, I_0) \xrightarrow{\operatorname{Hom}_A(-, f)} \operatorname{Hom}_A(-, I_1) \longrightarrow F \longrightarrow 0$$

where $F = \operatorname{Coker} \operatorname{Hom}_A(-, f)$. Because T is a generator–cogenerator, I_0, I_1 belong to add T. Evaluating this sequence of functors on T, and using that $F(T)$ is a B-module, we have

$$\operatorname{pd} \operatorname{Hom}_A(T, M) = \operatorname{pd} F(T) - 2 \leq s - 2.$$

Let

$$0 \longrightarrow \mathrm{Hom}_A(T, T_{s-2}) \longrightarrow \ldots \longrightarrow \mathrm{Hom}_A(T, T_0) \longrightarrow \mathrm{Hom}_A(T, M) \longrightarrow 0$$

be a projective resolution with the T_i in add T. Corresponding to it, there is an exact sequence of functors from add T to mod B

$$0 \longrightarrow \mathrm{Hom}_A(-, T_{s-2}) \longrightarrow \ldots \longrightarrow \mathrm{Hom}_A(-, T_0) \longrightarrow \mathrm{Hom}_A(-, M) \longrightarrow 0.$$

Evaluating this sequence on A_A, which lies in add T, we get an exact sequence

$$0 \longrightarrow T_{s-2} \longrightarrow \ldots \longrightarrow T_0 \longrightarrow M \longrightarrow 0$$

in mod A. This completes the proof. □

Example VI.3.14. Let A be a hereditary nonsemisimple algebra and $T = A \oplus DA$. Let M be any module. Without loss of generality, we may assume that M is not in add T. But then M is noninjective and so $\mathrm{Hom}_A(DA, M) = 0$, because of Lemma IV.2.5(b). Therefore, $\mathrm{Hom}_A(T, M) \cong \mathrm{Hom}_A(A, M)$ and every projective resolution $0 \longrightarrow P_1 \longrightarrow P_0 \longrightarrow M \longrightarrow 0$ induces an exact sequence

$$0 \longrightarrow \mathrm{Hom}_A(T, P_1) \longrightarrow \mathrm{Hom}_A(T, P_0) \longrightarrow \mathrm{Hom}_A(T, M) \longrightarrow 0.$$

Theorem VI.3.13 gives rep. dim. $A \leq 3$. In addition, if A is representation-infinite, then it follows from Theorem VI.3.12 that rep. dim. $A \geq 3$. Therefore, if A is a representation-infinite hereditary algebra, then we have rep. dim. $A = 3$.

Exercises for Section VI.3

Exercise VI.3.1. Prove directly that $\mathscr{N}_0$ is an ideal in $m(\mathscr{I})$.

Exercise VI.3.2. Let A be the path algebra of the quiver

$$1 \longleftarrow 2 \longleftarrow 3$$

(a) Prove that the Auslander algebra of A is given by the quiver

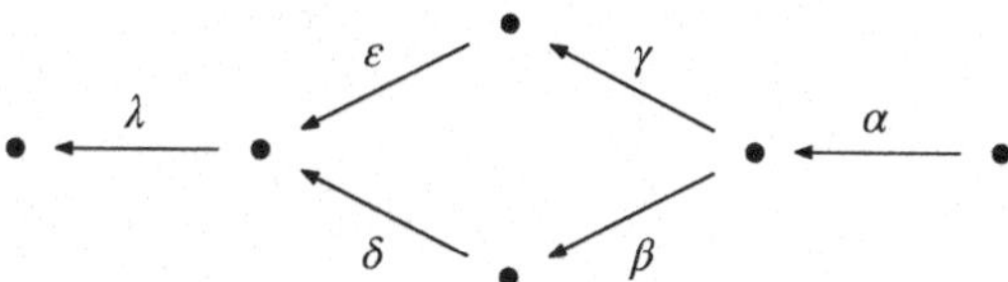

bound by $\alpha\beta = 0$, $\beta\delta + \gamma\epsilon = 0$, $\delta\lambda = 0$.

(b) Conversely, starting from the algebra given by the bound quiver of (a), prove that it satisfies the conditions of Theorem VI.3.6. Then verify that the endomorphism algebra of the direct sum of a complete set of representatives of the isoclasses of its indecomposable projective–injectives is precisely A.

Exercise VI.3.3. Let A be a representation-finite algebra, T the direct sum of a complete set of representatives of the isoclasses of indecomposable A-modules and $\mathscr{A} = \operatorname{End} T$ the Auslander algebra. Given an indecomposable A-module M, let S denote the top of $\operatorname{Hom}_A(T, M)$ in $\operatorname{mod} \mathscr{A}$. Prove the following statements:

(a) S is simple projective in $\operatorname{mod} \mathscr{A}$ if and only if M is simple projective in $\operatorname{mod} A$.
(b) $\operatorname{pd} S = 1$ if and only if M is projective but not simple.
(c) $\operatorname{pd} S = 2$ if and only if M is not projective.
(d) Assume that M is not projective and that

$$0 \longrightarrow \operatorname{Hom}_A(T, K) \longrightarrow \operatorname{Hom}_A(T, L) \longrightarrow \operatorname{Hom}_A(T, M) \longrightarrow S \longrightarrow 0$$

is a minimal projective resolution of S. Prove that

$$0 \longrightarrow K \longrightarrow L \longrightarrow M \longrightarrow 0$$

is an almost split sequence in $\operatorname{mod} A$.

Exercise VI.3.4. Prove that each of the following bound quiver algebras is the Auslander algebra of some representation-finite algebra. Compute the latter.

(a) 1 $\xleftarrow{\lambda}$ 3 $\xleftarrow{\beta}$ 4 $\xleftarrow{\alpha}$ 6; 2 $\xleftarrow{\mu}$ 3 $\xleftarrow{\delta}$ 5 $\xleftarrow{\gamma}$ 6 $\qquad \alpha\beta = \gamma\delta, \beta\lambda = 0, \delta\mu = 0$

(b) $1 \xleftarrow{\lambda} 2 \xleftarrow{\varepsilon} 3 \xleftarrow{\delta} 4 \xleftarrow{\gamma} 5 \xleftarrow{\beta} 6 \xleftarrow{\alpha} 7$ $\qquad \alpha\beta = 0, \gamma\delta = 0,$ $\varepsilon\lambda = 0$

(c) $1 \underset{\beta}{\overset{\alpha}{\rightleftarrows}} 2$ $\qquad \alpha\beta = 0$

(d) $1 \xleftarrow{\eta} 2 \xleftarrow{\nu} 3$; 3 $\xleftarrow{\lambda}$ 4 $\xleftarrow{\gamma}$ 6 $\xleftarrow{\alpha}$ 7; 3 $\xleftarrow{\mu}$ 5 $\xleftarrow{\delta}$ 6 $\xleftarrow{\beta}$ 8 $\qquad \alpha\gamma = 0, \beta\delta = 0,$ $\gamma\lambda = \delta\mu, \nu\eta = 0.$

Exercise VI.3.5. For each of the following hereditary algebras A, compute the endomorphism algebra of $A \oplus DA$ and prove directly that the global dimension equals 3.

(a) $1 \leftleftarrows 2$

(b) $$\begin{array}{ccccc} 1 & & & & 4 \\ & \nwarrow & & \swarrow & \\ & & 3 & & \\ & \swarrow & & \nwarrow & \\ 2 & & & & 5 \end{array}$$

Exercise VI.3.6. Let A be an algebra, T a generator of $\operatorname{mod} A$ and $B = \operatorname{End} T_A$. Let M be an A-module.

(a) Prove that every projective presentation

$$\operatorname{Hom}_A(T, T_1) \longrightarrow \operatorname{Hom}_A(T, T_0) \longrightarrow \operatorname{Hom}_A(T, M) \longrightarrow 0$$

in $\operatorname{mod} B$ is induced by an exact sequence $T_1 \longrightarrow T_0 \longrightarrow M \longrightarrow 0$.

(b) Prove that there exists an epimorphism $p : T_0 \longrightarrow M$ with T_0 in $\operatorname{add} T$ such that $\operatorname{Hom}_A(T, p) : \operatorname{Hom}_A(T, T_0) \longrightarrow \operatorname{Hom}_A(T, M)$ is a projective cover in $\operatorname{mod} B$.

Exercise VI.3.7. Let A be an algebra, T a generator–cogenerator of $\operatorname{mod} A$ and $B = \operatorname{End} T_A$. Prove that the following are equivalent:

(a) $\operatorname{gl.dim.} B \leq 3$,

(b) For every A-module M, there exists a short exact sequence

$$0 \longrightarrow T_1 \longrightarrow T_0 \xrightarrow{p} M \longrightarrow 0$$

with T_0, T_1 in $\operatorname{add} T$ and $\operatorname{Hom}_A(T, p) \colon \operatorname{Hom}_A(T, T_0) \longrightarrow \operatorname{Hom}_A(T, M)$ a projective cover in $\operatorname{mod} B$,

(c) For every A-module M, there exists a short exact sequence

$$0 \longrightarrow T_1 \longrightarrow T_0 \xrightarrow{p} M \longrightarrow 0$$

with T_0, T_1 in $\operatorname{add} T$ and $p : T_0 \longrightarrow M$ a right minimal $\operatorname{add} T$-approximation (see Exercise V.1.4).

In addition, prove that, if any of these conditions are satisfied, then $\operatorname{rep.dim.} A \leq 3$.

VI.4 The Four Terms in the Middle theorem

VI.4.1 Preparatory lemmata

Let A be a finite dimensional algebra and $0 \longrightarrow L \longrightarrow M \longrightarrow N \longrightarrow 0$ an almost split sequence in mod A. The middle term of the sequence may be decomposed into indecomposable direct summands as $M = \oplus_{i=1}^{t} M_i$. It is known that the integer t is uniquely determined by M whereas the M_i are unique up to isomorphism. However, to decompose a module M into indecomposable direct summands is in general a difficult problem. Therefore, a measure of the complexity of such an almost split sequence is the invariant t, called the number of middle terms. A reasonable question is then: how large may the integer t be? It is always finite, because we deal with finitely generated modules, but easy examples show that the integer t may be arbitrarily large, see Exercise VI.4.2 below.

It turns out that, if the algebra A is representation-finite, then $t \le 4$, and, if $t = 4$, then exactly one of the indecomposable summands of the middle term is projective–injective. This is called the Four Terms in the Middle theorem, or the **Bautista–Brenner theorem**. Our objective in this section is to prove this theorem. We mostly follow the neat proof given by Liu and Krause.

Throughout, we assume that A is a representation-finite algebra. In this case, because of Proposition IV.1.7, the set M^+ of all arrows leaving a point M in the Auslander–Reiten quiver $\Gamma(\text{mod}\, A)$ can and will be identified to the set of direct successors of M, no two of which are isomorphic. In particular, $|M^+|$ is the number of direct successors of M. Similarly, the set M^- of all arrows entering M can and will be identified with the set of all its direct predecessors, no two of which are isomorphic, and $|M^-|$ is the number of these direct predecessors. In this proof, we use essentially the notion of sectional path, see Definition IV.1.21.

Lemma VI.4.1. *Let M be an indecomposable A-module. Then, there exists $s \ge 0$ such that $\tau^s M$ has a projective sectional predecessor.*

Proof. There exists an indecomposable projective module P such that $\text{Hom}_A(P, M) \neq 0$. Because A is representation-finite, Corollary VI.1.3 yields a path $P = M_0 \longrightarrow M_1 \longrightarrow \dots \longrightarrow M_{m-1} \longrightarrow M_m = M$ of irreducible morphisms. Therefore, there exists $i \ge 0$ maximal with the property that $L = \tau^t M_i$ is projective for some $t \ge 0$. Then, we clearly get a sectional path for some $s \ge t$. □

Lemma VI.4.2. *Let $M_m \longrightarrow \dots \longrightarrow M_1 \longrightarrow M_0 = M$ be a sectional path. If*

$$\sum_{L \in M^-} l(L) - l(M_1) \ge l(M),$$

then none of the M_i, with $0 \le i \le m$, is projective.

Proof. Assume that M is projective, then $l(M) > \sum_{L\in M^-} l(L)$, contrary to the hypothesis. Therefore, M itself is not projective. In addition,

$$l(\tau M) = \sum_{L \in M^-} l(L) - l(M) \geq l(M_1).$$

Because $\tau M \in M_1^-$, this shows that M_1 is not projective. The given path is sectional; thus, $\tau M \neq M_2$ and we have

$$\sum_{L_1 \in M_1^-} l(L_1) - l(M_2) \geq l(\tau M) \geq l(M_1).$$

The same argument as before gives that M_2 is not projective. We finish the proof by induction. □

Corollary VI.4.3. *Let M be nonprojective such that $l(\tau M) \geq l(L)$ for all $L \in M^-$. Then, no sectional predecessor of M is projective.*

Proof. Let $X = M_m \longrightarrow \ldots \longrightarrow M_1 \longrightarrow M_0 = M$ be a sectional path. Because of the hypothesis, we have $l(\tau M) \geq l(M_1)$. Then,

$$\sum_{L \in M^-} l(L) = l(\tau M) + l(M)$$

yields

$$\sum_{L \in M^-} l(L) - l(M_1) = l(M) + l(\tau M) - l(M_1) \geq l(M).$$

Because of Lemma VI.4.2, X is not projective. □

Lemma VI.4.4. *Let N_1, N_2 be distinct elements of M^+. If $l(M) \geq l(N_1) + l(N_2)$, then*

(a) *M, N_1, N_2 are not projective; and*
(b) $l(\tau M) \geq \sum_{L \in M^-} l(L) - l(\tau N_1) - l(\tau N_2)$.

Proof. For $i \in \{1, 2\}$, we have $l(N_i) \leq l(N_1) + l(N_2) \leq l(M)$; hence, neither N_1 nor N_2 is projective. The two inequalities $l(N_i) + l(\tau N_i) \geq l(M)$ for $i = 1, 2$ give that

$$l(\tau N_1) + l(\tau N_2) \geq 2l(M) - l(N_1) - l(N_2) \geq l(M)$$

where we used the inequality in the hypothesis. Therefore, M is not projective and we have proven (a).

In addition, we have

$$l(\tau M) = \sum_{L \in M^-} l(L) - l(M) \geq \sum_{L \in M^-} l(L) - l(\tau N_1) - l(\tau N_2)$$

where we used the inequality that we just proved. This proves (b). □

Lemma VI.4.5. *Assume that $f : M \longrightarrow \oplus_{i=1}^4 N_i$ is irreducible, with M and the N_i indecomposable. If no N_i is projective, then f is a monomorphism and $l(M) < l(\tau^{-1}M)$.*

Proof. We assume that f is an epimorphism, or that $l(M) \geq l(\tau^{-1}M)$ with M noninjective, and try to reach a contradiction.

We first claim that $2l(M) \geq \sum_{i=1}^4 l(N_i)$. This is clear if f is an epimorphism. On the other hand, if M is noninjective and $l(M) \geq l(\tau^{-1}M)$, then $2l(M) \geq l(M) + l(\tau^{-1}M) \geq \sum_{i=1}^4 l(N_i)$, as required.

This inequality implies that $l(M) \geq l(N_1) + l(N_2)$ or $l(M) \geq l(N_3) + l(N_4)$. We may, without loss of generality, assume the former. Because of Lemma VI.4.4 above, M is not projective. We claim that, for any $L \in M^-$, we have $l(\tau M) \geq l(L)$.

There are two cases to consider. If $L \neq \tau N_i$ for every i, then

$$\begin{aligned}
l(\tau M) &\geq l(L) + \sum_{i=1}^4 l(\tau N_i) - l(M) \\
&\geq l(L) + \sum_{i=1}^4 [l(M) - l(N_i)] - l(M) \\
&= l(L) + 3l(M) - \sum_{i=1}^4 l(N_i) \\
&\geq l(L) + l(M) \\
&\geq l(L)
\end{aligned}$$

because $2l(M) \geq \sum_{i=1}^4 l(N_i)$. On the other hand, if $L \cong \tau N_i$ for some i, say, for $i = 1$, we have

$$\begin{aligned}
l(\tau M) &\geq \sum_{i=1}^4 l(\tau N_i) - l(M) \\
&\geq l(L) + \sum_{i=2}^4 [l(M) - l(N_i)] - l(M) \\
&= l(L) + 2l(M) - \sum_{i=2}^4 l(N_i) \\
&\geq l(L).
\end{aligned}$$

This establishes our claim. Because of Corollary VI.4.3, no sectional predecessor of M is projective. In particular, no τN_i is projective. Then we have

$$\begin{aligned} l(\tau M) &\geq \sum_{i=1}^{4} l(\tau N_i) - l(M) \\ &\geq \sum_{i=1}^{4} [l(M) - l(N_i)] - l(M) \\ &= 3l(M) - \sum_{i=1}^{4} l(N_i) \\ &\geq l(M). \end{aligned}$$

Because τN_i is not projective for every i, and τM is noninjective with $l(\tau M) \geq l(M) = l(\tau^{-1}(\tau M))$, we may repeat this reasoning for τM, getting that no sectional predecessor of τM is projective. Inductively, for every $s \geq 0$, $\tau^s M$ has no projective sectional predecessor. This contradicts Lemma VI.4.1. □

In the statement, the facts that f is an irreducible monomorphism and that M is indecomposable imply that M cannot be injective; thus, $\tau^{-1} M$ exists.

We need to state the dual of Lemma VI.4.5: assume that $g\colon \oplus_{i=1}^{4} L_i \longrightarrow M$ is irreducible, with M and the L_i indecomposable. If no L_i is injective, then g is an epimorphism and $l(\tau M) > l(M)$. Now, a last lemma.

Lemma VI.4.6. *Assume that M has an injective sectional successor and let $f\colon M \longrightarrow \oplus_{i=1}^{t} N_i$ be left minimal almost split, with the N_i indecomposable. Then:*

(a) $t \leq 4$;
(b) *If $t = 4$, then one of the N_i is projective.*

Proof. Let $M = M_0 \longrightarrow M_1 \longrightarrow \ldots \longrightarrow M_m = I$ be a sectional path of shortest length with I injective. First, if $m = 0$, then M itself is injective. Hence, f is an epimorphism. If $t \geq 4$, then, because of Lemma VI.4.5, f is a monomorphism, a contradiction. Therefore, $t < 4$ in this case.

Assume, thus, that $m \geq 1$. Because of the dual of Lemma VI.4.2, we have $\sum_{N \in M^+} l(N) - l(M_1) < l(M)$. Let $M^+ = \{M_1 = N_1, \ldots, N_t\}$. Then we have $l(M) > \sum_{i \geq 2} l(N_i)$. Assume $t > 4$, then the composition f' of the left minimal almost split morphism f with the projection $\oplus_{i=1}^{t} N_i \longrightarrow \oplus_{i \geq 2}^{t} N_i$ is irreducible, because of Corollary II.2.25, and the last inequality implies that f' is an epimorphism. But, because of Lemma VI.4.4, it is a monomorphism whenever $t > 4$, a contradiction. Therefore, $t \leq 4$ and we have proven (a).

Assume now $t = 4$, and no N_i projective. As before, we have $l(M) > \sum_{N \in M^+} l(N) - l(M_1)$. In particular, for any two indices $i, j \in \{2, 3, 4\}$, $l(M) > l(N_i) + l(N_j)$. Because of Lemma VI.4.4, M, N_i, N_j are not projective and we have

$$l(\tau M) \geq \sum_{L \in M^-} l(L) - l(\tau N_i) - l(\tau N_j)$$

We claim that $l(\tau M) \geq l(L)$ for every $L \in M^-$. Indeed, taking $i = 2$ and $j = 3$, we get $l(\tau M) \geq l(\tau N_1)$, $l(\tau M) \geq l(\tau N_4)$ and $l(\tau M) \geq l(L)$ for $L \neq \tau N_i$. Next, taking $i = 3$, $j = 4$, we get $l(\tau M) \geq l(\tau N_2)$. Finally, $i = 2$, $j = 4$ yields $l(\tau M) \geq l(\tau N_3)$. This establishes our claim. Invoking Corollary VI.4.3, we get that no sectional predecessor of M is projective. In particular, no τN_i is projective and clearly, no τN_i is injective either. Applying Lemma VI.4.5 to the irreducible morphism $\tau M \longrightarrow \oplus_{i=1}^4 \tau N_i$, we get $l(\tau M) < l(\tau^{-1}(\tau M)) = l(M)$. Applying its dual to the irreducible morphism $\oplus_{i=1}^4 \tau N_i \longrightarrow M$, we get $l(\tau M) > l(M)$, a contradiction. Therefore, one of the N_i is projective. This proves (b). □

Proposition VI.4.7. *Let M be an indecomposable A-module. Then*

(a) $|M^+| \leq 4$;
(b) *If* $|M^+| = 4$, *then M is noninjective and M^+ contains a projective.*

Proof. Because of the dual of Lemma VI.4.1, there exists a least $s \geq 0$ such that $N = \tau^{-s}M$ has an injective sectional successor. The minimality of s says that, for every t such that $0 \leq t < s$, there is no sectional path $\tau^{-t}M \rightsquigarrow I$, with I injective. In particular, the almost split sequence starting with $\tau^{-t}M$ for every $t < 5$ has no injective middle term. Therefore, $|M^+| \leq |N^+|$.

Because of Lemma VI.4.6, we have $|N^+| \leq 4$. Therefore, $|M^+| \leq 4$. Assume $|M^+| = 4$. Because $|M^+| \leq |N^+|$, we have $|N^+| = 4$ and actually, for every $t \leq s$, we have $|(\tau^{-t}M)^+| = 4$. Because of Lemma VI.4.6, one of the elements of N^+ is projective and this can only happen if $s = 0$. Therefore, M^+ contains a projective.

Because no projective has an injective as immediate predecessor in $\Gamma(\operatorname{mod} A)$, then M is noninjective. □

VI.4.2 The theorem

The Four Terms in the Middle theorem says that over a representation-finite algebra, an almost split sequence has at most four middle terms and, if it has four, then exactly one of these middle terms is projective–injective.

Theorem VI.4.8. *Let A be a representation-finite algebra and*

$$0 \longrightarrow L \longrightarrow \oplus_{i=1}^t M_i \longrightarrow N \longrightarrow 0$$

be an almost split sequence, with the M_i indecomposable. Then, $t \leq 4$ and, if $t = 4$, then one of the M_i is projective–injective, but the others are neither projective nor injective.

Proof. Because of Proposition VI.4.7, we have $|L^+| \leq 4$. In addition, if $|L^+| = 4$, then one of the M_i is projective. Dually, $|N^-| = 4$ implies that one of the M_j is injective. Because of Proposition III.3.2, we have $i = j$ and M_i is projective–injective, whereas all M_k with $k \neq i$ are neither projective nor injective. The proof is complete. □

Besides the intrinsic interest of the theorem, the proof shows how to use properties of the Auslander–Reiten quiver efficiently to obtain a new result. We end with an example.

Example VI.4.9. Let A be given by the quiver

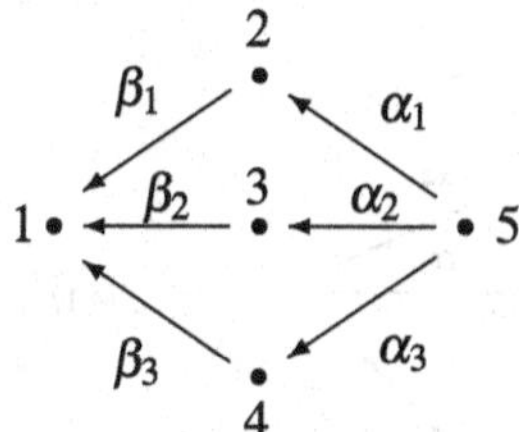

bound by all possible commutativity relations, that is, $\alpha_1\beta_1 = \alpha_2\beta_2 = \alpha_3\beta_3$. The projective–injective indecomposable module $P_5 = I_1$ is a direct summand of the middle term of the following almost split sequence with four middle terms:

$$0 \longrightarrow \operatorname{rad} P_5 \longrightarrow P_5 \oplus S_2 \oplus S_3 \oplus S_4 \longrightarrow \frac{P_5}{\operatorname{soc} P_5} \longrightarrow 0.$$

The algebra A is representation-finite and its Auslander–Reiten quiver is

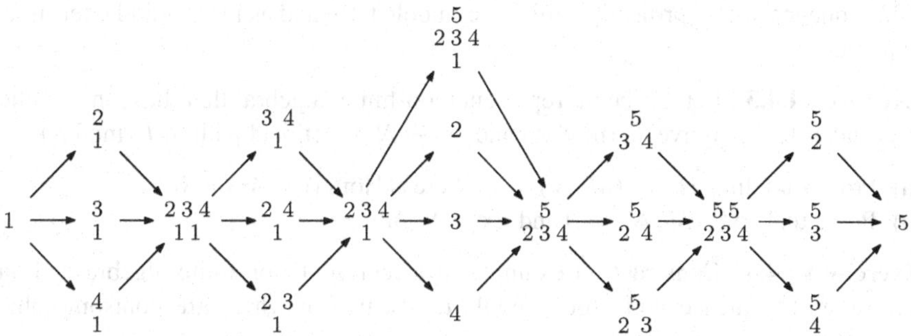

Exercises for Section VI.4

Exercise VI.4.1. For each of the following bound quiver algebras, construct an almost split sequence satisfying the conditions of the Four Terms in the Middle theorem. Then, show that the algebras of (a) and (b) are representation-finite, whereas that of (c) is representation-infinite.

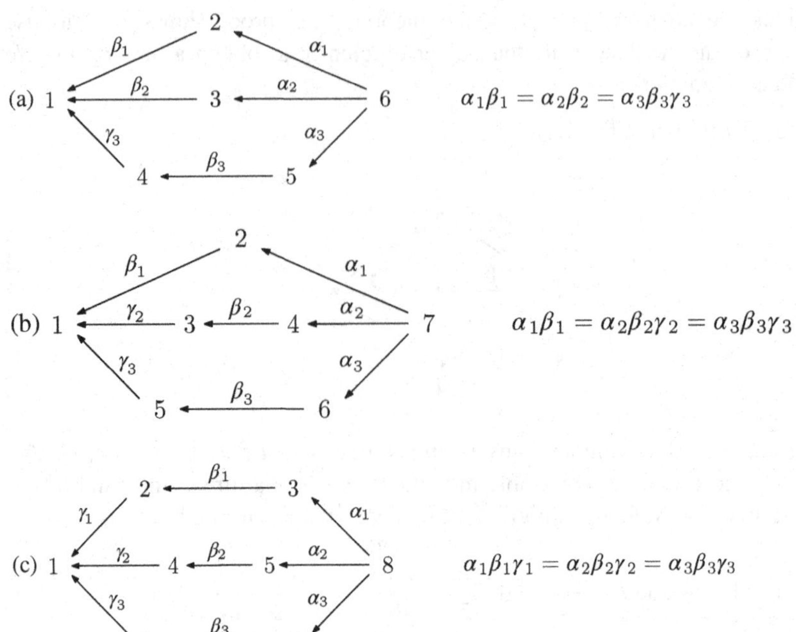

Exercise VI.4.2. Give an example of a bound quiver algebra that has an almost split sequence with a projective–injective middle term and as many middle terms as required.

Exercise VI.4.3. Let A be a representation-finite algebra that has an acyclic Auslander–Reiten quiver $\Gamma(\text{mod}\, A)$ and $M \rightsquigarrow N$ a sectional path in $\Gamma(\text{mod}\, A)$.

(a) Prove that $\dim_k \text{Hom}_A(M, N) \leq 1$ whereas $\text{Hom}_A(N, M) = 0$.
(b) Prove that $\text{Ext}^1_A(M, N) = 0$ and $\text{Ext}^1_A(N, M) = 0$.

Exercise VI.4.4. Construct an example of a representation-finite algebra and an almost split sequence with three middle terms, two of which are nonisomorphic projectives.

Bibliography

1. Textbooks on noncommutative and homological algebra

Anderson, F. W. and Fuller, K., Rings and categories of modules, Springer Verlag, Berlin, Heidelberg, New York (1973).
Assem, I., Algèbres et modules, Presses de l'université d'Ottawa, Ottawa, et Masson, Paris (1997).
Bourbaki, N., Algèbre Chapitre 10, Algèbre homologique, Masson, Paris (1980).
Cartan, H. and Eilenberg, S., Homological algebra, Princeton University Press, Princeton (1956).
Drozd, Yu. A. and Kiricenko, V. V., Finite dimensional algebras, Springer Verlag, Berlin, Heidelberg, New York (1994).
Gelfand, S. I. and Manin, Yu. I., Methods of homological algebra, Springer Verlag, Berlin, Heidelberg, New York (1996).
Rotman, J. J., An introduction to homological algebra, Academic Press (1979).

2. General texts on representations of algebras

Assem, I., Simson, D. and Skowronski, A., Elements of the Representation Theory of Associative Algebras, London Mathematical Society Student Texts 65, Cambridge University Press, Cambridge (2006).
Auslander, M., Reiten, I. and Smalø, S. O., Representation Theory of Artin Algebras, Cambridge Studies in Advanced Mathematics 36, Cambridge University Press, Cambridge (1995).
Barot, M., Introduction to the Representation Theory of Algebras, Springer Verlag, Berlin, Heidelberg, New York (2015).
Gabriel, P. and Roiter, A. V., Representations of Finite-Dimensional Algebras, Algebra VIII, Vol. 73 of the Encyclopedia of Mathematical Sciences, Springer Verlag, Berlin, Heidelberg, New York (1992).
Happel, D., Triangulated Categories in the Representation Theory of Finite Dimensional Algebras, London Mathematical Society Lecture Note Series 119, Cambridge University Press, Cambridge (1988).
Ringel, C. M., Tame algebras and integral quadratic forms, Lecture Notes in Mathematics 1099, Springer Verlag, Berlin, Heidelberg, New York (1984).

I. Assem, F. U. Coelho, *Basic Representation Theory of Algebras*, Graduate Texts in Mathematics 283, https://doi.org/10.1007/978-3-030-35118-2

Schiffler, R., Quiver Representations, CMS Books in Mathematics, Springer Verlag, Berlin, Heidelberg, New York (2014).

Simson, D. and Skowronski, A., Elements of the Representation Theory of Associative Algebras. 2. Tubes and concealed algebras of Euclidean type, London Mathematical Society Student Texts 71, (2007) Cambridge University Press.

Simson, D. and Skowronski, A., Elements of the Representation Theory of Associative Algebras. 3. Representation-infinite tilted algebras, London Mathematical Society Student Texts 72, (2007) Cambridge University Press.

Skowronski, A. and Yamagata, K., Frobenius Algebras I, Basic Representation Theory, European Mathematical Society, Zürich (2011).

Skowronski, A. and Yamagata, K., Frobenius Algebras II, Tilted and Hochschild Extension Algebras, European Mathematical Society, Zürich (2017).

3. Original papers or surveys related to the contents of the book

Assem, I., Tilting theory -an introduction, Topics in Algebra, Banach Center Publ. Vol. 26, Part 1, PWN (1990) 127–180.

Assem, I. and Marmaridis, N., Tilting modules over split-by-nilpotent extensions, Comm. Algebra, Vol 26 (1998), Issue 5, 1547–1555.

Auslander, Maurice, Reiten, Idun; Smalø, Sverre O.; Solberg, Øyvind, eds., Selected works of Maurice Auslander. Part 1, Providence, R.I.: American Mathematical Society (1999).

Auslander, Maurice, Reiten, Idun; Smalø, Sverre O.; Solberg, Øyvind, eds., Selected works of Maurice Auslander. Part 2, Providence, R.I.: American Mathematical Society(1999).

Bautista, R., Irreducible morphisms and the radical of a category, preprint Universidad Nacional Autonoma de Mexico No. 3 (1979).

Bautista, R. and Brenner, S., On the number of terms in the middle of an almost split sequence, Proc. ICRA III (1980) Springer Lecture Notes in Mathematics 903 (1981) 1–8.

Bautista, R. and Brenner, S., Replication numbers for non-Dynkin sectional subgraphs in finite Auslander-Reiten quivers and some properties of Weyl roots, Proc. London Math. Soc. (3) 47 (1983) 429–462.

Bautista, R. and Smalø, S. O., Nonexistent cycles, Comm. Algebra 11 (16) (1983) 1755–1767.

Bernstein, I.M., Gelfand, I.M. and Ponomarev, V.A., Coxeter functors and Gabriel's theorem, Uspechi Mat. Nauk 28 (1973) 19–23 (in Russian), English translation in Russian Math. Surveys 28 (1973) 17–32.

Bongartz, K., Tilted algebras, Proc ICRA III (1980) Springer Lecture Notes in Mathematics 903 (1981) 26–38.

Bongartz, K., On a result of Bautista and Smalø on cycles, Comm. Algebra 11 (18) (1983) 2123–2124.

Brenner, S., On the kernel of an irreducible map, Linear Algebra and its Applications Volume 365 (2003), 91–97.

Brenner, S. et Butler, M. C. R., Generalizations of the Bernstein-Gelfand-Ponomarev reflection functors, Proc. ICRA II (1979) Springer Lecture Notes in Mathematics 832 (1980) 103–170.

Butler, M. C. R., The construction of almost split sequences I, Proc. London Math. soc. 40 (1980) 72–86.

Chaio, C., Coelho, F. U. and Trepode, S., On the composite of two irreducible morphisms in radical cube, J. Algebra 312 (2007) 650–667.

Chaio, C. and Liu, S., A note on the radical of a module category, Comm. Algebra 41 (2013) 4419–4424.

Coelho, F. U. and Platzeck, M.-I., On the representation dimension of some classes of algebras, J. Algebra 275 (2004) 615–628.

Erdmann, K., Holm, T., Iyama, O. and Schröer, J., Radical embeddings and representation dimension, Advances in Mathematics 185 (2004), 159–177

Gabriel, P., Unzerlegbare Darstellungen I, Manuscripta Math. 6 (1972) 71–103.

Gabriel, P., Indecomposable representations II, Symposia Mat. Inst. Naz. Alta Mat. 11 (1973) 81–104.

Gabriel, P., Auslander-Reiten sequences and representation-finite algebras, Proc. ICRA II (1979) Springer Lecture Notes in Mathematics 831 (1980) 1–80.

Happel, D. and Ringel, C. M., Tilted algebras, Trans. Amer. Math. Soc. 274, No. 2 (1982) 399–443.

Krause, H., On the Four Terms in the Middle Theorem for almost split sequences, Arch. Math. Vol. 62 (1994) 501–505.

Liu, S., Almost split sequences for non-regular modules, Fund. Math. 143 (1993) 183–190.

Reiten, I., The use of almost split sequences in the representation theory of artin algebras, Proc. Workshop ICRA III, Springer Lecture Notes in Mathematics 944 (1981) 29–104.

Ringel, C. M., Report on the Brauer-Thrall conjectures, Proc. Workshop ICRA II (1979) Springer Lecture Notes in Mathematics 831 (1980) 104–136.

Index

T-presentation, 240
k-algebra, 1
k-category, 41
k-functor, 42
k-linear category, 42
τ-orbit, 194
$\mathscr{P}(T)$-presented functor, 237

Additive generator, 282
Algebra, 2
  basic, 6
  bound quiver algebra, 15
  connected, 6
  elementary, 18
  finite dimensional, 2
  hereditary, 31
  local, 4
  opposite algebra, 2
  representation-finite, 27
  representation-infinite, 27
  selfinjective, 35
  triangular matrix, 38
Almost split sequence, 68
Approximation
  left approximation, 245
  right approximation, 245
Associative product, 144
Auslander algebra, 282
Auslander–Reiten formulae, 114
Auslander–Reiten quiver, 161
Auslander–Reiten sequence, 68
Auslander–Reiten translate, 105
Auslander–Reiten translation, 105
Auslander's theorem, 275

Bautista–Brenner theorem, 298
Brauer–Thrall conjecture, 276
Brenner–Butler theorem, 262

Canonical sequence, 254
Category
  direct sum, 47
  injectively stable, 101
  linearisation, 53
  projectively stable, 101
  quotient, 43
  semisimple, 47
Component
  acyclic, 193
  postprojective, 193
  preinjective, 193
Composition series, 8
  composition factor, 9
  length, 8
Counit of adjunction, 242
Cycle, 13, 192

Depth of a morphism, 207
Dimension vector, 9
Dominant dimension, 289
Duality functor, 6

Endomorphism algebra, 3
Essential monomorphism, 64
Ext-injective, 267
Ext-projective, 267

I. Assem, F. U. Coelho, *Basic Representation Theory of Algebras*, Graduate Texts in Mathematics 283, https://doi.org/10.1007/978-3-030-35118-2

Four Terms in the Middle theorem, 302
Functor
  evaluation functor, 236
  finitely generated, 79
  kernel, 43
  Nakayama functor, 7
  projection functor, 44
  projective, 78
  simple, 80
  subfunctor, 43
  support, 277
  support-finite, 97

Generator–cogenerator, 290
Grothendieck group, 8

Harada–Sai lemma, 272

Ideal
  admissible, 15
  in a **k**-category, 43
Idempotent, 6
  central, 6
  complete set, 6
  orthogonal, 6
  primitive, 6
Injective envelope, 7
Injective points, 177

Knitting, 167
Kronecker algebra, 32
Kronecker quiver, 33

Loewy length, 23
Loop, 13

Maximal
  subfunctor, 48
  submodule, 3
Mesh, 161, 177
Modular law, 11
Module, 2
  composition length, 9
  finitely generated, 3
  indecomposable, 5
  injective, 7
  length, 9
  partial tilting, 246
  postprojective, 193
  preinjective, 193
  projective, 5
  regular, 223
  tilting, 246
  T-presented, 240
  uniserial, 27
Morphism, 3
  cancellable, 286
  irreducible, 55
  left almost split, 61
  left minimal, 64
  left minimal almost split, 65
  negligible, 102, 104
  radical morphism, 46
  retraction, 5
  right almost split, 61
  right minimal, 64
  right minimal almost split, 66
  section, 5
Morphism of algebras, 2

Nakayama algebra, 28
Nakayama's lemma, 11
Nilpotency index, 278

One-point extension, 38

Path, 13, 95
  length, 13, 95
  of irreducible morphisms, 95
  radical, 95
  source, 13
  stationary, 13
  target, 13
  trivial, 13
Path algebra, 14
Predecessor, 95
Projective cover, 7, 80
Projective points, 177

Quiver, 13
  acyclic, 13
  arrow, 13
  bound quiver, 15
  connected, 13
  finite, 13
  full subquiver, 13
  locally finite, 163, 177
  ordinary, 18

point, 13
repetitive quiver, 178
source, 13
subquiver, 13
target, 13
translation quiver, 177
underlying graph, 14

Radical, 3, 46
filtration, 23
infinite radical, 89
radical square, 55
Reciprocity theorem, 260
Regular representation, 223
Relation, 16
commutativity, 17
zero-relation, 17
Representation, 2, 39, 217
Representation dimension, 290

Sectional path, 179, 208
Sectional predecessor, 179
Sectional successor, 179
Space of irreducible morphisms, 158
Split
extension, 145
morphism, 5
torsion pair, 256
Split extension
split-by-nilpotent extension, 145
Stable tube, 179
rank, 179
Structural map, 217
Successor, 95
Superfluous epimorphism, 64
Support of a module, 137

Tilting theorem, 262
Torsion
class, 253
modules, 253
pair, 252
Torsion-free
class, 253
modules, 253
Trace, 251
Translation, 177
Transpose, 106
Transposition, 106
Trivial extension, 154

Unique decomposition theorem, 5
Unit of adjunction, 242

Walk, 13

Yoneda's lemma, 77

Zero object, 42

Printed in the United States
by Baker & Taylor Publisher Services

Printed in the United States
by Baker & Taylor Publisher Services